T0156089

UNITEXT for Physics

UNITEXT for Physics series, formerly UNITEXT Collana di Fisica e Astronomia, publishes textbooks and monographs in Physics and Astronomy, mainly in English language, characterized of a didactic style and comprehensiveness. The books published in UNITEXT for Physics series are addressed to upper undergraduate and graduate students, but also to scientists and researchers as important resources for their education, knowledge and teaching.

More information about this series at http://www.springer.com/series/13351

Carlo Alabiso · Ittay Weiss

A Primer on Hilbert Space Theory

Linear Spaces, Topological Spaces, Metric Spaces, Normed Spaces, and Topological Groups

Second Edition

 Springer

Carlo Alabiso
Department of Physics
University of Parma
Parma, Italy

Ittay Weiss
School of Mathematics and Physics
University of Portsmouth
Portsmouth, UK

ISSN 2198-7882   ISSN 2198-7890   (electronic)
UNITEXT for Physics
ISBN 978-3-030-67419-9   ISBN 978-3-030-67417-5   (eBook)
https://doi.org/10.1007/978-3-030-67417-5

This Springer imprint is published by the registered company Springer Nature Switzerland AG
The registered company address is: Gewerbestrasse 11, 6330 Cham, Switzerland

Ittay Weiss dedicates the book to his daughters, Eugenia and Esther, for allowing him to work on it.

Preface to the Second Edition

Other than correction of typos, stylistic improvements, and the addition of many new exercises, two new chapters have been added to the book offering a non-classical view of the classical spaces and a guided tour of some of the main attractions of Hilbert space land. The new material creates a more evenly spread experience across the text and interjects the abstract ambient theory with more concrete scenarios.

The classical sequence spaces ℓ_p and function spaces L_p are fundamental and go hand-in-hand with the general theory. However, the function spaces are technically demanding and so they were only superficially treated in the first edition. The new chapter devoted to the classical spaces builds on the metric machinery already in place and attempts to create a clear conceptual framework that emphasizes the similarities between the two families of spaces. This approach complements the standard measure-theoretic presentations found in the literature.

The chapter devoted to Hilbert spaces is the new closing chapter of the book and is written in the form of a discussion. It highlights the practical simplification borne by the presence of an inner product as it draws upon the material in the previous chapters to compare against the situation in Banach spaces. In this way, the chapter serves both as a guided tour for the reader's future journey to Hilbert space theory and a retrospective reflection on its preceding chapters. The topics covered are classical in Hilbert space theory but as they are handled here after a preliminary exposition of Banach spaces, the chapter takes the opportunity to create a richer historical context provided by a brief account of famous results in the development of the general theory.

Parma, Italy   Carlo Alabiso
Portsmouth, UK   Ittay Weiss
November 2020

Preface to the First Edition

The Structure of the Book

The book consists of eight chapters with an additional chapter of solved problems arranged by topic. Each chapter is composed of five sections, with each section accompanied by a set of exercises (with the exception of the shorter Chapter 7 with just a single batch of exercises). The total of 422 exercises and 50 solved problems comprise an integral part of the book designed to assist the reader and hone her intuition.

Chapter 1 is a general introduction to analysis and, in particular, to each of the subjects presented in the chapters that follow. Chapter 1 also contains a Preliminaries section, intended to quickly orient the reader as to the notation and concepts used throughout the book, starting with sets and ending with an axiomatic presentation of the real and complex numbers.

Chapter 2 is devoted to linear spaces. At the advanced undergraduate level, the reader is already familiar with at least some aspects of linear spaces, primarily finite-dimensional ones. The chapter does not rely on any previous knowledge though, and is in that sense self-contained. However, the material is somewhat advanced since the focus is the technically and conceptually demanding infinite-dimensional linear spaces.

Chapter 3 is an introduction to topology, a subject considered to be at a rather high level of abstraction. The main aim of the chapter is to familiarize the reader with the fundamentals of the theory, particularly the portions that are most directly relevant for analysis and Hilbert spaces. Care is taken to find a reasonable balance between the study of extreme topology, i.e., spaces or phenomena that one may consider pathological but that hone the topological intuition, and mundane topology, i.e., those spaces or phenomena one is most likely to find in nature but that may obscure the true nature of topology.

Chapter 4 is a study of metric spaces. Once the necessary fundamentals are covered, the main focus is complete metric spaces. In particular, the Banach Fixed

Point Theorem and Baire's Theorem are proved and completions are discussed, topics that are indispensable for Hilbert space theory.

Chapter 5 is a non-classical view of the classical sequence and function spaces. The ℓ_p and L_p spaces are presented and the classical facts regarding their separability properties and duality relationships are discussed. The presentation borrows ideas from category theory in order to stress the formal similarities between the two families of spaces.

Chapter 6 introduces and studies Banach spaces and their operators. Starting with semi-normed spaces the chapter establishes the fundamentals and goes on to introduce Banach spaces, treating the Open Mapping Theorem, the Hahn-Banach Theorem, the Closed Graph Theorem, and, toward the end, touches upon unboundedness.

Chapter 7 is a short introduction to topological groups, emphasizing their relation to Banach spaces. The chapter does not assume any knowledge of group theory, and thus, to remain self-contained, it presents all relevant group-theoretic notions. The chapter ends with a treatment of uniform spaces and a hint of their usefulness in the general theory.

Chapter 8 is a short excursion to Hilbert space land. Reflecting on the material developed in the preceding chapters, the chapter, written in discussion form, considers several problems that arise naturally in the context of Banach spaces, delves sufficiently deep to witness the inherent difficulty, and proceeds to display the enormous benefit of the geometry of Hilbert spaces toward a resolution. The topics covered are the closest point property, orthogonal complements, bases, Fourier series, and the Riesz representation theorem.

A Word About the Intended Audience

The book is aimed at the advanced undergraduate or beginning postgraduate level, with the general prerequisite of sufficient mathematical maturity as expected at that level of studies. The book should be of interest to the student knowing nothing of Hilbert space theory who wishes to master its prerequisites. The book should also be useful to the reader who is already familiar with some aspects of Hilbert space theory, linear spaces, topology, or metric spaces, as the book contains all the relevant definitions and pivotal theorems in each of the subjects it covers.

A Word About the Authors

The authors of the book, a physicist and a mathematician, by writing the entire book together, and through many arguments about notation and style, hope that this clash between the desires of a physicist to quickly yet intelligibly get to the point and the insistence of a mathematician on rigor and conciseness did not leave the pages of this book tainted with blood but rather that it resulted in a welcoming introduction for both physicists and mathematicians interested in Hilbert Space Theory.

Prof. Carlo Alabiso obtained his Degree in Physics at Milan University, and then taught for more than 40 years at Parma University, Parma, Italy (with a period spent as a research fellow at the Stanford Linear Accelerator Center and at Cern, Geneva). His teaching encompassed topics in quantum mechanics, special relativity, field theory, elementary particle physics, mathematical physics, and functional analysis. His research fields include mathematical physics (Padé approximants), elementary particle physics (symmetries and quark models), and statistical physics (ergodic problems), and he has published articles in a wide range of national and international journals as well as the previous Springer book (with Alessandro Chiesa), Problemi di Meccanica Quantistica non Relativistica.

Dr. Ittay Weiss earned his Ph.D. in mathematics from Utrecht University, The Netherlands, in 2007. He has published work on algebraic topology, general topology, metric space theory, and category theory, and taught undergraduate and postgraduate mathematics in The Netherlands, Fiji, and England. He is currently Senior Lecturer in Mathematics at the School of Mathematics and Physics in the University of Portsmouth, UK. Born in Israel, his first encounter with advanced mathematics was at the age of 16 while he was following computer science courses at the Israeli Open University. So profound was his fascination with the beauty and utility of mathematics that, despite the digital economic bloom of the time, he enrolled, as soon as he could, for the B.Sc. in mathematics at the Hebrew University and continued to pursue his M.Sc. immediately afterwards, receiving both degrees cum laude. Convinced that the distinction between pure and applied mathematics is illusory and that the abstract and the concrete form a symbiosis of endless mutual nourishment, he finds great joy in digging into the mathematical foundations of applied topics. Occasionally he finds an angle he particularly likes that he might communicate as an article in the online journal The Conversation.

A Word of Gratitude

First and foremost, the authors extend their gratitude to Prof. Adriano Tomassini and to Dr. Daniele Angella for their contribution to the Solved Problems chapter. The authors wish to thank Rahel Berman for her numerous stylistic suggestions and to Alveen Chand for useful comments that improved the Preliminaries chapter. The second named author wishes to thank Michał Gnacik for interesting discussions and suggestions and Levana Cohen for being a very engaging student.

Parma, Italy   Carlo Alabiso
Portsmouth, UK   Ittay Weiss

About This Book

This textbook is a treatment of the structure of abstract spaces, in particular, linear, topological, metric and normed spaces, as well as topological groups, in a rigorous and reader-friendly fashion. The assumed background knowledge on the part of the reader is modest, limited to basic concepts of finite-dimensional linear spaces and elementary analysis. The book's aim is to serve as an introduction toward the theory of Hilbert spaces.

The formalism of Hilbert spaces is fundamental to Physics and in particular to Quantum Mechanics, requiring a certain amount of fluency with the techniques of linear algebra, metric space theory, and topology. Typical introductory-level books devoted to Hilbert spaces assume a significant level of familiarity with the necessary background material and consequently present only a brief review of it. A reader who finds the overview insufficient is forced to consult other sources to fill the gap. Assuming only a rudimentary understanding of real analysis and linear algebra, this book offers an introduction to the mathematical prerequisites of Hilbert space theory in a single self-contained source. The text is suitable for advanced undergraduate or introductory graduate courses for both Physics and Mathematics students.

Contents

List of Symbols

$\mathbb{N}$	The natural numbers
$\mathbb{Z}$	The integer numbers
$\mathbb{Q}$	The rational numbers
$\mathbb{R}$	The real numbers
$\mathbb{C}$	The complex numbers
id	Identity function
id_X	Identity function on the set X
I	Either the identity operator on a linear space or an interval
I_n	The $n \times n$ identity matrix
K^B	Set of all functions from B to K
$|X|$	Cardinality of a set X
$\mathcal{P}(X)$	Power-set of the set X
$\in$	Set memberhood
$\subseteq$	Set inclusion
$\subset$	Proper set inclusion
$\cup$	Union of sets
$\cap$	Intersection of sets
$X - Y$	The relative complement of the set Y in the set X
$\sim$	Equivalence relation
$\sim_X$	Equivalence relation on the set X
$X/\!\sim$	Quotient set/space
$\cong$	Indicates isomorphic, isometric, or homeomorphic spaces
$\times$	The product of sets, linear spaces, topologies spaces, metric spaces, or Banach spaces
$\mathcal{B}$	Stands for a basis in a linear space or a topological space, or for a Banach space
$\mathcal{H}$	A Hilbert space
c	Space of convergent sequences
c_0	Space of sequences converging to 0
c_{00}	Space of eventually 0 sequences

ℓ_p	Space of absolutely p-power summable sequences
ℓ_∞	Space of bounded sequences
$C(I, \mathbb{R})$	Space of real-valued continuous functions on I
$C^k(I, \mathbb{R})$	Space of k-fold differentiable real-valued functions
$C^\infty(I, \mathbb{R})$	Space of infinitely differentiable real-valued functions
$C(I, \mathbb{C})$	Space of complex-valued continuous functions on I
$\mathbb{R}^n$	Space of n-tuples of real numbers
$\mathbb{R}^\infty$	Space of infinite sequences of real numbers
$\mathbb{C}^n$	Space of n-tuples of complex numbers
$\mathbb{C}^\infty$	Space of infinite sequences of complex numbers
$P_n, \mathbb{P}_{n+1}$	Space of polynomial functions of degree at most n
$P, \mathbb{P}$	Space of all polynomial functions
Y^X	Set of all functions from a set X to a set Y
$(K^B)_0$	Space of all functions from a set B to a field K that are almost always 0
$\oplus$	Direct sum
$\|x\|$	Norm of a vector x
$\|x\|_p$	The ℓ_p or L_p norm of x
$\|x\|_\infty$	The ℓ_∞ or L_∞ norm of x
$\langle x, y \rangle$	The inner product of x and y
$p(x)$	The semi-norm of a vector x
$\mathrm{Ker}(T)$	Kernel of a linear operator T
$\mathrm{Im}(T)$	Image of a linear operator T
$\dim(V)$	Dimension of a linear space V
τ	A topology
τ_X	A topology on the set X
$\mathbb{S}$	Sierpinski space
$\bar{S}$	Closure of a subset
$\partial(S)$	Boundary of a subset
$\mathrm{int}(S)$	Interior of a subset
$\mathrm{ext}(S)$	Exterior of subset
$B_\varepsilon(x)$	Open ball of radius ε and center x
$\bar{B}_\varepsilon(x)$	Closed ball of radius ε and center x
$\mathbb{R}_+$	The extended non-negative real numbers
d	Distance function in a metric space
d_X	Distance function on the set X
$\mathbf{B}(V, W)$	The set of bounded linear operators from V to W
$\mathbf{B}(\mathcal{B})$	The set of bounded linear operators on $\mathcal{B}$
V^*	The dual space of V
e	The identity element in a group

Chapter 1
Introduction and Preliminaries

1.1 Hilbert Space Theory—A Quick Overview

Hilbert space theory is the mathematical formalism for non-relativistic Quantum Mechanics. A Hilbert space is a linear space together with an inner product that endows the space with the structure of a complete metric space. Hilbert space theory is thus a fusion of algebra, topology, and geometry.

The reader is already familiar with the intermingling of algebra and geometry, namely in the linear space $\mathbb{R}^n$. Elements in $\mathbb{R}^n$ can be thought of as points in n-dimensional space but also as vectors. Typically, points have coordinates and vectors can be added and scaled. Moreover, in the presence of the standard inner product, given by $\langle x, y \rangle = \sum_{k=1}^{n} x_k y_k$, the length of a vector is given by the norm

$$\|x\| = \sqrt{\langle x, x \rangle}$$

and the angle between vectors can be computed by means of the formula

$$\theta = \arccos \frac{\langle x, y \rangle}{\|x\| \|y\|}.$$

At this level, the interaction between the algebra and the geometry is quite smooth. Things change though as soon as one considers infinite dimensional linear spaces, and this is also where topology comes into the picture. For finite dimensional linear spaces, bases (and thus coordinates) are readily available, all linear operators are continuous, convergence of operators has a single meaning, any linear space is naturally isomorphic to its double dual, and the closed unit ball is compact. These conveniences do not exist in the infinite-dimensional case. While bases do exist, the proof of their existence is non-constructive and more often than not no basis can be given explicitly. Consequently, techniques that rely on coordinates and matrices are generally unsuitable. Linear operators need not be continuous, and in fact many linear operators of interest are not continuous. The space consisting of all linear

© Springer Nature Switzerland AG 2021
C. Alabiso and I. Weiss, *A Primer on Hilbert Space Theory*, UNITEXT for Physics,
https://doi.org/10.1007/978-3-030-67417-5_1

operators between two linear spaces carries two different topologies, and thus two different notions of convergence. The correct notion of the dual space is that of all continuous linear operators into the ground field and even then the original space only embeds in its double dual. Lastly, the closed unit ball is not compact.

In a sense, the aim of Hilbert space theory is to develop adequate machinery to be able to reason about infinite-dimensional linear spaces subject to these inherent difficulties and topological subtleties. The mathematics background required for the study of Hilbert spaces thus includes linear algebra, topology, the theory of metric spaces, and the theory of normed spaces. The theory of topological groups, a fusion between group theory and topology, arises naturally as well. Before we proceed to elaborate on each of these topics we visit the most fundamental space of them all, that of the real numbers.

1.1.1 The Real Numbers—Where it All Begins

Most fundamental to human observation of the external world are the real numbers. The outcome of a measurement is almost always assumed to be a real number (at least in some ideal sense). That this idealization is deeply ingrained in the scientific lore is witnessed by the very name of the *real* numbers; among all numbers, these are the real ones. The reasons for science's strong preference for the real numbers is a debate for philosophers. We merely observe, whether the use of real numbers is justified or not, that the success of the real numbers in science is unquestionable.

Mathematically though, the real numbers pose several non-trivial challenges. One of those challenges is the very definition of the real numbers, or, in other words, answering the question: what is a real number? For the ancient Greeks, roughly speaking until the Pythagoreans, the real numbers were taken for granted as being the same as what we would today call the rationals. That was the commonly held belief until the discovery that $\sqrt{2}$ is an irrational number (the precise circumstances of that discovery are unclear). The construction of precise models for the reals had to wait numerous centuries, and so did the discovery of transcendental numbers and the understanding of just how many real numbers there are.

1.1.1.1 Constructing the Real Numbers

The reader is likely to have her own personal idea(lization) regarding what the real numbers really are. Any precise interpretation or elucidation of the true entities which are the real numbers belongs to philosophy. Choosing to stay firmly afoot in mathematics, we present one of many precise constructions of a model of the real numbers. As with the construction of any model, one can not construct something from nothing. We thus assume the reader accepts the existence of (a model of) the rational numbers. In other words, we pretend to agree on what the rational numbers are, but to have no knowledge of the real number system. We will give the precise construction but, to remain in the spirit of an introduction, we avoid the details of the proofs.

To motivate the following construction of the real numbers from the rational numbers, recall that the rational numbers are dense in the real numbers, meaning that between any two real numbers there exists a rational number. This simple observation gives rise to the crucial fact that every real number is the limit of a sequence of rational numbers. Thus, the convergent sequences of rational numbers give us access to all of the real numbers. Since many different sequences of rational numbers may converge to the same real number, the set of all convergent sequences of rational numbers is too wasteful to be taken as the definition of the real numbers. One needs to introduce an equivalence relation on it, one that identifies two sequences of rational numbers if they converge to the same real number.

If S denotes the set of all convergent sequences of rational numbers, then we expect to be able to identify an equivalence relation $\sim$ on S such that $S/\sim$ forms a model of the real numbers. However, to even take the first step in this plan without committing the sin of a circular argument, one must first be able to identify the convergent sequences among all sequences of rational numbers without any a-priori knowledge of the real numbers. That task is achieved by appealing to the familiar notion of a Cauchy sequence, a condition which, for sequences of real numbers, is well known to be equivalent to convergence. The key observation is that the Cauchy condition for sequences of rational numbers can be stated without mentioning real numbers at all. In this way the convergent sequences of rational numbers are carved from the set of all sequences of rationals by means of an inherently rational criterion.

The construction we give is due to Georg Cantor, the creator of the theory of sets. On the set $\mathbb{Q}$ of rational numbers, which we assume is endowed with the familiar notions of addition and multiplication, and thus also with the notions of subtraction and division, consider the function $d : \mathbb{Q} \times \mathbb{Q} \to \mathbb{Q}$ given by $d(x, y) = |x - y|$. We define a Cauchy sequence to be a sequence (x_n) of rational numbers which satisfies the condition that for every rational number $\varepsilon > 0$ there exists $N \in \mathbb{N}$ such that $d(x_n, x_m) < \varepsilon$ for all $n, m > N$. Let now S be the set of all Cauchy sequences and define the relation $\sim$ on S as follows. Given Cauchy sequences $(x_n), (y_n) \in S$, declare that $(x_n) \sim (y_n)$ if $d(x_n, y_n) \to 0$. Note that convergence to 0 (or any other rational number) of a sequence of rational numbers can be stated without assuming the existence of real numbers. Indeed, the meaning of $(x_n) \sim (y_n)$ is that for every rational number $\varepsilon > 0$ there exists an $N \in \mathbb{N}$ such that $d(x_n, y_n) < \varepsilon$ for all $n > N$.

It is a straightforward matter to show that $\sim$ is an equivalence relation on S. We now define the set of real numbers to be $\mathbb{R} = S/\sim$, namely the quotient set determined by the equivalence relation. One immediate thing to notice is that $\mathbb{R}$ contains a copy of $\mathbb{Q}$. Indeed, for any rational number $q \in \mathbb{Q}$ consider the constant sequence $x_q = (x_n = q)$. It is clearly a Cauchy sequence, and thus $x_q \in S$, and its equivalence class $[x_q]$ is thus, per definition, a real number. Further, if $q' \neq q$ is another rational, then $[x_q] \neq [x_{q'}]$ since the respective constant sequences are not equivalent, and thus not identified in the quotient set. In other words, the function $q \mapsto [x_q]$ is an injection from $\mathbb{Q}$ to $\mathbb{R}$.

The familiar algebraic structure of $\mathbb{Q}$ carries over to $\mathbb{R}$ as follows. Given real numbers x, y, represented by Cauchy sequences (x_n) and (y_n) respectively, the sum $x + y$ is defined to be represented by $(x_n + y_n)$ and the product xy to be represented by $(x_n y_n)$. Of course, one needs to check these are well-defined notions, i.e., that the proposed sequences are Cauchy and that the resulting sum and product are independent of the choice of representatives. This verification, especially for the sum, is straightforward. Once that is done, the verification of the familiar algebraic properties of addition and multiplication is easily performed, proving that $\mathbb{R}$ is a field. Similarly, the order structure of $\mathbb{Q}$ carries over to $\mathbb{R}$ by defining, for real numbers x and y as above, that $x < y$ if there exists a rational number $\varepsilon > 0$ such that $x_n \leq y_n - \varepsilon$ for all but finitely many n. It is quite straightforward to verify that $\mathbb{R}$ is then an ordered field. Finally, and by far least trivially, $\mathbb{R}$ satisfies the least upper bound property, proving that it is Dedekind complete.

The embedding of $\mathbb{Q}$ in $\mathbb{R}$ described above is easily seen to respect the order and algebraic structure of $\mathbb{Q}$. Therefore the field $\mathbb{Q}$ of rational numbers can be identified with its isomorphic copy in $\mathbb{R}$, namely the image of the embedding $\mathbb{Q} \to \mathbb{R}$, and one then considers $\mathbb{R}$ as an extension of $\mathbb{Q}$.

1.1.1.2 Transcendentals and Uncountability

The divide of the real numbers into rational and irrational numbers represents the first fundamental indication of the complexity of the real numbers. Together with the realization that $\sqrt{2}$ is an irrational number comes the necessity, if only out of curiosity, of asking about the nature of other real numbers, for instance π and e. The number e was introduced by Jacob Bernoulli in 1683 but it was only some 50 years later that Leonard Euler established its irrationality. In sharp contrast, the number π, known since antiquity, was only shown to be irrational in 1761 by Johann Heinrich Lambert. If this is not indication enough that the real number system holds more secrets than one would superficially expect, the fact that to date it is unknown whether or not $\pi + e$ is irrational should remove any doubt.

Some real numbers, while irrational, may still be nearly rational, in the following sense. Consider a rational number $r = p/q$. Clearly, r satisfies the equation

$$qt - p = 0,$$

and thus r is a root of a polynomial with integer coefficients. More generally, any real number that is a root of a polynomial with integer coefficients is called an algebraic number. For instance, $\sqrt{2}$ is algebraic since it is a root of the polynomial $t^2 - 2$. A real number that is not algebraic is called transcendental. The existence of transcendental numbers was not established until 1844 by Joseph Liouville who, shortly afterwards, constructed the first explicit example of a transcendental number (albeit a somewhat artificial one). More prominent members of the transcendental class are the numbers e, proved transcendental by Charles Hermite in 1873, and the number π, proved transcendental by Ferdinand von Lindemann in 1882. While much is known about transcendental numbers, it is still unknown if $\pi + e$ is algebraic or not.

In 1884 Georg Cantor, following shortly after Liouville's proof of the existence of transcendental numbers, proved not only that transcendental numbers exist but that in some sense the vast majority of all real numbers are transcendental. The technique employed by Cantor was that of transfinite counting. With his notion of cardinality of sets, Cantor counted how many algebraic numbers there are and how many real numbers there are, and demonstrated that there are strictly more real numbers than algebraic ones. The inevitable conclusion that transcendental numbers exist was met with some resistance with objections to the set-theoretic proof technique.

We end this short historical excursion and turn to quickly address the main topics covered in this book.

1.1.2 Linear Spaces

A linear space, also known as a vector space, embodies what is perhaps the simplest notion of a mathematical space. When thought of as modeling actual physical space, a linear space appears to be the same in all directions and is completely devoid of any curvature. Linear spaces, and the closely related notion of linear transformation or linear operator, are fundamental objects. For instance, the derivative of a function at a point is best understood as a linear operator on tangent spaces, particularly for functions of several variables. Differentiable manifolds are spaces which may be very complicated but, at each point, they are locally linear. It is perhaps somewhat of an exaggeration, but it is famously held to be true that if a problem can be linearized, then it is as good as solved.

1.1.3 Topological Spaces

Topology is easy to define but hard to explain. The ideas that led to the development of topology were lurking beneath the surface for some time and it is difficult to pinpoint the exact time in history when topology was born. What is clear is that after its birth the advancement of topology in the first half of the 20th century was rapid. The unifying power of topology is immense and its explanatory ability is a powerful one. For instance, familiar theorems of first year real analysis, such as the existence of maxima and minima for continuous functions on a closed interval, or the uniform continuity of a continuous function on a closed interval, may make the closed interval appear to have special significance. Topology is able to clarify the situation, identifying a particular topological property of the closed interval, namely that it is compact, as the key ingredient enabling the proof to carry over into much more general situations.

1.1.4   Metric Spaces

Metric spaces appeared in 1906 in the PhD dissertation of Maurice Fréchet. A metric space is a set where one can measure distances between points, and in a strong sense a metric space carries much more geometric information than can be described by a topological space. With stronger axioms come stronger theorems but also less examples. However, the axioms of a metric space allow for a vast and varied array of examples and the theorems one can prove are very strong. In particular, complete metric spaces, i.e., those that, intuitively, have no holes, admit two very strong theorems. One is the Banach Fixed Point Theorem and another is Baire's Theorem. The former can be used to solve, among other things, differential equations, while the latter has deep consequences to the structure of complete metric spaces and to continuous functions.

1.1.5   Classical Sequence and Function Spaces

Spaces of infinite sequences and spaces of functions are of great importance in modern analysis. They arise often in applications and provide a rich source of problems both in historical and contemporary contexts. A rigorous presentation of the sequence spaces is certainly within grasp with just a modest amount of undergraduate analysis. Their elementary structure can be appreciated as well without too much difficulty. However, the function spaces are well known to present a far greater challenge. The approach taken in the chapter presents the families of sequence and function spaces as metric completions using ideas from category theory in order to extract a strong similarity between the two families of spaces. This approach allows the reader to export understanding of the technically simpler sequence spaces to the more elusive function spaces.

1.1.6   Banach Spaces

A very basic attribute one can associate with a vector is its norm, i.e., its length. The abstract formalism is given by the axioms of a normed space. The presence of a norm allows one to define the distance between any two vectors, by the so-called induced metric, and one obtains a metric space, as well as the topology it induces. Thus any normed space immediately incorporates both algebra and geometry. A Banach space is a normed space which, as a metric space, is complete. The interaction between the algebra and the geometry is then particularly powerful, allowing for very strong results.

1.1.7 Topological Groups

Banach spaces are a fusion of algebra and geometry. Similarly, topological groups are a fusion between algebra and topology. Unlike a Banach space, the algebraic structure is that of a group while the geometry is reduced from a metric to a topology. Thus, a topological group is a much weaker structure than a Banach space, yet the interaction between the algebra and the topology still gives rise to a very rich and interesting theory.

1.1.8 Hilbert Spaces

In a sense the whole book is geared toward this chapter. The content is a discussion of the closest point property, orthogonal complements, bases, Fourier series, and the Riesz representation theorem. The location of this chapter as the final one, after the rudiments of Banach space theory, allows for a presentation that relates to broader concepts and illustrates the differences between Banach spaces and Hilbert spaces.

1.2 Preliminaries

The aim of this section is to recount some of the fundamental notions of set theory, the common language for rigorous mathematical discussions. It serves to quickly orient the reader with respect to the notation used throughout the book as well as to present fundamental results on cardinals and demonstrate the technique of Zorn's Lemma. This section is designed to be skimmed through and only consulted for a more detailed reading if needed further on. For that reason, and unlike the rest of the book, the style of presentation in this section is succinct.

1.2.1 Sets

The notion of a *set* is taken as a primitive notion, left undefined. In an axiomatic approach to sets one lists certain well formulated and carefully chosen axioms, while in a naive approach one relies on a common understanding of what sets are, avoiding the technical difficulties of precise definitions at the cost of some rigor. We adopt the naive approach and thus for us a set is a *collection* of *elements*, where no repetitions are allowed and no ordering of the elements of the set is assumed.

1.2.2 Common Sets

Among the sets we will encounter are the sets $\mathbb{N} = \{1, 2, 3, \ldots\}$ of natural numbers, $\mathbb{Z} = \{\ldots, -3, -2, -1, 0, 1, 2, 3, \ldots\}$ of integers, $\mathbb{Q} = \{p/q \mid p, q \in \mathbb{Z}, q \neq 0\}$ of rational numbers, $\mathbb{R}$ of real numbers, and $\mathbb{C} = \{x + iy \mid x, y \in \mathbb{R}\}$ of complex numbers. The notation illustrates two commonly occurring ways for introducing sets, the informal ... which implies that the reader knows what the author had in mind, and the precise form $\{x \in X \mid P(x)\}$, to be interpreted as the set of all those elements x in X satisfying property P. The *empty set* is denoted by $\emptyset$, and it simply has no elements in it. A set $\{x\}$, consisting of just a single element, is called a *singleton* set.

1.2.3 Relations Between Sets

The notation $x \in X$ was already used above to indicate that x is a *member* of the set X. To indicate that x is not a member of X we write $x \notin X$. Two sets X and Y may have the property that any element $x \in X$ is also a member of Y, a situation denoted by $X \subseteq Y$, meaning that X is *contained* in Y, X is a *subset* of Y, Y *contains* X, and that Y is a *superset* of X. If, moreover, $X \neq Y$, then X is said to be a *proper subset* of Y, denoted by $X \subset Y$. In particular, $X \subset Y$ means that every element $x \in X$ satisfies $x \in Y$ and that for at least one element $y \in Y$ it holds that $y \notin X$. An often useful observation is that two sets X and Y are equal, denoted by $X = Y$, if, and only if, $X \subseteq Y$ and $Y \subseteq X$.

1.2.4 Families of Sets; Union and Intersection

Any set I may serve as an indexing set for a set of sets. In that case it is customary to speak of a *family of sets*, a *collection of sets*, or an *indexed set*. For instance, consider $I = [0, 1] = \{t \in \mathbb{R} \mid 0 \leq t \leq 1\}$, and for each $t \in I$ let $X_t = (0, t) = \{x \in \mathbb{R} \mid 0 < x < t\}$. Then $\{X_t\}_{t \in I}$ is a family of sets indexed by I. If $X_i \cap X_j = \emptyset$ for all $i, j \in I$ with $i \neq j$, then the collection is said to be *pairwise disjoint*. For any indexed collection, its *union* is the set

$$\bigcup_{i \in I} X_i$$

which is the set of all those elements x for which there exists an index $i \in I$ (which may depend on x) such that $x \in X_i$. The *intersection* of the collection is the set

$$\bigcap_{i \in I} X_i$$

which is the set of all those elements x that belong to X_i for all $i \in I$. For the indexed collection given above one may verify that

$$\bigcup_{t \in I} X_t = (0, 1)$$

and that

$$\bigcap_{t \in I} X_t = \emptyset.$$

When the indexing set I is finite, one typically takes $I = \{1, 2, \ldots, n\}$ and writes

$$X_1 \cup X_2 \cup \cdots \cup X_n$$

and

$$X_1 \cap X_2 \cap \cdots \cap X_n$$

for the union and intersection, respectively.

1.2.5 Set Difference, Complementation, and De Morgan's Laws

The *set difference* $X \setminus Y$ is the set $\{x \in X \mid x \notin Y\}$. In the presence of a *universal set* or a *set of discourse*, that is a set U which is the relevant ambient set for a given context, the *complement* of a set $X \subseteq U$ is the set $X^c = \{y \in U \mid y \notin X\}$. In other words, $X^c = U \setminus X$. Needless to say, the notation X^c pre-supposes knowledge of U, and different choices of U yield different complements. In this context we mention that the set difference $X \setminus Y$ is also known as the *relative complement* of Y in X.

Among the many properties of sets and the operations of union, intersection, and complementation we only mention *De Morgan's laws*: Given any indexed collection $\{X_i\}_{i \in I}$ of subsets of a set X, the equalities

$$\left(\bigcup_{i \in I} X_i \right)^c = \bigcap_{i \in I} X_i^c$$

and

$$\left(\bigcap_{i \in I} X_i \right)^c = \bigcup_{i \in I} X_i^c$$

both hold. Here the complements are relative to the set X.

1.2.6   Finite Cartesian Products

Given m sets $(m \geq 1)$, $X_1, \ldots, X_m$, their *cartesian product* is the set

$$X_1 \times \cdots \times X_m = \{(x_1, \ldots, x_m) \mid x_k \in X_k, \text{ for all } 1 \leq k \leq m\}.$$

In other words, the cartesian product of m sets $X_1, \ldots, X_m$ (in that order) is the set consisting of all m-tuples where the k-th component (for $1 \leq k \leq m$) is taken from the k-th set. In case $X_k = X$ for all $1 \leq k \leq m$, the cartesian product is denoted by X^m and is called the *m-fold cartesian product* of X with itself.

From a set-theoretic perspective cartesian products present some subtleties involving the issue of the definition itself and the question of coherence with regard to repeated application of cartesian products. These aspects are investigated in the exercises.

1.2.7   Functions

Given any two sets X and Y, a *function* or a *mapping*, denoted by $f : X \to Y$ or $X \xrightarrow{f} Y$ is, informally, a rule that associates with each element $x \in X$ an element $f(x) \in Y$. More formally, a function $f : X \to Y$ is a subset $f \subseteq X \times Y$ satisfying the property that for all $x \in X$ and $y_1, y_2 \in Y$, if $(x, y_1) \in f$ and $(x, y_2) \in f$, then $y_1 = y_2$, and that for all $x \in X$ there exists some $y \in Y$ with $(x, y) \in f$. The set X is the *domain* of the function while Y is the *codomain*. For any set X there is always the *identity function* $\mathrm{id}_X : X \to X$ given by $\mathrm{id}_X(x) = x$. Further, given functions $f : X \to Y$ and $g : Y \to Z$, their *composition* is defined to be the function $g \circ f : X \to Z$ given by $(g \circ f)(x) = g(f(x))$. One easily sees that composition of functions is associative, that is, if

$$X_1 \xrightarrow{f} X_2 \xrightarrow{g} X_3 \xrightarrow{h} X_4,$$

then

$$h \circ (g \circ f) = (h \circ g) \circ f.$$

A function $f : X \to Y$ is *injective*, or *one-to-one*, if $f(x_1) = f(x_2)$ implies $x_1 = x_2$, for all $x_1, x_2 \in X$. A function $f : X \to Y$ is *surjective*, or *onto*, if for all $y \in Y$ there exists an $x \in X$ such that $f(x) = y$. A function $f : X \to Y$ is called *bijective* if it is both injective and surjective. A function $f : X \to Y$ is *invertible* if there exists a function $f^{-1} : Y \to X$, called the *inverse* of f, such that $f \circ f^{-1} = \mathrm{id}_Y$ and $f^{-1} \circ f = \mathrm{id}_X$. Such an inverse, if it exists, is easily seen to be unique. Lastly, we note that a function f is invertible if, and only if, it is bijective.

Given a function $f : X \to Y$ and a subset $S \subseteq X$, the *restriction* of f to S is the function $g : S \to Y$ given by $g(s) = f(s)$. This restriction function is denoted by $f|_S$. We also mention that any two sets X and Y with $X \subseteq Y$ give rise to the *inclusion function* $\iota : X \to Y$, defined by $\iota(x) = x$. Note that the inclusion function differs from the identity function only by the codomain, not by the functional assignments. The set of all functions $f : X \to Y$ is denoted by Y^X.

1.2.8 Arbitrary Cartesian Products

The *cartesian product* of any indexed family $\{X_i\}_{i \in I}$ of sets is the set

$$\{f : I \to \bigcup_{i \in I} X_i \mid f(i) \in X_i, \forall i \in I\}.$$

To justify the definition, observe that when $I = \{1, 2, 3, \ldots, m\}$, each function f as above can be uniquely identified with the m-tuple $(f(1), f(2), \ldots, f(m))$, in other words, with an element of $X_1 \times \cdots \times X_m$, so this definition recovers the definition of the cartesian product of finitely many sets as given above. When $X_i = X$ for all $i \in I$, the cartesian product is denoted by X^I. Notice that X^I is simply the set of all functions $f : I \to X$.

1.2.9 Direct and Inverse Images

The *power set* of a set X is the set of all subsets of X, and is denoted by $\mathcal{P}(X)$. Every function $f : X \to Y$ induces the *direct image* function $f : \mathcal{P}(X) \to \mathcal{P}(Y)$, denoted again by f, given by $f(A) = \{f(a) \mid a \in A\}$, for all $A \subseteq X$ (notice that this definition introduces a new way to construct sets). Similarly, any function $f : X \to Y$ induces the *inverse image* function $f^{-1} : \mathcal{P}(Y) \to \mathcal{P}(X)$, which, for every $B \subseteq Y$, is given by $f^{-1}(B) = \{x \in X \mid f(x) \in B\}$. Notice that $f^{-1}(Y) = X$ but that the inclusion $f(X) \subseteq Y$ may be proper. The set $f(X)$ is called the *image* of f. Note that $f(X) = Y$ if, and only if, f is surjective. We remark that the notation f^{-1} for the inverse image conflicts with the notation for the inverse function of f (when it exists) and thus some care should be exercised. At times, the notation $f_\to$ is used for the direct image and $f^\leftarrow$ for the inverse image function.

The following properties of functions are easily verified. For all functions

$$X \xrightarrow{f} Y \xrightarrow{g} Z$$ and $S \subseteq Z$ a subset, we have that

$$(g \circ f)^{-1}(S) = f^{-1}(g^{-1}(S)).$$

Further, for all subsets $S_1, S_2 \subseteq Y$,

$$f^{-1}(S_1 \cap S_2) = f^{-1}(S_1) \cap f^{-1}(S_2)$$

and

$$f^{-1}(S_1 \cup S_2) = f^{-1}(S_1) \cup f^{-1}(S_2).$$

More generally, given any collection $\{S_i\}_{i \in I}$ of subsets of Y,

$$f^{-1}\left(\bigcap_{i \in I} S_i\right) = \bigcap_{i \in I} f^{-1}(S_i)$$

and

$$f^{-1}\left(\bigcup_{i \in I} S_i\right) = \bigcup_{i \in I} f^{-1}(S_i).$$

If the domain X of a function contains an element x that is also a subset of X, namely both $x \in X$ and $x \subseteq X$ hold, then the meaning of $f(x)$ is ambiguous; it may refer to the value of f at x or to the image of the subset x under f. This unfortunate scenario results from overloading the notation f to mean the function and its direct image.

When considering the inverse image of a singleton set $\{y\}$, namely $f^{-1}(\{y\})$, it is convenient to shorten the notation and write $f^{-1}(y)$, remembering that the outcome is a subset of the domain that may well be empty.

1.2.10   Indicator Functions

Let $\mathbb{S}$ be the set $\{0, 1\}$ and fix some set X. The subsets of X are easily seen to be in bijective correspondence with the functions $X \to \mathbb{S}$. The correspondence is given as follows. Given a subset $S \subseteq X$ let $F(S) : X \to \mathbb{S}$ be given by

$$F(S)(x) = \begin{cases} 1 & \text{if } x \in S \\ 0 & \text{if } x \notin S. \end{cases}$$

The function $F(S)$ is called the *indicator function* of the subset $S \subseteq X$. Conversely, given any function $f : X \to \mathbb{S}$, the subset of X it determines is $G(f) = f^{-1}(\{1\})$. It is easily verified that F and G are each other's inverses, setting up the stated bijective correspondence. Note that $f^{-1}(\{0\})$ is the complement $X \setminus f^{-1}(\{1\})$.

1.2.11 Cardinality

Two sets X and Y are said to have the same *cardinality* if there exists a bijection $f : X \to Y$, in which case we write $|X| = |Y|$. If there is an injection $f : X \to Y$, then X is said to have cardinality less than or equal to that of Y, and we write $|X| \leq |Y|$.

Two finite sets have the same cardinality if, and only if, they have the same number of elements. It is thus common to write $|X|$ for a finite set X to indicate the number of elements in X.

A set X is said to be *countable* if it is empty or if there exists a surjective function $f : \mathbb{N} \to X$. Equivalently, a set is countable if it has the same cardinality as some subset of $\mathbb{N}$. Equivalently still, a set X is countable precisely when there exists an injective function $X \to \mathbb{N}$. A set that is both countable and infinite is said to be *countably infinite*. A set that is not countable is called *uncountable*. A set which has the same cardinality as $\mathbb{R}$ is said to have the cardinality of the *continuum*. Examples of countable sets include every finite set, the set $\mathbb{N}$, as well as the set $\mathbb{Z}$ of integers and the set $\mathbb{Q}$ of rational numbers. Sets of the cardinality of the continuum include the set $\mathbb{R}$ of real numbers as well as $\mathbb{C}$ of complex numbers, the n-fold products $\mathbb{R}^n$ and $\mathbb{C}^n$ for any $n \in \mathbb{N}$, as well as the power set $\mathcal{P}(\mathbb{N})$, namely the set of all subsets of natural numbers, and the set of all continuous functions $f : \mathbb{R} \to \mathbb{R}$. Uncountable sets of cardinality larger than that of the continuum include the power set $\mathcal{P}(\mathbb{R})$ of all subsets of real numbers, as well as the set of all functions $f : \mathbb{R} \to \mathbb{R}$. Establishing such cardinality claims requires set-theoretic tools of which the one presented next is of fundamental use.

1.2.12 The Cantor-Shröder-Bernstein Theorem

The result we present now is a very convenient tool in establishing that two sets have the same cardinality.

Theorem 1.1 (Cantor-Shröder-Bernstein) *For all sets X and Y, if $|X| \leq |Y|$ and $|Y| \leq |X|$, then $|X| = |Y|$.*

Proof By the condition in the assertion, there exists an injective function $f : X \to Y$ and an injective function $g : Y \to X$. To construct a bijection $h : X \to Y$, we consider the behaviour of elements in both X and Y with respect to the given functions f and g. Notice that since g is injective, the inverse image $g^{-1}(x_0)$, for any $x_0 \in X$, is either empty or is of the form $\{y_0\}$, where $g(y_0) = x_0$. Similarly, $f^{-1}(y_1)$, for any $y_1 \in Y$, is either empty or is of the form $\{x_1\}$ with $f(x_1) = y_1$. We thus write $f^{-1}(y_1) = x_1$ and $g^{-1}(x_0) = y_0$, when these sets are not empty. Starting with $x_0 \in X$ we may successively consider the sequence

$$x_0 \mapsto g^{-1}(x_0) = y_0 \mapsto f^{-1}(y_0) = x_1 \mapsto g^{-1}(x_1) = y_1 \mapsto \cdots$$

where at each step we apply f^{-1} or g^{-1} alternatively, if they exist. Such a sequence exhibits precisely one of three possible behaviours. It may be infinite, in which case x_0 is said to have type ∞, or it may terminate since $g^{-1}(x_k)$ was empty, in which case x_0 is said to have type X, or the sequence may stop due to the fact that $f^{-1}(y_k)$ was empty, in which case x_0 is said to have type Y. In exactly the same way, each element of Y can be classified as having precisely one type, either type ∞, or type X or type Y.

Let us now denote by X_∞ all type ∞ elements in X, by X_X all type X elements in X, and by X_Y all type Y elements in X. Similarly, introduce the notation Y_∞, Y_X, and Y_Y. Thus

$$X = X_\infty \cup X_X \cup X_Y$$

and the union is pairwise disjoint. Similarly we obtain the pairwise disjoint union

$$Y = Y_\infty \cup Y_X \cup Y_Y.$$

The function given by

$$h(x) = \begin{cases} f(x) & \text{if } x \in X_X \\ g^{-1}(x) & \text{if } x \in X_Y \\ f(x) & \text{if } x \in X_\infty \end{cases}$$

is now easily seen to be a bijection, for instance by constructing an inverse function for it. $\square$

For two non-trivial intervals (a, b) and (c, d) in $\mathbb{R}$ the function $f(x) = \frac{d-c}{b-a} \cdot (x - a) + c$ is a bijection from (a, b) to (c, d). Thus all open intervals in $\mathbb{R}$ have the same cardinality. Other explicit bijections are provided by $\arctan : (-\pi/2, \pi/2) \to \mathbb{R}$ and $\exp : \mathbb{R} \to (0, \infty)$, showing that $|\mathbb{R}| = |(-\pi/2, \pi/2)| = |(0, \infty)|$. Generally, finding explicit bijections can be very difficult and then the Cantor-Shröder-Bernstein Theorem is handy. For instance, for the intervals $(0, 1)$ and $[0, 1]$ the inclusion $x \mapsto x$ is an injection from $(0, 1) \to [0, 1]$ while $x \mapsto (x + 1)/4$ is an injection in the other direction. It follows that $|(0, 1)| = |[0, 1]|$ but an explicit bijection is illusive (and certainly no continuous bijection exists).

1.2.13 Countable Arithmetic

Some aspects of the arithmetic of the natural numbers is a reflection of properties of finite sets. Passing to infinite sets quickly leads to axiomatic subtleties. In this section we present useful results pertaining to the arithmetic of countable sets where the axiomatic difficulties are less pronounce. The countable case suffices for most applications in the context of this book.

Theorem 1.2 *A subset of a countable set is countable.*

Proof Suppose a set X is countable and $S \subseteq X$ is a subset. If $S = \emptyset$, then it is countable, so we may assume $S \neq \emptyset$ and let us fix some $s_0 \in S$. Since X is countable, there exists a surjection $f : \mathbb{N} \to X$. Clearly, there is a surjection $g : X \to S$ (for instance, $g(x) = x$ if $x \in S$, and $g(x) = s_0$ if $x \notin S$) and, since the composition of surjective functions is surjective, the function $g \circ f : \mathbb{N} \to S$ is surjective, showing that S is countable.   □

Theorem 1.3 *A cartesian product of finitely many countable sets is countable.*

Proof Let $X_1, \ldots, X_m$ be countable sets and choose, for each $1 \leq k \leq m$, an injective function $f_k : X_k \to \mathbb{N}$. These functions together give rise to

$$f : X_1 \times \cdots \times X_m \to \mathbb{N}^m$$

given by

$$f(x_1, \ldots, x_m) = (f_1(x_1), \ldots, f_m(x_m)),$$

which is clearly injective. Now fix m distinct prime numbers $p_1, \ldots, p_m$ and let

$$g : \mathbb{N}^m \to \mathbb{N}$$

be given by

$$g(n_1, \ldots, n_m) = p_1^{n_1} \cdots \cdot p_m^{n_m}.$$

The fact that every natural number has an essentially unique decomposition into prime factors implies that g is injective. To conclude, the composition

$$X_1 \times \cdots \times X_m \xrightarrow{f} \mathbb{N}^m \xrightarrow{g} \mathbb{N}$$

is injective, and thus $X_1 \times \cdots \times X_m$ is countable.   □

Theorem 1.4 *A countable union of countable sets is itself countable.*

Proof Let $\{X_m\}_{m \in \mathbb{N}}$ be a countable family of countable sets. We may assume, without loss of generality, that the collection is pairwise disjoint and that $X_m \neq \emptyset$ for all $m \in \mathbb{N}$. Since each X_m is countable, there exists a surjective function $f_m : \mathbb{N} \to X_m$. Let

$$X = \bigcup_{m \in \mathbb{N}} X_m$$

and define now the function

$$g : \mathbb{N} \times \mathbb{N} \to X$$

by

$$g(m, n) = f_m(n),$$

which is clearly surjective. To finish the proof, note that $\mathbb{N} \times \mathbb{N}$, being the product of two countable sets, is countable, and thus there is a surjection $h : \mathbb{N} \to \mathbb{N} \times \mathbb{N}$. The composition

$$\mathbb{N} \xrightarrow{\ h\ } \mathbb{N} \times \mathbb{N} \xrightarrow{\ f\ } X$$

is thus a surjection, and so X is countable. $\square$

We may now conclude, for instance, that the set $\mathbb{Z}$ of integers is countable since $\mathbb{Z} = \{0, 1, 2, 3, \dots\} \cup \{-1, -2, -3, \dots\}$, a union of two countable sets. The set $\mathbb{Z} \times \mathbb{Z}$ is thus countable as well. The set $\mathbb{Q}$ of rational numbers is also countable since fixing any choice for writing each rational number as a ratio of two integers yields an injection $\mathbb{Q} \to \mathbb{Z} \times Z$. It now follows that $\mathbb{Q}^n$, for any $n \geq 1$, is countable as well.

1.2.14   Relations

A *relation* R on a set X is a subset $R \subseteq X \times X$, but we write $x R y$ instead of $(x, y) \in R$. The relation R is said to be *reflexive* if $x R x$ holds for all $x \in X$. The relation R is *symmetric* if $x R y$ implies $y R x$, for all $x, y \in X$. The relation R is *anti-symmetric* if, for all $x, y \in X$, the assertions $x R y$ and $y R x$ together imply that $x = y$. Finally, the relation R is said to be *transitive* if $x R y$ and $y R z$ together imply $x R z$, for all $x, y, z \in X$.

1.2.15   Equivalence Relations

A relation R on a set X is called an *equivalence relation* if it is reflexive, symmetric, and transitive, in which case we denote R by $\sim$, and write $x \sim y$ instead of $x R y$. If $x \sim y$ we say that x and y are *equivalent*. A *partition* of X is a collection $\{X_i\}_{i \in I}$ of non-empty pairwise disjoint subsets of X such that $\bigcup_{i \in I} X_i = X$.

Given a partition $\{X_i\}_{i \in I}$ of X, defining $x \sim y$ precisely when there exists an $i \in I$ with $x, y \in X_i$ is easily seen to define an equivalence relation. If we denote by $\mathrm{Par}(X)$ the set of all partitions of X and by $\mathrm{Equ}(X)$ the set of all equivalence relations on X, then the process just described defines a function $E : \mathrm{Par}(X) \to \mathrm{Equ}(X)$. Conversely, given an equivalence relation $\sim$ on X we may define, for each $x \in X$, the set

$$[x] = \{y \in X \mid x \sim y\}$$

of all those elements $y \in X$ which are equivalent to x. This set is known as the *equivalence class* of x, and x is called a *representative* of the equivalence class $[x]$ (if it is important to emphasize the relation $\sim$, then we write $[x]_\sim$). It can easily be verified that two elements $x, y \in X$ represent the same equivalence class, that is $[x] = [y]$, if, and only if, $x \sim y$. It is completely straightforward to demonstrate that the set $\{[x] \mid x \in X\}$ of all equivalence classes is a partition of X, and we thus obtain a function $P : \mathrm{Equ}(X) \to \mathrm{Par}(X)$. It is quite easy to verify that in fact P is the inverse function of E and thus we have established a bijective correspondence between equivalence relations on X and partitions of X.

Given an equivalence relation $\sim$ on a set X, the set $\{[x] \mid x \in X\}$ is denoted by $X/\sim$ and is called the *quotient set* of X *modulo* $\sim$. There is also the corresponding function $\pi : X \to X/\sim$, given by $\pi(x) = [x]$, called the *canonical projection*.

1.2.16   Ordered Sets

A *poset* (short for *partially ordered set*) is a pair $(X, \leq)$ where X is a set and $\leq$ is a reflexive, transitive, and anti-symmetric relation on X. Any one of the sets $\mathbb{N}$, $\mathbb{Z}$, $\mathbb{Q}$, and $\mathbb{R}$ has a familiar ordering making each one a poset. There is an abundance of other useful posets.

Example 1.1 Given any set X, let $\mathcal{P}(X)$ be the set of all subsets of X. Defining, for subsets $Y_1, Y_2 \subseteq X$, that $Y_1 \leq Y_2$ precisely when $Y_1 \subseteq Y_2$, is easily seen to endow $\mathcal{P}(X)$ with the structure of a poset. In this case we say that $\mathcal{P}(X)$ is ordered by *set inclusion*. More generally, if A is some collection of subsets of X, then it too can be endowed with the ordering induced by set inclusion, in precisely the same manner. A related construction is to consider the set P of all pairs (Y, f) where $Y \subseteq X$ and $f : Y \to Z$ is a function to some fixed set Z. Defining $(Y_1, f_1) \leq (Y_2, f_2)$ when $Y_1 \subseteq Y_2$ and when f_2 extends f_1 (namely, $f_2(y) = f_1(y)$ holds for all $y \in Y_1$) is again easily seen to endow P with a poset structure. Again, one may consider variants of this construction, for instance by demanding extra conditions on the sets Y or on the functions f.

In the context of a general poset, the meaning of $x < y$ is taken to be: $x \leq y$ and $x \neq y$. The relations $x > y$ and $x \geq y$ are similarly derived from the given relation $\leq$. Since not all elements x and y in a poset need be *comparable* (x and y are comparable when either $x \leq y$ or $y \leq x$), the correct interpretation of, for instance, the negation of $x \leq y$ is not that $x > y$ but rather that either x and y are incomparable, or $x > y$.

Among all posets, some of which may be quite wild, a tamer sub-variety is formed by the complete lattices.

Definition 1.1 Consider a subset $S \subseteq X$ of a poset X. The *join* of S is an element $\bigvee S \in X$ characterised by the property

$$\bigvee S \leq x \iff s \leq x \text{ for all } s \in S$$

for all $x \in X$. Similarly, the *meet* of S is an element $\bigwedge S \in X$ characterised by the property

$$x \le \bigwedge S \iff x \le s \text{ for all } s \in S$$

for all $x \in X$.

Note that by taking $x = \bigvee S$ the condition on the left is trivially satisfied and so $s \le \bigvee S$ holds for all $s \in S$. In other words, $\bigvee S$ is an upper bound of S. Conversely, if $t \in X$ is some upper bound of S, namely if $s \le t$ for all $s \in S$, then taking $x = t$ the condition on the right is satisfied and so $\bigvee S \le t$. Stated differently, $\bigvee S$ can be interpreted as the least upper bound of S. Similarly, $\bigwedge S$ can be seen as the greatest lower bound of S.

When the set S consists of just two elements, say $S = \{a, b\}$, we write $a \vee b$ and $a \wedge b$ instead of, respectively, $\bigvee\{a, b\}$ and $\bigwedge\{a, b\}$. Similarly, the join of finitely many elements is denoted by $a_1 \vee \cdots \vee a_n$ and their meet by $a_1 \wedge \cdots \wedge a_n$.

For instance, in the poset $\mathcal{P}(X)$ ordered by set inclusion given $S, T \in \mathcal{P}(X)$ the fact that, for any subset $W \subseteq X$, the inclusion $S \cup T \subseteq W$ is equivalent to the inclusions $S \subseteq W$ and $T \subseteq W$ makes the union $S \cup T$ the join of S and T in the poset $\mathcal{P}(X)$. Similarly, one verifies that the meet is given by $S \cap T$. More generally, for any collection $\mathscr{C} \subseteq \mathcal{P}(X)$ of subsets of X one has

$$\bigvee \mathscr{C} = \bigcup_{S \in \mathscr{C}} S \text{ and } \bigwedge \mathscr{C} = \bigcap_{S \in \mathscr{C}} S$$

computed in the poset $\mathcal{P}(X)$. In particular, $\mathcal{P}(X)$ admits all joins and all meets. Posets with this property deserve special attention.

Definition 1.2 A poset X in which every subset $S \subseteq X$ admits a join $\bigvee S$ and a meet $\bigwedge S$ is called a *complete lattice*.

The discussion preceding the definition shows that $\mathcal{P}(X)$, for any set X, is a complete lattice. Another naturally arising family of posets is, for each $n \ge 1$, the set D_n consisting of the positive divisors of n. Declaring $k \le m$ precisely when k divides m introduces a poset structure on D_n. The join $a \vee b$ is then the least common multiple of a and b and the meet $a \wedge b$ is their greatest common divisor. In fact, D_n admits all joins and meets and is thus a complete lattice.

In any poset X one can single out two special elements.

Definition 1.3 A *bottom element* in a poset X is an element $\bot \in X$ satisfying $\bot \le x$ for all $x \in X$. Similarly, a *top element* is an element $\top \in X$ with $x \le \top$ for all $x \in X$.

In $\mathcal{P}(X)$ the bottom element is $\bot = \emptyset$ and the top element is $\top = X$ while in D_n we have $\bot = 1$ and $\top = n$. Expectedly, top and bottom elements, joins and meets, need not exist in an arbitrary poset. For instance, $\mathbb{N}$ with its usual ordering is a poset with $\bot = 1$ but no top element, and where every non-empty subset $S \subseteq \mathbb{N}$ has a meet given by the smallest element in S, and only finite subsets S admit a join given by the largest element in S.

In the complete lattices we encountered above it was the case that top and bottom elements were present. In fact, any complete lattice has a top element and a bottom element due to the following observation.

Proposition 1.1 *The empty join $\bigvee \emptyset$ in a poset X, if it exists, is precisely a bottom element in X. Similarly, the empty meet $\bigwedge \emptyset$ in X, if it exists, is a top element.*

Proof This is an exercise in carefully following definitions and correctly establishing vacuous validity. We demonstrate that $\bigvee \emptyset$ is a bottom element and leave the rest to the reader. Assuming $\bigvee \emptyset$ exists, its defining property is that

$$\bigvee \emptyset \leq x \iff s \leq x \text{ for all } s \in \emptyset$$

for all $x \in X$. But for any given $x \in X$ the condition on the right is vacuously true, and so $\bigvee \emptyset \leq x$. As $x \in X$ was arbitrary this makes $\bigvee \emptyset$ the bottom element.

Implicit in the linguistic use of the definite article when referring to the join and the top element, and similarly to the meet and the bottom element, is the following uniqueness clause.

Proposition 1.2 *Let X be a poset and $S \subseteq X$ a subset. If the join $\bigvee S$ exists, then it is unique. Similarly, if the meet $\bigwedge S$ exists, then it is unique.*

Proof We establish the uniqueness of joins, so assume $a, b \in X$ are so that each satisfies the defining property of $\bigvee S$. If we can now show that $a \leq b$, then due to the symmetry between the two, $b \leq a$ would follow as well and thus the claim that $a = b$. So, to show $a \leq b$, since a satisfies the defining property of $\bigvee S$, it suffices to show that $s \leq b$ for all $s \in S$. Recall that we already established that $\bigvee S$ is an upper bound of S, and so, since b satisfies the properties of $\bigvee S$, it is indeed the case that $s \leq b$ for all $s \in S$. This completes the argument.

The feeling one gets when translating claims and proofs about joins to their counterparts about meets is of repeating the same argument while standing upside down. The redundancy of adapting claims and proofs in the obvious manner can be avoided through a formal process of turning a poset X to its *opposite*. In more detail, if X is a poset, then its opposite X^{op} is the same set but where the ordering has been reversed, namely $x \leq y$ holds in X^{op} precisely when $y \leq x$ holds in X. It is routine to verify that X^{op} is a poset. Every concept of posets now acquires a dual by interpreting the same concept in the opposite poset. In this way, top element is dual to bottom element and, more generally, joins are dual to meets.

1.2.17 Zorn's Lemma

A poset P is said to be *linearly ordered* or a *total order* if for all $x, y \in P$ either $x \leq y$ or $y \leq x$ holds. The poset $\mathcal{P}(X)$ discussed in Example 1.1 is linearly ordered if, and only if, $|X| \leq 1$.

Any subset $S \subseteq P$ in a poset $(P, \leq)$ inherits a poset structure from P. We also say that P induces a poset structure on S. In more detail, the poset S is given by the ordering $\preceq$ defined, for all $x, y \in S$ by $x \preceq y$ precisely when $x \leq y$ in P. We usually do not make any notational distinction between $\leq$ and the *induced order* $\preceq$ and simply write $x \leq y$ when referring to the ordering in S.

Definition 1.4 A *chain* in a poset P is a subset $S \subseteq P$ that, with the induced ordering, is linearly ordered. An *upper bound* of a set $S \subseteq P$ (whether a chain or not) is an element $y \in P$ for which $x \leq y$ holds for all $x \in S$. A *maximal* element in a poset P is an element y_M such that $y_M < x$ does not hold for any $x \in P$.

Notice that upper and lower bounds were encountered in the discussion on lattices above where we remarked that the join is the least upper bound and the meet is the greatest lower bound.

Lemma 1.1 (Zorn's Lemma) *Let P be a non-empty poset. If every chain in P has an upper bound, then P has a maximal element.*

Zorn's Lemma is a powerful tool that is more of a proof technique than a lemma in much the same way that proof by induction is a proof technique rather than a lemma. A proof of Zorn's Lemma is subtle and in fact it is well-known that Zorn's Lemma is equivalent to the Axiom of Choice.

1.2.18 A Typical Application of Zorn's Lemma

Given any two sets X and Y, it is natural to wonder if their cardinalities compare. That is, is it always the case that either $|X| \leq |Y|$ or $|Y| \leq |X|$? A positive answer to this question turns out to be equivalent to the Axiom of Choice. We will only prove half of this equivalence, primarily to illustrate a typical application of Zorn's Lemma.

Theorem 1.5 *For all sets X and Y either $|X| \leq |Y|$ or $|Y| \leq |X|$.*

Proof The case where either X or Y is empty is easily dispensed with, and so we proceed under the assumption that both are non-empty. As a first step we construct a suitable poset, similar to the one given in Example 1.1.

Let P be the set of triples (X', f, Y') where $X' \subseteq X$, $Y' \subseteq Y$, and $f : X' \to Y'$ is an injective function. The partial order structure on P is given by

$$(X', f, Y') \leq (X'', g, Y'')$$

precisely when

$$X' \subseteq X'', \quad Y' \subseteq Y'', \quad and\ g(x) = f(x)\ \text{for all } x \in X'.$$

It is straightforward to verify that this is indeed a partial order.

Next, we show that $(P, \leq)$ satisfies the conditions of Zorn's Lemma. Firstly, $P \neq \emptyset$ since the triple $(\emptyset, \emptyset, \emptyset)$ is in P. Next, suppose that $\{(X_i, f_i, Y_i)\}_{i \in I}$ is a chain in P, and we shall construct an upper bound for it. Let

$$X_0 = \bigcup_{i \in I} X_i$$

$$Y_0 = \bigcup_{i \in I} Y_i$$

and $f : X_0 \to Y_0$ given by

$$f(x) = f_{i_x}(x).$$

Let us explain the definition of the function f. Given any $x \in X_0$, there is an $i_x \in I$ such that $x \in X_{i_x}$. We may thus consider the value $f_{i_x}(x) \in Y_{i_x} \subseteq Y_0$. However, there may be another index $i'_x \in I$ with $x \in X_{i'_x}$, and, a-priori,

$$f_{i'_x}(x) \neq f_{i_x}(x)$$

is a possibility. However, since $\{(X_i, f_i, Y_i)\}_{i \in I}$ is a chain, we may assume, without loss of generality, that $(X_{i_x}, f_{i_x}, Y_{i_x}) \leq (X_{i'_x}, f_{i'_x}, Y_{i'_x})$. But then, the definition of $\leq$ in P implies that

$$f_{i'_x}(x) = f_{i_x}(x)$$

and so the function $f : X_0 \to Y_0$ above is well defined. A similar argument shows that f is injective.

Now that we have verified the conditions of Zorn's Lemma we are guaranteed of the existence of a maximal element $(X_M, f_M, Y_M) \in P$. That is, $f_M : X_M \to Y_M$ is an injective function for some $X_M \subseteq X$ and $Y_M \subseteq Y$. Let us now entertain the possibility that both X_M and Y_M are proper subsets, namely, that there exist elements $x_! \in X \setminus X_M$ and $y_! \in Y \setminus Y_M$. But then we may define

$$X_! = X_M \cup \{x_!\}$$
$$Y_! = Y_M \cup \{y_!\}$$

and $f_! : X_! \to Y_!$ by

$$f_!(x) = \begin{cases} f_M(x) & \text{if } x \in X_M \\ y_! & \text{if } x = x_! \end{cases}$$

giving rise to the element $(X_!, f_!, Y_!) \in P$ (the reader is invited to verify membership in P) with

$$(X_M, f_M, Y_M) < (X_!, f_!, Y_!),$$

contradicting the maximality of (X_M, f_M, Y_M).

We thus conclude that either $X_M = X$ or $Y_M = Y$. If $X_M = X$, then the composition $X \xrightarrow{f_M} Y_M \xrightarrow{\text{incl.}} Y$ with the inclusion function yields an injection $X \to Y$, thus showing that $|X| \leq |Y|$. If $Y_M = Y$, then the composition $Y \xrightarrow{f_M^{-1}} X_M \xrightarrow{\text{incl.}} X$ with the inclusion function is an injection $Y \to X$ (where viewing f_M^{-1} as a function $Y \to X_M$, necessarily injective, is justified by the injectivity of f_M), showing that $|Y| \leq |X|$, completing the proof.   □

1.2.19   The Real Numbers

The set $\mathbb{R}$ of real numbers may be constructed in numerous different ways and can also be characterized axiomatically in different ways. We will not concern ourselves here with the construction of a model of the reals, and thus accept its existence and only state the governing axioms, namely that the reals form a Dedekind complete totally ordered field.

The statement that $\mathbb{R}$, with addition and multiplication, is a field is the claim that the following axioms hold:

Associativity of addition:	$(a + b) + c = a + (b + c)$
Commutativity of addition:	$a + b = b + a$
Neutrality of 0:	$a + 0 = a$
Existence of additive inverses:	$a + x = 0$ admits a solution
Associativity of multiplication:	$(a \cdot b) \cdot c = a \cdot (b \cdot c)$
Commutativity of multiplication:	$a \cdot b = b \cdot a$
Neutrality of 1:	$1 \cdot a = a$
Existence of reciprocals	$a \cdot x = 1$ admits a solution if $a \neq 0$
Distributivity:	$a \cdot (b + c) = a \cdot b + a \cdot c$

for all $a, b, c \in \mathbb{R}$. In general, any set K with addition and multiplication operations, and distinguished elements $0 \neq 1$ in K, that satisfy these axioms is said to be a *field*. For instance, both $\mathbb{C}$ and $\mathbb{Q}$, with the usual notion of addition and multiplication, are fields. In contrast, $\mathbb{Z}$ with the same familiar addition and multiplication is not a field.

The statement that the reals form an ordered field is the claim that, with the usual notion of $a \leq b$ for the real numbers, $\mathbb{R}$ is a totally ordered set, and that the following axioms hold:

Translation invariance:	$a \leq b \implies a + c \leq b + c$
Scale invariance:	$a \leq b \implies a \cdot c \leq b \cdot c$, if $c \geq 0$

for all $a, b, c \in \mathbb{R}$. In general, a field K with a total order on it such that these two axioms hold is called an *ordered field*. For instance, $\mathbb{Q}$ with the usual ordering is an ordered field. For the field $\mathbb{C}$, however, it can be shown that no ordering exists turning it into an ordered field.

Finally, perhaps the most important property of the field of real numbers, and certainly one that sets it apart among all ordered fields, is its Dedekind completeness. Using the language of complete lattices, first we note that the ordering of the reals makes $\mathbb{R}$ a poset and we may wonder wether it is a complete lattice or not. The question is thus whether all subsets admit a join and a meet. Certainly, a set that is not bounded above, for instance $\mathbb{N}$ cannot have a join in $\mathbb{R}$. For a similar reason, the empty set does not have a meet since the empty meet is the top element and clearly $\mathbb{R}$ lacks a top element. Similarly, sets not bounded below fail to have a meet and the empty join does not exist. An ordered field in which unboundedness and emptiness are the only obstructions for possessing meets and joins is said to be Dedekind complete. The field $\mathbb{R}$ of real numbers does have this property.

We phrase it in more detail using the synonym 'supremum' instead of 'join'. Recall that an upper bound for a subset $S \subseteq \mathbb{R}$ is a real number a such that $s \leq a$ for all $s \in S$. A *supremum* (or *least upper bound*) of S is an upper bound a with the property that if b is any upper bound of S, then $a \leq b$. Similarly, one defines the *infimum* (or *greatest lower bound*) of S to be a lower bound that is not smaller than any other lower bound. The statement that $\mathbb{R}$ is *Dedekind complete* is the claim: any non-empty bounded above subset $S \subseteq \mathbb{R}$ has a supremum. It can then be shown that this property is equivalent to: any non-empty bounded below subset $S \subseteq \mathbb{R}$ has an infimum.

In general, any Dedekind complete ordered field can be taken as essentially the field of the real numbers. In other words, the axioms listed above of a Dedekind complete ordered field determine the reals up to an isomorphism, meaning that the only difference between any two Dedekind complete ordered fields is the naming of the elements.

1.2.20 The Complex Numbers

The field $\mathbb{C}$ of complex numbers is an important extension of the field of real numbers. We give, in condensed form, the basics of the algebraic and analytical properties of $\mathbb{C}$.

A *complex number* $z \in \mathbb{C}$ consists of a *real part* $\mathrm{Re}(z)$ and an *imaginary part* $\mathrm{Im}(z)$, each a real number. This is indicated in the *cartesian* notation

$$z = \mathrm{Re}(z) + i\mathrm{Im}(z)$$

where i is a formal symbol. We often write $z = x + iy$ where it is understood that x is then the real part and y the imaginary part of z. The set

$$\mathbb{C} = \{x + iy \mid x, y \in \mathbb{R}\}$$

is thus the set of complex numbers. By identifying a real number x with the complex number $x + i \cdot 0$ we may write $\mathbb{R} \subset \mathbb{C}$. Addition and multiplication of complex numbers is given by

$$(x_1 + iy_1) + (x_2 + iy_2) = (x_1 + x_2) + i(y_1 + y_2)$$
$$(x_1 + iy_1) \cdot (x_2 + iy_2) = (x_1 x_2 - y_1 y_2) + i(x_1 y_2 + x_2 y_1)$$

which arise by preservation of the algebraic structure of $\mathbb{R}$ and setting $i^2 = -1$. It is a lengthy yet largely uneventful exercise to verify that with these operations $\mathbb{C}$ is a field.

The real numbers are often depicted as forming the real line. The complex numbers, since each is a pair of real numbers, form a plane and consequently enjoy a rich geometry. The *modulus* of $z = x + iy$ is the real number

$$|z| = \sqrt{x^2 + y^2}$$

and it measures the length of the segment from the origin to z. The *conjugate* of z is

$$\bar{z} = x - iy$$

obtained by reflection through the real line. The two operations are related by the equality

$$|z|^2 = z\bar{z}$$

as a straightforward computation shows. Given an angle θ we write

$$e^{i\theta} = \cos\theta + i\sin\theta,$$

and elementary trigonometry reveals it to be a complex number of modulus 1 forming an angle of θ radians from the positive real axis to it. Moreover, the formula

$$e^{i(\theta+\rho)} = e^{i\theta} \cdot e^{i\rho}$$

is readily verified. In other words, the set

$$\{e^{i\theta} \mid 0 \leq \theta < 2\pi\}$$

corresponds precisely to the unit circle in the plane. Now suppose $z \neq 0$. The number $z/|z|$ is then a point on the unit circle and so we can write $z/|z| = e^{i\theta}$ for a suitable angle. Denoting $r = |z|$ we obtain the *polar decomposition*

$$z = re^{i\theta}$$

where, in this form, r is the modulus and θ is the *argument* of the complex number. It is worth noting that addition of complex numbers is best carried out in cartesian form while multiplication is simplified in polar form.

Finally we mention the striking analytical difference between $\mathbb{R}$ and $\mathbb{C}$. Clearly, not every polynomial with real coefficients has a root (e.g., $x^2 + 1$). This same polynomial does have a root in $\mathbb{C}$, namely i. The field of complex numbers can be viewed as an extension of the reals obtained by adjoining just this one new root of a single polynomial that failed to have a root in $\mathbb{R}$. Quite profoundly this process cannot be repeated with polynomials with coefficients in $\mathbb{C}$. This is the content of the fundamental theorem of algebra which we state without proof.

Theorem 1.6 *For any polynomial function $f(z)$ with complex coefficients there exists a complex number w with $f(w) = 0$.*

Exercises

Exercise 1.1 For sets X, Y, and Z prove that:

1. $X \subseteq X$.
2. If $X \subseteq Y$ and $Y \subseteq X$, then $X = Y$.
3. If $X \subseteq Y$ and $Y \subseteq Z$, then $X \subseteq Z$.

Exercise 1.2 For sets X and Y show that either one of $X \cap Y = X$ and $X \cup Y = Y$ is equivalent to $X \subseteq Y$.

Exercise 1.3 For sets X and Y show that:

1. The intersection satisfies $X \cap Y \subseteq X$ and $X \cap Y \subseteq Y$, and it is the largest set with this property.
2. The union satisfies $X \subseteq X \cup Y$ and $Y \subseteq X \cup Y$, and it is the smallest set with this property.

Exercise 1.4 Assuming a universal set of discourse, show for sets X and Y that $(X^c)^c = X$ and that $X \subseteq Y$ if, and only if, $Y^c \subseteq X^c$.

Exercise 1.5 Prove the following distributivity properties of intersection over union of sets:

1. $X \cap (Y_1 \cup Y_2) = (X \cap Y_1) \cup (X \cap Y_2)$.
2. $X \cap (Y_1 \cup Y_2 \cup \cdots \cup Y_n) = (X \cap Y_1) \cup (X \cap Y_2) \cup \cdots \cup (X \cap Y_n)$ for all $n \geq 2$.
3. $X \cap \bigcup_{i \in I} Y_i = \bigcup_{i \in I} (X \cap Y_i)$.
4. $\bigcap_{i \in I} \bigcup_{j \in J_i} Y_{i,j} = \bigcup_{f \in C} \bigcap_{i \in I} Y_{i,f(i)}$ for all indexed collections $\{Y_{i,j}\}$. Here the index i ranges over a set I, and for each $i \in I$ the index j ranges over a set J_i. The set C is the set of choice functions, namely those functions $f : I \to \bigcup_{i \in I} J_i$ satisfying $f(i) \in J_i$ for all $i \in I$.

Exercise 1.6 State and prove distributivity properties of union over intersection in the spirit of the previous exercise. Also, show the last item there implies the others.

Exercise 1.7 Assume a set of discourse S. For a set $X \subseteq S$ show $X \cap X^c = \emptyset$. In fact, prove that among all subsets $Y \subseteq S$ with $X \cap Y = \emptyset$, the complement of X is the largest one.

Exercise 1.8 Fix a set X and recall the correspondence F from $\mathcal{P}(X)$ to the set $\mathcal{I}(X)$ of indicator functions $X \to \{0, 1\}$ given by $F(S)(x) = 1$ if, and only if, $x \in S$. Any function $\psi \colon \{0, 1\} \times \{0, 1\} \to \{0, 1\}$ defines an operation $*_\psi$ on $\mathcal{I}(X)$ by $(f *_\psi g)(x) = \psi(f(x), g(x))$ and thus an operation on subsets $S, T \subseteq X$ by $S \star_\psi T = F^{-1}(F(S) *_\psi F(T))$. Enumerate all functions ψ and express each corresponding set operation $\star_\psi$ in terms of set operations. Extending this idea, express, for a function $f \colon X \to Y$, the inverse image function $f^{-1} \colon \mathcal{P}(Y) \to \mathcal{P}(X)$ in terms of indicator functions. Can you do the same for the direct image function?

Exercise 1.9 Show that if a set X has n elements, then the power set $\mathcal{P}(X)$ has 2^n elements.

Exercise 1.10 Show that a set X satisfies $|X| = 1$ if, and only if, $|X \times Y| = |Y|$ holds for all sets Y.

Exercise 1.11 Let X be a set. Prove that:

1. For every set Y there exists a unique function $X \to Y$ if, and only if, $X = \emptyset$.
2. For every set Y there exists a unique function $Y \to X$ if, and only if, X is a singleton set.

Exercise 1.12 Consider sets X, Y and the diagram

$$X \xleftarrow{\ \pi_X\ } X \times Y \xrightarrow{\ \pi_Y\ } Y$$
$$\nwarrow_{\forall f_X} \quad \uparrow_{\exists! g} \quad \nearrow_{\forall f_Y}$$
$$T$$

in which $\pi_X(x, y) = x$ and $\pi_Y(x, y) = y$ are the projection functions. The diagram reads as: for all sets T and functions f_X, f_Y there exists a unique function g satisfying $\pi_X \circ g = f_X$ and $\pi_Y \circ g = f_Y$. Prove this is so.

Exercise 1.13 Consider sets X, Y and the diagram

$$X \xrightarrow{\ \iota_X\ } X \coprod Y \xleftarrow{\ \iota_Y\ } Y$$
$$\searrow_{\forall f_X} \quad \downarrow_{\exists! g} \quad \swarrow_{\forall f_Y}$$
$$T$$

in which $X \coprod Y = (X \times \{1\}) \cup (Y \times \{2\})$, $\iota_X(x) = (x, 1)$, and $\iota_Y(y) = (y, 2)$. The diagram reads as: for all sets T and functions f_X, f_Y there exists a unique function g satisfying $g \circ \iota_X = f_X$ and $g \circ \iota_Y = f_Y$. Prove this is so. Note the formal duality between this diagram and the one in the previous exercise.

Exercise 1.14 For sets X, Y, Z construct explicit bijections showing:

1. $|Z^{X \cup Y}| = |Z^X \times Z^Y|$ provided $X \cap Y = \emptyset$.
2. $|(X \times Y)^Z| = |X^Z \times Y^Z|$.
3. $|Z^{X \times Y}| = |(Z^Y)^X| = |(Z^X)^Y|$.

In the first item, what happens if X and Y are not disjoint? How can you fix the problem?

Exercise 1.15 Under what conditions on sets X, Y, Z is the cartesian product an associative operation? In general show that $|(X \times Y) \times Z| = |X \times (Y \times Z)|$ by using the *associator function* $a_{X,Y,Z} \colon (X \times Y) \times Z \to X \times (Y \times Z)$ given by $a_{X,Y,Z}((x, y), z) = (x, (y, z))$.

Exercise 1.16 Consider four sets W, X, Y, Z and the diagram

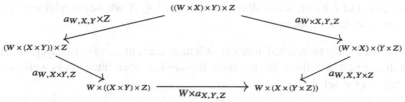

where the functions are obtained from the family of associator functions from the previous exercise. For instance, the function at the bottom is given by $W \times a_{X,Y,Z}(w, ((x, y), z)) = (w, (x, (y, z)))$. Verify that the diagram commutes in the sense that the two compositions starting at the top of the diagram are equal functions. Mac Lane's coherence theorem states that the commutativity of this pentagon implies the commutativity of any diagram built from associators. It is the mathematical reason why it is safe to pretend that the cartesian product of sets is associative.

Exercise 1.17 The sets $\mathbb{R} \times \mathbb{R}^2$ and $\mathbb{R}^3$ are different but in bijective correspondence and are often identified. The previous exercise gives a precise and safe such identification. Prior to any identification consider the set T whose elements are all possible tuples that one can construct from the real numbers, including tuples of tuples to any finite degree of nesting (e.g., $(1, (3, 2)), (1, (1, 2)), (1, 2, 3)$) is a possible element in T). For any $t \in T$ let $-t$ be defined by negating all of the numbers appearing in t. For $X \subseteq T$ let $-X = \{-x \mid x \in X\}$. For any three such sets X, Y, Z let $\beta_{X,Y,Z} \colon (X \times Y) \times Z \to X \times (Y \times Z)$ be given by $\beta_{X,Y,Z}((x, y), z) = (-x, (-y, -z))$. Show that even though $\beta_{X,Y,Z}$ is a bijection the pentagon from the previous exercise does not generally commute. Contemplate the headache that would ensue if one carelessly used this family of functions to make the familiar identifications.

Exercise 1.18 Let X be a set. Show the existence of an injective function $X \to \mathcal{P}(X)$ from X to its power set. Show that no such function can be surjective as follows. Suppose that such a function f exists and use it to define the set $S = \{x \in X \mid x \notin f(x)\}$. Exploit the surjectivity of f to obtain $s \in X$ with $S = f(s)$ and reach a contradiction.

Exercise 1.19 For sets X and Y write $|X| < |Y|$ to mean that $|X| \leq |Y|$ yet $|X| \neq |Y|$. State the meaning of $|X| < |Y|$ in terms of the existence and non-existence of suitable functions. Use the previous exercise to prove Cantor's Theorem: $|X| < |\mathcal{P}(X)|$ for all sets X. Deduce that a set of largest cardinality does not exist, nor does the set of all sets.

Exercise 1.20 Let $f: X \to Y$ and $g: Y \to Z$ be functions and consider the direct and inverse image functions. Show that $(g \circ f)_\to = g_\to \circ f_\to$ and $(g \circ f)^\leftarrow = f^\leftarrow \circ g^\leftarrow$. Show further that $S \subseteq f^\leftarrow(f_\to(S))$ for all $S \subseteq X$ and $T \supseteq f_\to(f^\leftarrow(T))$ for all $T \subseteq Y$. Exhibit cases of strict inclusion.

Exercise 1.21 Let X be a countable set. Show that the set $\mathcal{P}_{\text{fin}}(X)$ of all finite subsets $S \subseteq X$ is countable. Establish a bijection between $\mathcal{P}_{\text{fin}}(X)$ and $\mathcal{P}_{\text{cof}}(X)$ of all co-finite subsets $S \subseteq X$, namely those for which $X \setminus S$ is finite. Conclude that $\mathcal{P}_{\text{cof}}(X)$ is countable. Let T be the set of all infinite subsets $S \subseteq X$ whose complement $X \setminus S$ is also infinite. Is T countable?

Exercise 1.22 Any two closed intervals in $\mathbb{R}$ have the same cardinality. Demonstrate that with explicit bijections. Do the same for any two open intervals as well as any open ray and the set $\mathbb{R}$.

Exercise 1.23 For a set X let $\text{Bij}(X)$ be the set of all bijections $f: X \to X$. Show for finite X with $|X| = n$ that $|\text{Bij}(X)| = n!$. What is the cardinality of $\text{Bij}(\mathbb{N})$?

Exercise 1.24 For a function $f: X \to Y$ and $y \in Y$ consider the equation $f(x) = y$. Show that:

1. The function f is injective if, and only if, for each $y \in Y$ the equation has at most one solution.
2. The function f is surjective if, and only if, for each $y \in Y$ the equation has at least one solution.
3. The function f is bijective if, and only if, for each $y \in Y$ the equation has a unique solution.

Exercise 1.25 Let $f: X \to Y$ be a function. Say that f is left-cancelable if for all sets T and functions $g, h: T \to X$, if $f \circ g = f \circ h$, then $g = h$. Similarly, f is right-cancelable if for all sets T and functions $g, h: Y \to T$, if $g \circ f = h \circ f$, then $g = h$. Show that:

1. f is injective if, and only if, f is left-cancelable.
2. f is surjective if, and only if, f is right-cancelable.
3. f is bijective if, and only if, f is and left and right-cancelable.

Exercise 1.26 Injectivity of a function cannot be lost by restricting the domain nor can it be gained by extending the function to a larger domain. Give a mathematically rigorous formulation of this claim and then prove it. Similarly, formalise the claim that surjectivity of a function can be gained by restricting the codomain or by extending the function to a larger domain.

Exercise 1.27 Prove that for finite sets X, Y of equal cardinality injectivity, surjectivity, and bijectivity of a function $f : X \to Y$ coincide. Show that the finiteness assumption is crucial.

Exercise 1.28 If functions $X \xrightarrow{f} Y \xrightarrow{g} X$ satisfy $g \circ f = \mathrm{id}_X$, then f is said to be a left inverse of g and g a right inverse of f.

1. Show that $h : S \to T$ with $S \neq \emptyset$ is injective if, and only if, h has a left inverse.
2. Show that any $h : S \to T$ is surjective if, and only if, h has a right inverse. (Caution: this claim is equivalent to the Axiom of Choice).

Exercise 1.29 With notation as in the previous exercise, for X and Y finite show that if f has a left inverse, then it is bijective. For a set S denote by S^∞ the set of all infinite sequences of elements of S. By considering a suitable shift operation on S^∞ show that the finiteness assumption is crucial. What happens if one replaces left inverse by the evident notion of right inverse?

Exercise 1.30 Show that injectivity is preserved under composition in the sense that the composition of two injective functions is injective. Show that surjectivity and bijectivity are also preserved under composition.

Exercise 1.31 Construct a set X that has an element $x \in X$ satisfying $x \subseteq X$. Use X to obtain a function $f : X \to Y$ for which $f(x)$ is ambiguous (in the context of the direct image function). For each $n \geq 1$ find a set X_n and a function $f : X_n \to Y$ such that the expression $f(S) = \{f(s) \mid s \in S\}$ for the direct image function has n distinct meanings. Are infinitely many meanings possible?

Exercise 1.32 Consider the set $\mathcal{P}(X)$ of all subsets of a given set X. With union playing the role of addition and intersection as multiplication, which of the field axioms are satisfied?

Exercise 1.33 Consider the set $\mathbb{S}$ of all sequences $s = (s_1, s_2, \dots)$ of real numbers. With addition and multiplication defined component wise, which of the field axioms hold true for $\mathbb{S}$?

Exercise 1.34 A *filter* on a set X is a collection $\mathcal{F}$ of subsets of X with the properties that $\emptyset \notin \mathcal{F}$, $X \in \mathcal{F}$, $F \in \mathcal{F}$ and $F \subseteq S \subseteq X$ imply $S \in \mathcal{F}$, and $F_1, F_2 \in \mathcal{F}$ implies $F_1 \cap F_2 \in \mathcal{F}$. It is convenient to think of the sets in $\mathcal{F}$ as large portions of X. Show that if X is an infinite set, then the collection $\mathcal{F}$ consisting of those subsets $F \subseteq X$ whose complement $X \setminus F$ is finite forms a filter (known as the Fréchet filter).

A filter $\mathcal{F}$ is an *ultrafilter* if for any subset $S \subseteq X$ either $S \in \mathcal{F}$ or $X \setminus S \in \mathcal{F}$. Use Zorn's Lemma to prove that every filter $\mathcal{F}$ on a set X is contained in an ultrafilter on X.

Exercise 1.35 With reference to the previous exercise let $\mathcal{F}$ be the Fréchet filter on the set $\mathbb{N}$ of natural numbers and let $\mathcal{G}$ be an ultrafilter containing $\mathcal{F}$. Introduce the relation $\sim$ on $\mathbb{S}$ by declaring $s \sim t$ when the set $\{n \in \mathbb{N} \mid s_n = t_n\}$ belongs to the ultrafilter $\mathcal{G}$. Show that $\sim$ is an equivalence relation. Considering the quotient set $\mathbb{S}/\sim$, show that defining addition by $[s] + [t] = [s + t]$ and multiplication by $[s] \cdot [t] = [s \cdot t]$ is well defined. Establish (not for the faint of heart) that $\mathbb{S}/\sim$ is then a field and identify in it the field $\mathbb{R}$ by considering constant sequences. Finally, declare that a property P holds for $[s] \in \mathbb{S}/\sim$ precisely when the set $\{n \in \mathbb{N} \mid s_n$ has property $P\}$ belongs to $\mathcal{G}$. Show independence of the representative s.

Exercise 1.36 An *infinitesimal* is an element t satisfying $t > 0$ and $t < 1/n$ for all natural numbers n. Prove that $\mathbb{R}$ does not contain any infinitesimals. Building on the previous exercise show that in the field $\mathbb{S}/\sim$ the element $[s]$ where $s_n = 1/n$ is an infinitesimal.

Exercise 1.37 Let X be a set and $O(X)$ the set of all partial order relations on X. Declare, for $\leq_1, \leq_2$ in $O(X)$, that $\leq_1 \preceq \leq_2$ if $x \leq_2 y$ whenever $x \leq_1 y$. Prove the following:

1. The relation $\preceq$ is a partial order on $O(X)$.
2. Any chain in $O(X)$ has an upper bound.
3. There exist maximal elements in $O(X)$ and each is a linear ordering.
4. Any partial ordering on X can be extended to a linear ordering. Hint: Use Zorn's Lemma on a slightly adapted $O(X)$.
5. Any partial ordering on X is the intersection of all linear orderings containing it.
6. The intersection of all linear orders on X is the equality relation.

Exercise 1.38 Prove that in any ordered field if $x \neq 0$, then $x^2 > 0$. Deduce that there is no ordering on $\mathbb{C}$ making the complex numbers an ordered field.

Exercise 1.39 Fix a set R and for any set X write X^* for the set of all functions $f : X \to R$. Show that $G(x)(f) = f(x)$ defines a function $G : X \to (X^*)^*$. Is G injective? Is it surjective? Could the answer change under certain conditions on R or on X?

Exercise 1.40 Show that every function $f : X \to Y$ can be written as a composition $f = g \circ h$ where g is injective and h is surjective. Show that f can also be written as such a composition but with g surjective and h injective.

Exercise 1.41 Let X, Y be sets. A *set operator* is a function $O : \mathcal{P}(X) \to \mathcal{P}(Y)$ that is monotone, i.e., $O(S) \subseteq O(T)$ whenever $S \subseteq T \subseteq X$. Define an order relation on the set $\text{Ope}(X, Y)$ of all set operators $O : X \to Y$ by setting $O_1 \leq O_2$ when $O_1(S) \subseteq O_2(S)$ for all $S \subseteq X$. Show that $\text{Ope}(X, Y)$ with $\leq$ is a complete lattice.

Complementation in the domain and codomain enables the definition of the complement of O, i.e., $O^c(S) = (O(S^c))^c$. Show that $\mathcal{O} \to \mathcal{O}^c$ defines a function $\text{Ope}(X, Y) \to \text{Ope}(X, Y)$ that reverses the ordering, namely if $O_1 \leq O_2$, then $(O_2)^c \leq (O_1)^c$. Show further that O is expanding, i.e., $S \subseteq O(S)$, if, and only if, O^c is shrinking, i.e., $O^c(S) \subseteq S$.

Exercise 1.42 With terminology from the previous exercise, consider a function $f : X \to Y$. Show that the inverse image $f^{\leftarrow} : \mathcal{P}(Y) \to \mathcal{P}(X)$ and the direct image $f_{\to} : \mathcal{P}(X) \to \mathcal{P}(Y)$ are operators.

1. Show that $f^\leftarrow$ is its own complement.
2. Show that the complement of $f_\rightarrow$, denoted by $f^\rightarrow$, need not be $f_\rightarrow$.
3. For all $S \subseteq X$ and $T \subseteq Y$ show that $f_\rightarrow(S) \subseteq T$ if, and only if, $S \subseteq f^\leftarrow(T)$
4. For all $T \subseteq Y$ and $S \subseteq X$ show that $f^\leftarrow(T) \subseteq S$ if, and only if, $T \subseteq f^\rightarrow(S)$.

Exercise 1.43 Let X be a set and $R \subseteq X \times X$ a relation on X.

1. Show that $R \cup \{(x, x) \mid x \in X\}$ is a reflexive relation on X and that among all reflexive relations on X that contain R it is the smallest one. This relation is called the reflexive closure of R.
2. Show that $R \cup \{(y, x) \mid (x, y) \in R\}$ is a symmetric relation on X and that among all symmetric relations on X that contain R it is the smallest one. This relation is called the symmetric closure of R.

Exercise 1.44 Let X be a set and $R \subseteq X \times X$ a relation on X. Consider the set T of all transitive relations on X that contain R.

1. Prove that the intersection $S = \bigcap T$ of all the relations in T is a transitive relation on X that contains R.
2. Among all transitive relations on X that contain R show that S is the smallest one. The relation S is called the transitive closure of R.

Exercise 1.45 A relation from a set X to a set Y is any subset $R \subseteq X \times Y$. Given also a relation S from Y to a set Z the set $S \circ R = \{(x, z) \mid$ there is $y \in Y$ with $(x, y) \in R, (y, z) \in S\}$ is called the composition of S and R. Show that, much like for functions, composition of relations is associative.

Exercise 1.46 With relation to the previous two exercises, given a relation R on X define R^n to be the n-fold composition of R with itself. Prove that $\bigcup_{n=1}^\infty R^n$ is the transitive closure of R.

Exercise 1.47 Verify that multiplication of complex numbers is associative.

Exercise 1.48 For complex numbers $z, w \in \mathbb{C}$ show that $|z + w| \leq |z| + |w|, |zw| = |z||w|$, and that $|z| = 0$ if, and only if, $z = 0$.

Exercise 1.49 Compute $(1 + i)^{107}$.

Exercise 1.50 Let $z \in \mathbb{C}$ be a non-zero complex number. Show that in polar form $z = re^{i\theta}$ the inverse is given by $z^{-1} = r^{-1}e^{-i\theta}$. In cartesian form $z = x + iy$ show that the inverse is given by $z^{-1} = (x/|z|^2) - i(y/|z|^2)$.

Exercise 1.51 Show that $\overline{z + w} = \overline{z} + \overline{w}$ and that $\overline{zw} = \overline{z}\,\overline{w}$ for all complex numbers $z, w \in \mathbb{C}$. Prove that if $f(z)$ is a polynomial function with real coefficients and w is a root of f, then so is $\overline{w}$. Conclude that if f has odd degree, then it must have a real root.

Exercise 1.52 For an arbitrary complex number $z = x + iy$ define $e^z = e^x \cdot e^{iy}$. Show that $|e^z| = e^{\text{Re}z}$.

Exercise 1.53 For a non-zero complex number z show that there exists a unique complex number w of modulus 1 with zw a positive real number.

Further Reading

The closing section of this chapter (and of each of the forthcoming chapters) consists of suggestions for further reading. The list of sources is deliberately kept short and is thus by necessity not comprehensive. It aims to be a starting point for the reader interested in reading more about particular aspects of the chapter that, for whatever reason, were not elaborated upon in the text.

For a broad historical perspective on the development of modern mathematics, including a detailed discussion of the birth of modern mathematical analysis, see [11]. The reader interested in the interplay between mathematics as a formal system and the real world, and the student baffled by the usefulness of mathematics in physics, is referred to [8]. For a discussion of the internal forces governing and influencing the development of mathematics see the classic text [7]. Along these lines a modern, comprehensive, and sharp philosophical discussion is offered by [3].

For a more in-depth treatment of set theory the reader may consult the book [9] which is also an introduction to logic and proof theory. This book also includes a detailed and elementary presentation of the Axiom of Choice and some of its equivalents, i.e., Zorn's Lemma, Zermelo's well-ordering principle, and the principle of cardinal comparability, as well as an elementary treatment of cardinal arithmetic. Various books (e.g., [6, 10]) address the Axiom of Choice from a historical point-of-view, discussing its influence on and within mathematics.

For more on the various possibilities of constructing the real numbers see [12]. Some constructions refer to techniques of nonstandard analysis, i.e., to models of real numbers that contain actual infinitesimals. The reader interested in this somewhat unorthodox approach to analysis may consult the books [1, 4, 5] as well as [2] giving an eightfold path to the subject.

References

1. Albeverio, S., Høegh-Krohn, R., Fenstad, J.E., Lindstrøm, T.: Nonstandard Methods in Stochastic Analysis and Mathematical Physics. Pure and Applied Mathematics, vol. 122, p. xii+514. Academic Press Inc, Orlando, FL (1986)
2. Benci, V., Forti, M., Di Nasso, M.: The eightfold path to nonstandard analysis. Nonstandard Methods and Applications in Mathematics. Lecture Notes Log., vol. 25, pp. 3–44. Assoc. Symbol. Logic, La Jolla, CA (2006)
3. Corfield, D.: Towards a Philosophy of Real Mathematics. Cambridge University Press, Cambridge (2003), 300 pp
4. Davis, M.: Applied Nonstandard Analysis. Pure and Applied Mathematics, p. xii+181. Wiley-Interscience [John Wiley & Sons], New York-London-Sydney (1977)
5. Goldblatt, R.: Lectures on the Hyperreals. Graduate Texts in Mathematics, vol. 188, p. xiv+289. Springer, New York (1998)
6. Jech, T.J.: The Axiom of Choice. Dover Books on Mathematics, p. 224. Dover Publications (2008)
7. Lakatos, I.: Proofs and Refutations: The Logic of Mathematical Discovery. Cambridge University Press, Cambridge (1976), 188 pp

8. Mac Lane, S.: Mathematics, Form and Function, p. xi+476. Springer, New York (1986)
9. Moerdijk, I., van Oosten, J.: Sets, Models and Proofs, p. 156. Springer (2018)
10. Moore, G.H.: Zermelo's Axiom of Choice. Dover Books on Mathematics, p. 446. Dover Publications (2013)
11. Smoryński, C.: Adventures in Formalism. Texts in Mathematics, vol. 2, p. xii+606. College Publications, London (2012)
12. Weiss, I.: The real numbers - a survey of constructions. Rocky Mt. J. Math. **45**(3), 737–762 (2015)

Chapter 2
Linear Spaces

This chapter assumes a rudimentary understanding of the linear structure of $\mathbb{R}^n$, a very rich structure, both algebraically and geometrically. Elements in $\mathbb{R}^n$, when thought of as vectors, that is as entities representing direction and magnitude, can be used to form parallelograms, can be scaled, the angle between two vectors can be computed, and the length of a vector can be found. These geometric features are given algebraically by means of, respectively, vector addition, scalar multiplication, the inner product of two vectors, and the norm of a vector. In this chapter these notions are abstracted to give rise to the concepts of linear space, inner product space, and normed space.

The chapter gives a detailed presentation of all of the relevant notions of linear spaces, provides examples, and contains rigorous proofs of all of the results therein. Exploiting the assumption of a rudimentary understanding of the linear structure of $\mathbb{R}^n$, and thus of finite-dimensional linear spaces, the chapter has as secondary goal the elucidation of the subtleties of infinite-dimensional linear spaces. This is an absolute necessity with Hilbert spaces in mind since the infinite-dimensional ones are paramount.

A consequence of this infinite-dimensional theme of the chapter is that some proofs are considerably more involved than their finite-dimensional counterparts. In particular, the theorems establishing the existence of bases and the concept of dimension are sophisticated and resort to an application of Zorn's Lemma. Also, cardinality considerations are important, since one needs to be able to compute, at least a little bit, with infinite quantities. To facilitate an easier reading of this chapter, the Preliminaries contain an account of Zorn's Lemma (Sect. 1.2.17) and basic cardinal arithmetic (Sects. 1.2.12 and 1.2.13). Moreover, the text below indicates which proofs can safely be skipped on a first reading, leaving it to the discretion of the reader when, and whether, to tackle the technicalities involved in mastering the more advanced techniques.

Section 2.1 introduces the axioms of linear spaces in full generality, establishes basic properties and explores examples, most of which will be revisited throughout

© Springer Nature Switzerland AG 2021
C. Alabiso and I. Weiss, *A Primer on Hilbert Space Theory*, UNITEXT for Physics,
https://doi.org/10.1007/978-3-030-67417-5_2

the book. Section 2.2 is concerned with establishing the notion of dimension for an arbitrary linear space. In particular, the dimension need not be finite, which, at times, necessitates some more intricate proofs. Section 2.3 discusses linear operators, the natural choice of structure preserving functions between linear spaces, studies their basic properties, and discusses the notion of isomorphic linear spaces. Section 2.4 introduces standard constructions producing new spaces from given ones, and in particular the kernel and image of a linear operator are discussed. The final section is devoted to inner product spaces and normed spaces and presents several important examples such as function spaces and sequence spaces.

2.1 Linear Spaces—Elementary Properties and Examples

In $\mathbb{R}^2$ the scalar product $\alpha \cdot x$ is between a real number $\alpha \in \mathbb{R}$ and a vector $x \in \mathbb{R}^2$, and yields again a vector in $\mathbb{R}^2$. However, in similar situations, such as in $\mathbb{C}^2$, the scalar multiplication $\alpha \cdot x$ is now defined between any complex number $\alpha \in \mathbb{C}$ and an arbitrary vector x. The most general situation is when α is allowed to vary over the elements of an arbitrary (but fixed) *field* K. Prominent examples of fields are the field $\mathbb{R}$ of real numbers and the field $\mathbb{C}$ of complex numbers (for a more detailed discussion of fields the reader is referred to Sect. 1.2.19 of the Preliminaries). The definition of linear space given below is the result of the distillation of key properties of vector addition and scalar multiplication in $\mathbb{R}^2$ or $\mathbb{R}^3$, and is formalized in the most general form, namely with arbitrary fields. If the reader is not familiar with any fields other than $\mathbb{R}$ and $\mathbb{C}$ (and $\mathbb{Q}$, the field of rational numbers), then it is perfectly safe to proceed and replace any occurrence of an arbitrary field K by either $\mathbb{R}$ or $\mathbb{C}$. In the context of this book these are in any case the most important fields.

Definition 2.1 (*Linear Space*) Let K be an arbitrary field (such as the field of real numbers or of complex numbers, for concreteness). A set V of elements $x, y, z, \ldots$ of arbitrary nature, together with an operation, called *vector addition*, or simply *addition*, associating with any two elements $x, y \in V$ an element $z \in V$, called the *sum* of x and y, and denoted by $z = x + y$, as well as an operation associating with any $x \in V$ and $\alpha \in K$ an element $w \in V$, called the *product* or *scalar product* of α and x, and denoted by $w = \alpha \cdot x$, is called a *linear space* if:

1. For all $x, y, z \in V$, the operation of vector addition satisfies:

Associativity:	$(x + y) + z = x + (y + z))$
Commutativity:	$x + y = y + x$
Existence of a *neutral* element:	there exists $0 \in V$ with $x + 0 = x = 0 + x$
Existence of *additive inverses* :	there is $x' \in V$ with $x + x' = 0 = x' + x$.

2. For all $x \in V$ and $\alpha, \beta \in K$, the scalar product operation satisfies:

    Associativity: $\qquad\qquad\qquad\qquad\quad \alpha \cdot (\beta \cdot x) = (\alpha\beta) \cdot x$

    Neutrality of field unit: $\qquad\qquad\quad 1 \cdot x = x.$

3. For all $x, y \in V$ and $\alpha, \beta \in K$, the scalar product and vector addition operations are compatible in the sense that

    Scalar product distributes over vector addition: $\quad \alpha \cdot (x + y) = \alpha \cdot x + \alpha \cdot y$

    Scalar product distributes over scalar addition: $\quad (\alpha + \beta) \cdot x = \alpha \cdot x + \beta \cdot x.$

A linear space is also called a *vector space*, its elements are called *vectors*, and the elements of the field K are called *scalars*. If we wish to emphasize the relevant field, then we say that V is a *linear space over* K. Otherwise, the assertion that V is a linear space includes the implicit introduction of a field K serving as the field of scalars for V.

Scalars will typically be denoted by lowercase greek letters from the beginning of the alphabet, namely α, β, γ, and so on, while vectors will be denoted by x, y, z, etc. Subscripts or superscripts may be used to enhance readability.

2.1.1 Elementary Properties of Linear Spaces

We establish several properties of linear spaces that follow immediately from the axioms.

Proposition 2.1 *In any linear space V the following statements hold.*

1. *The neutral element $0 \in V$ is unique.*
2. *For all $x \in V$ the additive inverse x' is unique.*
3. *For all $\alpha \in K$ and $x \in V$ the equation $\alpha \cdot x = 0$ holds if, and only if, $\alpha = 0$ or $x = 0$.*

Proof 1. Suppose that $0' \in V$ is, like $0 \in V$, a neutral element. That is

$$x + 0' = x$$

for all $x \in V$, and thus using the neutrality of $0'$, commutativity, and the neutrality of 0 lead to
$$0 = 0 + 0' = 0' + 0 = 0'.$$

2. Suppose that $x'' \in V$ satisfies that $x + x'' = 0$, then

$$x' = x' + 0 = x' + (x + x'') = (x' + x) + x'' = 0 + x'' = x''.$$

3. For $\alpha = 0$

$$\alpha \cdot x = 0 \cdot x = (0+0) \cdot x = 0 \cdot x + 0 \cdot x \implies 0 \cdot x = 0.$$

For $x = 0$

$$\alpha \cdot x = \alpha \cdot 0 = \alpha \cdot (0+0) = \alpha \cdot 0 + \alpha \cdot 0 \implies \alpha \cdot 0 = 0.$$

In the other direction, if $\alpha \cdot x = 0$ and $\alpha \neq 0$, then upon multiplication by α^{-1} one obtains

$$x = 1 \cdot x = (\alpha^{-1} \cdot \alpha) \cdot x = \alpha^{-1} \cdot (\alpha \cdot x) = \alpha^{-1} \cdot 0 = 0.$$

$\square$

Remark 2.1 It similarly follows that for any vector x, the additive inverse x' is given by $x' = (-1) \cdot x$. It is further common to neglect the $\cdot$ denoting the scalar product, write $x' = -x$ and resort to the familiar conventions for algebraic manipulations on vectors and scalars commonly used for addition and multiplication, e.g., we write $x - y$ for $x + (-y)$ or $x + y + z$ for $(x + y) + z$, and so on. This convention, of course, considerably shortens proofs such as those given above and will be silently used throughout the text below.

Remark 2.2 Most of the linear spaces we are concerned with will be over the field $\mathbb{R}$ of real numbers, in which case they are called *real linear spaces*, or over the field $\mathbb{C}$ of complex numbers, in which case they are called *complex linear spaces*. In the absence of further specification, or as implied by context, a linear space is assumed to be either a real or a complex linear space.

Remark 2.3 We make no notational distinction between the *zero vector*, i.e., the neutral element with respect to vector addition, and the element 0 in the field K. This is generally a safe practice since context typically points to the correct interpretation. For instance, in the equation $0 \cdot x = 0$, context dictates that the 0 on the left-hand side is the scalar $0 \in K$ while on the right-hand side it is the zero vector $0 \in V$.

2.1.2 Examples of Linear Spaces

Since linear spaces are very common in mathematics, presenting an exhaustive list of linear spaces is a daunting task. The chosen examples below are meant to present some commonly occurring linear spaces, to explore some less common possibilities, and to familiarize the reader with some linear spaces of great importance in the context of this book. The latter refers to linear spaces of sequences and functions, which will be revisited throughout the book.

Example 2.1 In the familiar spaces $\mathbb{R}^2$ and $\mathbb{R}^3$, the mathematical models of the physical plane and space, vector addition has the geometric interpretation known as the parallelogram law, while scalar multiplication αx has a scaling effect on the vector x determined by the magnitude and sign of α.

Example 2.2 Given a natural number $n \geq 1$, let $\mathbb{R}^n$ be the set of n-tuples of real numbers. Thus if $x \in \mathbb{R}^n$, then it is of the form $x = (x_1, \ldots, x_n)$ where (for every $1 \leq k \leq n$) x_k, the k-th *component of x*, is a real number.

For all $x, y \in \mathbb{R}^n$ and $\alpha \in \mathbb{R}$ defining

$$x + y = (x_1 + y_1, x_2 + y_2, \ldots, x_n + y_n)$$
$$\alpha x = (\alpha x_1, \alpha x_2, \ldots, \alpha x_n),$$

endows $\mathbb{R}^n$ with the structure of a linear space over the field $\mathbb{R}$, as is easy to verify. Obviously, the cases $n = 2$ and $n = 3$ recover the familiar linear spaces $\mathbb{R}^2$ and $\mathbb{R}^3$. Analogously, the set $\mathbb{C}^n$ of all n-tuples of complex numbers, with similar coordinate-wise operations, is a linear space over the field $\mathbb{C}$.

Example 2.3 Let $\mathbb{R}^\infty$ be the set of infinite sequences of real numbers. Thus, a typical element $x \in \mathbb{R}^\infty$ is of the form $x = (x_1, x_2, \ldots, x_k, \ldots)$ where $x_k \in \mathbb{R}$, for each $k \geq 1$, is a real number called the k-th *component* of x. For all $x, y \in \mathbb{R}^\infty$ and $\alpha \in \mathbb{R}$ setting

$$x + y = (x_1 + y_1, x_2 + y_2, \ldots, x_k + y_k, \ldots)$$
$$\alpha x = (\alpha x_1, \alpha x_2, \ldots, \alpha x_k, \ldots)$$

endows $\mathbb{R}^\infty$ with the structure of a linear space over the field $\mathbb{R}$, as is easily seen. Similarly, the set $\mathbb{C}^\infty$ of all infinite sequences of complex numbers, with similar coordinate-wise operations, is a linear space over the field $\mathbb{C}$.

Remark 2.4 The convention that for an element x in either $\mathbb{R}^n$, $\mathbb{C}^n$, $\mathbb{R}^\infty$, or $\mathbb{C}^\infty$ its k-th *component* is denoted by x_k (as illustrated in the preceding examples) will be used throughout this text. To refer to a sequence of such vectors we may thus use super scripts for the different vectors, e.g., $\{x^{(m)}\}_{m \geq 1}$, and then $x_k^{(m)}$ refers to the k-th component of the m-th vector.

Example 2.4 Consider the subset $c \subseteq \mathbb{C}^\infty$ consisting of all convergent sequences (here convergence is in the usual sense of convergence of sequences of complex numbers), let $c_0 \subseteq c$ be the subset consisting of all sequences that converge to 0, and let $c_{00} \subseteq c_0$ be the set

$$c_{00} = \{(x_1, \ldots, x_m, 0, 0, 0, \ldots) \mid m \geq 1, \quad x_1, \ldots, x_m \in \mathbb{C}\}$$

of all sequences that are eventually 0. With the same definition of addition and multiplication as in Example 2.3, each of these sets is easily seen to be a linear space over $\mathbb{C}$. Obviously, replacing $\mathbb{C}$ throughout by $\mathbb{R}$ yields similar linear spaces over $\mathbb{R}$.

Example 2.5 The constructions given above of the linear spaces $\mathbb{R}^n$, $\mathbb{C}^n$, $\mathbb{R}^\infty$, and $\mathbb{C}^\infty$ easily generalize to any field K and to any cardinality. Indeed, consider an arbitrary set B and an arbitrary field K. Recall from the Preliminaries (Sect. 1.2.8) that the set K^B is the set of all functions $x : B \to K$. For all such functions

$$x : B \to K, \quad y : B \to K$$

and

$$\alpha \in K$$

define the functions $x + y : B \to K$ and $\alpha x : B \to K$ by

$$(x + y)(b) = x(b) + y(b), \quad (\alpha x)(b) = \alpha \cdot x(b).$$

With these notions of addition and scalar multiplication the set K^B is easily seen to be a linear space over K. In particular, we obtain the linear spaces $\mathbb{R}^{\{1,2,\dots,n\}} = \mathbb{R}^n$, $\mathbb{C}^{\{1,2,\dots,n\}} = \mathbb{C}^n$, $\mathbb{R}^{\mathbb{N}} = \mathbb{R}^\infty$, and $\mathbb{C}^{\mathbb{N}} = \mathbb{C}^\infty$. Further, restricting attention to the subset $(K^B)_0$ consisting only of those functions $x : B \to K$ for which $x(b) = 0$ for all but finitely many $b \in B$ also yields a linear space, with exactly the same definition for addition and scalar multiplication as in K^B. In particular the linear space c_{00} from Example 2.4 is recovered by noticing that $c_{00} = (\mathbb{C}^{\mathbb{N}})_0$. Note that the spaces c and c_0 rely on the analytical notion of convergence and so their analogues are not available for arbitrary B.

Example 2.6 For $n \geq 0$, let P_n be the set of all polynomial functions, that is functions of the form $p(t) = a_n t^n + \cdots + a_1 t + a_0$, with real coefficients and degree at most n. With the ordinary operations

$$(p + q)(t) = p(t) + q(t), \quad (\alpha p)(t) = \alpha \cdot p(t)$$

taken as addition and scalar multiplication, it is immediate to verify that P_n is a linear space over $\mathbb{R}$. Removing the restriction on the degrees of the polynomials one obtains the set P of all polynomial functions with real coefficients which, again with the obvious notions of addition and scalar multiplication, forms a linear space over $\mathbb{R}$. Considering complex coefficients yields similar linear spaces over $\mathbb{C}$.

Example 2.7 Let I be a subset of $\mathbb{R}$ that is either an open interval (a, b), a closed interval $[a, b]$, or the entire real line $\mathbb{R}$. Consider the set $C(I, \mathbb{R})$ of all continuous real-valued functions $x : I \to \mathbb{R}$. The familiar definitions

$$(x + y)(t) = x(t) + y(t), \quad (\alpha x)(t) = \alpha \cdot x(t),$$

when applied to continuous functions $x, y : I \to \mathbb{R}$, are well known to produce continuous functions again, and it is easy to see that when these operations are taken as addition and scalar multiplication the set $C(I, \mathbb{R})$ is a linear space over $\mathbb{R}$. One may also consider the set $C(I, \mathbb{C})$ of all continuous complex-valued functions $x : I \to \mathbb{C}$

to similarly obtain a linear space over $\mathbb{C}$. One may also consider, for each $k \geq 1$, the set $C^k(I, \mathbb{R})$ of all functions $x : I \to \mathbb{R}$ with a continuous k-th derivative, which is similarly a linear space over $\mathbb{R}$. It is then customary to equate $C^0(I, \mathbb{R})$ with $C(I, \mathbb{R})$. We may also allow $k = \infty$ so as to obtain $C^\infty(I, \mathbb{R})$, the linear space of all infinitely differentiable functions.

Example 2.8 Let K be any field and F a proper subfield of K, namely F is a proper subset of K, and with the induced operations from K it is itself a field. For instance, $\mathbb{Q}$ is a subfield of $\mathbb{R}$ which in turn is a subfield of $\mathbb{C}$. In such a situation any linear space V over the larger field K is also a *different* linear space over the smaller field F. The reason is that, of course, the additive structure of V is unaffected by the choice of field of scalars, whereas the scalar product axioms, if they hold for scalars ranging over K, certainly continue to hold for scalars ranging over the smaller field F (since these axioms are universal equational quantifications). The process of considering a linear space over K as a linear space over F is named *restriction of scalars*.

In particular, each of the examples above of a linear space over $\mathbb{C}$ is also a linear space over $\mathbb{R}$, and also a linear space over $\mathbb{Q}$. Any linear space obtained by restriction of scalars $F \subset K$ is (except for trivial cases) very different than the original space. Another particular instance of restriction of scalars is the observation that since $\mathbb{R}$ is a linear space over itself (if this is confusing pause for a minute to think about the linear space $\mathbb{R}^n$ and take $n = 1$), restriction of scalars implies that $\mathbb{R}$ is also a linear space over $\mathbb{Q}$.

Exercises

Exercise 2.1 Prove that the cancelation law

$$x + y = x + z \implies y = z$$

holds for all vectors x, y, z in any linear space V.

Exercise 2.2 Let V be a linear space. Prove that

$$-x = (-1) \cdot x$$

for all vectors $x \in V$.

Exercise 2.3 Let V be a linear space over K and $f : V \to X$ a bijection where X is a set with no a-priori extra structure. Let $f^{-1} : X \to V$ be the inverse function of f. Prove that the operations

$$x + y = f(f^{-1}(x) + f^{-1}(y))$$

and

$$\alpha x = f(\alpha f^{-1}(x)),$$

defined for all $x, y \in X$ and $\alpha \in K$, turn X into a linear space over K.

Exercise 2.4 For any $\alpha \in \mathbb{C}$ let c_α be the set of all sequences $(x_1, x_2, \ldots) \in \mathbb{C}^\infty$ that converge to α. Prove that the linear structure of $\mathbb{C}^\infty$ restricts to a linear structure on c_α if, and only if, if $\alpha = 0$.

Exercise 2.5 Show that the set of solutions of the differential equation $y'' + y = 0$ is a linear space (with the usual operations).

Exercise 2.6 Show that the set of solutions of the integral equation $y(s) = \int_a^b dt\, K(s,t) y(s)$ is a linear space (with the usual operations).

Exercise 2.7 Let K be a field (such as $\mathbb{R}$ or $\mathbb{C}$ for familiarity) and let $M_{n,m}(K)$ be the set of $n \times m$ matrices with entries in the field K. Prove that with ordinary matrix addition and scalar product the set $M_{n,m}(K)$ is a linear space over K.

Exercise 2.8 With the usual operations of addition and multiplication, is the set $(\mathbb{R} \setminus \mathbb{Q}) \cup \{0\}$ a linear space over $\mathbb{Q}$?

Exercise 2.9 Let V be a linear space over K and fix some vector $w_0 \in V$. Define on the set V an addition operation by $x \oplus y = x + y - w_0$ and a scalar product operation by $\alpha \odot x = \alpha(x - w_0)$. Prove that with these operations V is a linear space over K whose zero vector is w_0.

Exercise 2.10 Prove that the set $\mathbb{R}$ of real numbers with addition given by $x \oplus y = xy$ and scalar multiplication given by $\alpha \odot y = y^\alpha$ is a linear space over the field $\mathbb{R}$.

Exercise 2.11 Let X be a singleton set. Prove that for any field K there is a unique choice of operations that turns X into a linear space over K.

Exercise 2.12 Show that the cardinality of a linear space V over a field K is either 1 or at least $|K|$.

Exercise 2.13 Let K be a field and $\mathcal{B}$ an arbitrary set. Consider the set $< \mathcal{B} >_K$ of all formal expressions of the form

$$\sum_{b \in \mathcal{B}} \alpha_b \cdot b$$

where $\alpha_b \in K$ for each $b \in \mathcal{B}$ and at most finitely many of the α_b are non-zero. Verify that the obvious way to define addition and scalar multiplication on the set $< \mathcal{B} >_K$ turns it into a linear space over K (known as the *free linear space generated by* $\mathcal{B}$).

2.2 The Dimension of a Linear Space

The familiar linear spaces $\mathbb{R}^n$ are all infinite sets (in fact, they all have the same cardinality, the cardinality c of the continuum). However, it is intuitively clear that

$\mathbb{R}^3$ is larger than, say, $\mathbb{R}^2$. This fact is usually expressed by the claim that $\mathbb{R}^3$ has dimension 3 while $\mathbb{R}^2$ has dimension 2. The notion of dimension in general linear spaces is rather subtle, especially for the infinite dimensional ones. We investigate this notion in full generality, starting with the related concepts of linear independence and spanning sets.

2.2.1 Linear Independence, Spanning Sets, and Bases

By forming linear combinations of vectors from any given set of vectors in a linear space V, one obtains a (potentially) large collection of vectors. Intuitively, the original set is spanning if this process yields every vector in the linear space V while it is linearly independent if no duplicate vectors are produced. Precise definitions follow.

Definition 2.2 Given any set $S \subseteq V$ of vectors in a linear space V (S may be finite or infinite), a *linear combination* of elements of S is any vector of the form

$$x = \sum_{k=1}^{m} \alpha_k x_k$$

with $x_1, \ldots, x_m$ vectors from S and $\alpha_1, \ldots, \alpha_m \in K$ arbitrary scalars. Equivalently,

$$x = \sum_{s \in S} \alpha_s \cdot s$$

with $\alpha_s = 0$ for all but finitely many $s \in S$. The two forms are essentially the same, differing only notationally. The *span* of S is then the set of all linear combinations of elements from S, and S is said to be a *spanning set* if its span is the entire linear space V. A spanning set S is a *minimal spanning set* if it is itself a spanning set but no proper subset of it is a spanning set. Further, S is said to be a set of *linearly independent* vectors if the only possibility of expressing the zero vector as a linear combination of elements from S is the trivial linear combination, i.e., where all the coefficients are 0. That is, S is linearly independent if whenever one has

$$0 = \sum_{k=1}^{m} \alpha_k x_k,$$

with $x_1, \ldots, x_m$ vectors in S, then necessarily $\alpha_k = 0$ for all $1 \le k \le m$. The set S is said to be a *maximal linearly independent* set if it is itself linearly independent but any set that properly contains it is not linearly independent. A set that is not linearly independent is also referred to as a *linearly dependent* set. Finally, a set that is both a spanning set and linearly independent is called a *basis* of the linear space.

Remark 2.5 We speak of vectors $x_1, \ldots, x_m \in V$ as being either spanning or linearly independent, if the set $\{x_1, \ldots, x_m\}$ is spanning or linearly independent. Of course, we may also consider countably infinitely many vectors $x_1, x_2, \ldots$ as being spanning or linearly independent, in a similar fashion.

Example 2.9 The situation in $\mathbb{R}^n$ is probably very familiar to the reader. Any m vectors $x_1, \ldots, x_m$ in $\mathbb{R}^n$ are linearly independent if, and only if, the equation

$$\sum_{k=1}^{m} \alpha_k x_k = 0$$

admits the unique solution $\alpha_1 = \alpha_2 = \cdots = \alpha_m = 0$. It is well known that linear independence implies $m \leq n$. Similarly, the given vectors are spanning if, and only if, for every vector $b \in \mathbb{R}^n$ the equation

$$\sum_{k=1}^{m} \alpha_k x_k = b$$

admits a solution. It is again a familiar fact that if the given vectors are spanning, then $m \geq n$. It thus follows that a basis for $\mathbb{R}^n$ must consist of precisely n vectors. In particular, all bases have the same size, n, which is referred to as the dimension of $\mathbb{R}^n$. Below we prove that every linear space has a dimension, provided we allow infinite cardinalities into the picture. The result in that generality subsumes the properties of $\mathbb{R}^n$ just mentioned.

Example 2.10 In the space $\mathbb{C}^2$, considered as a linear space over $\mathbb{C}$, the vectors $(1, 0)$ and $(0, 1)$ are immediately seen to form a basis. However, if $\mathbb{C}^2$ is considered as a linear space over $\mathbb{R}$ (by the procedure of restriction of scalars from Example 2.8), then these two vectors are (of course) still linearly independent but they fail to span $\mathbb{C}^2$. Indeed, since only real scalars may now be used to form linear combinations of these vectors, the span will only be $\mathbb{R}^2$. To obtain a basis, the two vectors need to be augmented, for instance, by the vectors $(i, 0)$ and $(0, i)$. The four vectors together do form a basis of $\mathbb{C}^2$.

Example 2.11 In the linear space $\mathbb{R}^n$ or $\mathbb{C}^n$, the vectors $e_1, \ldots, e_n$ where

$$e_k = (0, \ldots, 0, 1, 0, \ldots, 0)$$

with 1 in the k-th position, are easily seen to be spanning and linearly independent, and thus form a basis, called the *standard basis* of $\mathbb{R}^n$, respectively $\mathbb{C}^n$. It is obvious that $\mathbb{R}^n$ and $\mathbb{C}^n$ have infinitely many bases.

In the examples presented so far it was quite straightforward to obtain a basis. The following example shows that this is not always the case. In fact, it is not even clear that the next linear space even has a basis.

Example 2.12 Let $\mathbb{C}^\infty$ be the linear space from Example 2.3. The vectors $e_1, e_2, \ldots$, given by

$$e_k = (0, \ldots, 0, 1, 0, \ldots)$$

with 1 in the k-th position are easily seen to be linearly independent, but they do not form a spanning set. Indeed, since the span consists only of *finite* linear combinations of vectors from the set, the span in this case is the set of all vectors of the form $(a_1, \ldots, a_k, 0, \ldots, 0, 0, \ldots)$, namely those infinite sequences of complex numbers that are eventually 0. In other words, the span in $\mathbb{C}^\infty$ of the vectors $e_1, e_2, \ldots$ is the space c_{00} from Example 2.4, and thus we incidentally found a basis for c_{00}. It is now tempting to proceed as follows. Taking any vector $y_1 \in \mathbb{R}^\infty$ not spanned by $e_1, e_2, \ldots$, for instance the vector $y_1 = (1, 1, 1, \ldots, 1, \ldots)$, forming the set $\{y_1, e_1, e_2, \ldots\}$ must get us closer to obtaining a basis. Indeed, the new set is still linearly independent precisely because y_1 was not spanned by the rest of the vectors. But, this new set is still not a basis as there are still many vectors it fails to span, for instance the vector $y_2 = (1, 0, 1, 0, 1, 0, \ldots)$. Of course, we may now consider the larger set $\{y_1, y_2, e_1, e_2, \ldots\}$, but it too fails to be a basis. One may attempt to resolve the argument once and for all by claiming that proceeding in this way to infinity will eventually result in a basis. However, this is a very vague statement, and even if this process can be carried out mathematically (which it can, as will be shown below), it is entirely unclear as to which vectors will end up in the basis and which will not. In other words, even if this linear space has a basis, it is unlikely we can ever present one.

Naturally, similar observations hold true for $\mathbb{R}^\infty$ instead of $\mathbb{C}^\infty$, and in fact to most of the linear spaces in this book, and in analysis in general.

Example 2.13 Recall (see Example 2.8) that one can consider the space $\mathbb{R}$ as a linear space over the field $\mathbb{Q}$ of rational numbers (as a particular case of restriction of scalars). A real number α is said to be *transcendental* if it is not the root of a polynomial with rational coefficients. Examples of transcendental numbers include e and π, though the proofs are far from trivial. For any transcendental number x the set $\{1, x, x^2, x^3, \ldots\}$ is linearly independent (this is basically the definition of x being transcendental), but it is not a spanning set.

Example 2.14 Recall the space P_n from Example 2.6 of all polynomial functions with real coefficients of degree at most n. For every $k \geq 0$ let p_k be the vector $p_k(t) = t^k$. Clearly, the vectors $p_0, p_1, p_2, \ldots, p_n$ form a basis for P_n, but again there are infinitely many other choices of bases. The space P of all polynomial functions with real coefficients has a countable basis given by $p_0, p_1, p_2, \ldots$, a fact that is easily verified. Noticing that polynomials are continuous functions, no matter on which interval they are defined, we see by the above that in the linear space $C(I, \mathbb{R})$ of all continuous functions $x : I \to \mathbb{R}$, where I is a non-degenerate interval (i.e., not reducing to a point) the vectors $p_0, p_1, p_2, \ldots$ are linearly independent. However, they do not form a basis since any linear combination of these vectors is again a polynomial function, but not all continuous functions are polynomial functions.

Once more, it is not at all clear that $C(I, \mathbb{R})$ even has a basis (what would one look like?).

From the discussion above we see that in some linear spaces, such as $\mathbb{R}^n$, P_n, or P, it is quite easy to find bases while in other linear spaces, such as $\mathbb{R}^\infty$, $\mathbb{R}$ over $\mathbb{Q}$, or $C(I, \mathbb{R})$, it is a highly non-trivial task. In more detail, we saw that it is rather simple to exhibit a large set of linearly independent vectors, but it is not so easy to have these vectors also span the entire space (the spaces c_{00} and P, as discussed above, are the exception to the rule). In fact, it is not at all clear that one can find bases for V in the case of $\mathbb{R}^\infty$ or $\mathbb{C}^\infty$, as well as for $\mathbb{R}$ as a linear space over $\mathbb{Q}$, or for $C(I, \mathbb{R})$.

2.2.2   Existence of Bases

The existence of a basis for an arbitrary vector space, over an arbitrary field, relies essentially on Zorn's Lemma. Consequently, the existence proof is not constructive and in many cases an explicit basis cannot be exhibited.

Proposition 2.2 *Let $S \subseteq V$ be a set of vectors in a linear space V. The following conditions are equivalent.*

1. *S is a maximal linearly independent set.*
2. *S is a minimal spanning set.*
3. *S is a basis.*

Proof First we show that if S is a maximal linearly independent set, then it is a basis. All that is needed is to show that S is a spanning set. To that end, let $x \in V$ be a vector in the ambient linear space. If $x \in S$ then it is certainly spanned by S. If $x \notin S$ then by virtue of S being a maximal linearly independent set, the set $S \cup \{x\}$ is linearly dependent. Thus, there exist vectors $x_1, \ldots, x_m \in S \cup \{x\}$ and non-zero scalars $\alpha_1, \ldots, \alpha_m \in K$ with

$$0 = \sum_{k=1}^{m} \alpha_k x_k.$$

However, in the expression above it must be that x itself appears as a summand since otherwise we would have expressed 0 as a non-trivial linear combination of vectors from the linearly independent set S. We may thus isolate x in the expression above to obtain it as a linear combination of elements from S. As x was arbitrary, we conclude that S is a spanning set.

Next we show that if S is a minimal spanning set, then it is a basis. All that is needed is to show that S is linearly independent, and indeed, if it were linearly dependent, then we would have some expression as above, giving 0 as a non-trivial linear combination of vectors from S. Using that expression we are able to isolate some vector $x \in S$ and exhibit it as a linear combination of other vectors from S. It is then easy to see that $S \setminus \{x\}$ is still a spanning set (simply since any linear combination containing x can be replaced by one that does not). But this contradicts S being a minimal spanning set, and thus S must be linearly independent.

So far we have shown that each of conditions 1 and 2 implies condition 3. The proof will be completed by showing the converse of these implications. The details are very similar in spirit to those given so far, and thus the rest of the proof is left for the reader.   □

We are now ready to establish that every linear space has a basis. The proof makes essential use of Zorn's Lemma and the reader may safely choose to skip the proof on a first reading. To the reader interested in the technique of Zorn's Lemma we remark that the proof below is actually a straightforward application with little technical difficulties, and is thus a fortunate first encounter with this proof technique.

Theorem 2.1 *Every linearly independent set $A \subseteq V$ in a linear space V can be extended to a maximal linearly independent set. In particular, by considering the linearly independent set $\emptyset \subseteq V$ and by Proposition 2.2, it follows that every linear space V has a basis.*

Proof Consider the set P of all linearly independent subsets S of V that contain A, and order P by set inclusion. Evidently P is a poset, and to find a maximal linearly independent set that extends A amounts to finding a maximal element in P, and so we apply Zorn's Lemma (Sect. 1.2.17 of the Preliminaries). First we note that P is certainly not empty since clearly $A \in P$. Now, assume that $\{S_i\}_{i \in I}$ is a chain in P. We will show that $S = \bigcup_{i \in I} S_i$ is an upper bound for the chain. Clearly $S_i \subseteq S$ for all $i \in I$, thus the only thing to show is that $S \in P$, namely that S contains A (which is immediate) and that it is linearly independent. For that, assume that

$$0 = \sum_{k=1}^{m} \alpha_k x_k$$

is a non-trivial linear combination of the vectors $x_1, \ldots, x_m \in S$. Since S is the union of the S_i, it follows that $x_m \in S_{f(m)}$ for a suitable $f(m) \in I$, but since $\{S_i\}_{i \in I}$ is a chain it follows that there is a single index $i_0 \in I$ such that $x_1, \ldots, x_m \in S_{i_0}$. But then the equality above expresses the zero vector as a non-trivial linear combination of vectors from S_{i_0}, contradicting the fact that S_{i_0} is linearly independent. With that the conditions of Zorn's Lemma are satisfied, and so the existence of a maximal element in P is guaranteed. This maximal element is a set $S_M \in P$, namely S_M contains A and S_M is a maximal set of linearly independent vectors, as required.   □

2.2.3 Existence of Dimension

Now that we know that every linear space has at least one basis, it is tempting to define the dimension of a linear space to be the cardinality of its basis. However, for this

to make sense we need to know that all bases have the same cardinality. While this
is a very plausible assertion (certainly for such familiar spaces as $\mathbb{R}^n$), it requires a
careful proof, particularly in the infinite-dimensional case. The important ingredient
is the following lemma, stating that the cardinality of any linearly independent set
is not greater than the cardinality of any spanning set. This result again uses Zorn's
Lemma and its proof may safely be skipped on a first reading. A word of caution
to the ambitious reader that this proof, compared to the proof of Theorem 2.1, is
technically more demanding.

Lemma 2.1 *If V is a linear space, $I \subseteq V$ a linearly independent set of vectors, and
$S \subseteq V$ a spanning set, then there exists an injective function $f : I \to S$.*

Proof Consider a pair (J, f) where $J \subseteq I$ and $f : J \to S$ is an injection. The idea
will be (using Zorn's Lemma) to keep on extending the domain of f until the entire
set I is exhausted. Conceptually, we think of the injective function $f : J \to S$ as
an instruction to replace the vectors in J by their images in S, so we only consider
such pairs (J, f) for which $(I \setminus J) \cup f(J)$ is still a linearly independent set. Let
us now form the set P of all such pairs and introduce an ordering on it declaring
that $(J_1, f_1) \le (J_2, f_2)$ precisely when $J_1 \subseteq J_2$ and f_2 extends f_1 (the latter means
that $f_1(x) = f_2(x)$ holds for all $x \in J_1$). It is immediate that P is a poset, and that
a maximal element in it will furnish us with an injective function $f : J_M \to S$ for a
very large subset $J_M \subseteq I$. We will then show that necessarily $J_M = I$, and the result
will be established.

To verify the conditions of Zorn's Lemma, notice first that P is not empty. Indeed,
the pair $(\emptyset, \emptyset \to S)$ is in P. Next, let $\{(J_t, f_t)\}_{t \in T}$ be a chain in P, for which we
must now find an upper bound. Let

$$J = \bigcup_{t \in T} J_t$$

and notice that we may define $f : J \to S$ as follows. Given $x \in J$ there is some
$t \in T$ such that $x \in J_t$, and so let $f(x) = f_t(x)$. To see that this is a well-defined
function, i.e., that it is independent of the choice of $t \in T$, note that if $x \in J_{t'}$ then
either $J_t \subseteq J_{t'}$ or $J_{t'} \subseteq J_t$ and then either $f_{t'}$ extends f_t or f_t extends $f_{t'}$, and in either
case $f_t(x) = f_{t'}(x)$. It is clear that $J_t \subseteq J$ and that f extends f_t, for all $t \in T$, so all
we need to do in order to show that (J, f) is an upper bound for the chain is establish
that $(J, f) \in P$. Clearly, $J \subseteq I$, and $f : J \to S$ is injective (since we saw that in
fact for any two $x_1, x_2 \in J$ the function f agrees with some f_t, which is injective, so
$f(x_1) = f(x_2) \implies f_t(x_1) = f_t(x_2) \implies x_1 = x_2$). So it remains to observe that
$(I \setminus J) \cup f(J)$ is a linearly independent set of vectors. Indeed, if 0 can be obtained
as a non-trivial linear combination of the vectors $x_1, \ldots, x_m \in (I \setminus J) \cup f(J)$, then,
using the chain condition again, it follows that there exists an element $t \in T$ such
that $x_1, \ldots, x_m \in (I \setminus J_t) \cup f_t(J_t)$, which is a linearly independent set, yielding a
contradiction.

With the conditions of Zorn's Lemma now verified, it follows that there exists a
maximal element (J_M, f_M), with $J_M \subseteq I$ and $f_M : J_M \to S$ an injection. If $J_M = I$,

then we are done, since $f_M : I \to S$ is the required injection. Assuming this is not the case, let $x_! \in I \setminus J_M$. If we can find a vector $y_! \in S \setminus f_M(J_M)$ such that

$$(I \setminus (J_M \cup \{x_!\})) \cup (f_M(J_M) \cup \{y_!\})$$

is linearly independent, then we will have that $(J_M \cup \{x_!\}, f_!) \in P$, where $f_!$ is the extension of f_M given by $f_!(x_!) = y_!$. But that would contradict the maximality of (J_M, f_M), and we will have our contradiction. So, we proceed to prove the existence of such a vector $y_!$. If no such $y_!$ exists, then that means that the set

$$(I \setminus (J_M \cup \{x_!\})) \cup (f_M(J_M) \cup \{y\})$$

is a linearly dependent set for every $y \in S \setminus f_M(J_M)$. Now, since S is a spanning set we may write

$$x_! = \sum_{s \in S} \alpha_s \cdot s = \sum_{s \in f_M(J_M)} \alpha_s \cdot s + \sum_{s \in S \setminus f_M(J_M)} \alpha_s \cdot s = x_1 + x_2$$

where the sum is a finite sum, so that $\alpha_s = 0$ for all but finitely many s, and we simply split the sum according to whether or not $s \in f_M(J_M)$. By our assumption, the set

$$(I \setminus (J_M \cup \{x_!\})) \cup f_M(J_M)$$

is linearly dependent if any of the vectors $s \in S \setminus f_M(J_M)$ is added to it. Thus, this set is linearly dependent if any linear combination of such vectors, such as x_2, is added to it. Further, since x_1 is in the span of $f_M(J_M)$ it follows that the set above is linearly dependent if $x = x_1 + x_2$ is added to it. But the latter is the set

$$(I \setminus (J_M \cup \{x_!\})) \cup f_M(J_M) \cup \{x_!\} = (I \setminus J_M) \cup f_M(J_M)$$

which is linearly independent. The proof is now complete. □

We are now in position to establish the following important result.

Theorem 2.2 *Let V be a linear space. If B_1 and B_2 are any two bases for V, then they have the same cardinality.*

Proof By definition of basis, B_1 is linearly independent and B_2 is a spanning set. By Lemma 2.1, there is an injective function $f : B_1 \to B_2$. By the same argument there is also an injective function $g : B_2 \to B_1$. It now follows from the Cantor-Schröder-Bernstein Theorem (Theorem 1.1 in the Preliminaries) that the cardinalities of B_1 and B_2 are equal. □

Definition 2.3 The *dimension* of a linear space V, denoted by $\dim(V)$, is equal to the cardinality of a basis for it. By Theorem 2.1 at least one basis exists, and by Theorem 2.2 all bases have the same cardinality. Thus, the notion of dimension is well defined. The linear space V is said to be *finite dimensional* if its dimension is finite, and *infinite dimensional* otherwise. A basis for an infinite-dimensional linear space is called a *Hamel basis*.

Example 2.15 Finite-dimensional linear spaces, by Examples 2.11 and 2.14, include the spaces $\mathbb{R}^n$ and $\mathbb{C}^n$, having dimension n, and the space P_n of polynomials (see Example 2.6), which is of dimension $n + 1$. Infinite-dimensional linear spaces include, due to Examples 2.12 and 2.14, the space c_{00} of sequences that are eventually 0, the space P of all polynomials with real coefficients, and the space $C(I, \mathbb{R})$ of continuous functions $x : I \to \mathbb{R}$, as long as I does not reduce to a point.

We have already encountered linear spaces of very large dimension. To argue about the dimension of such spaces the following result, interesting on its own, is useful. Recall from Example 2.5 the linear space $(K^B)_0$ of all functions $x : B \to K$ satisfying $x(b) = 0$ for all but finitely many $b \in B$.

Proposition 2.3 *Let V be a linear space and B a basis for it. Then every vector $x \in V$ can be expressed uniquely as a linear combination of elements from B. In other words, there is a bijective correspondence between the vectors $x \in V$ and the elements of $(K^B)_0$.*

Proof Since a typical element in $(K^B)_0$ is nothing but a function $x : B \to K$ for which $x(b) = 0$ for all but finitely many $b \in B$, we may associate with each such x the vector $\sum_{b \in B} x(b) \cdot b$ (as the summation is finite). We claim that this correspondence is the desired bijection. Indeed, it is a tautology that the surjectivity of this process is the claim that B is a spanning set. It is almost a tautology that the injectivity of the process is the claim that B is linearly independent.   □

We thus see that a basis B endows a linear space with a notion of coordinates. This is certainly a useful thing to have, but for practical reasons its usefulness hangs on the ability to explicitly describe the basis B. In finite-dimensional linear spaces it is very common to work with bases but, as we remarked earlier, in infinite-dimensional linear spaces being able to explicitly describe a Hamel basis is the exception rather than the rule. Consequently, Hamel bases are used primarily for theoretical rather then practical purposes.

Example 2.16 From Example 2.8 the field $\mathbb{R}$ may be viewed as a linear space over $\mathbb{Q}$ so let B be a Hamel basis for $\mathbb{R}$ over $\mathbb{Q}$. By Proposition 2.3, there is then a bijection between $\mathbb{R}$ and the set $(\mathbb{Q}^B)_0$ of all functions $B \to \mathbb{Q}$ that attain 0 at all but finitely many arguments. In other words, $|\mathbb{R}| = |(\mathbb{Q}^B)_0|$ (see Sect. 1.2.11 of the Preliminaries for the basics of cardinalities). If B is countable, then it is easy to write $(\mathbb{Q}^B)_0$ as a countable union of countable sets (Sect. 1.2.13) and thus the set $(\mathbb{Q}^B)_0$ itself would be countable. But that would imply that $\mathbb{R}$ is a countable set while the reals are well known to be uncountable. We conclude that $\mathbb{R}$, as a linear space over $\mathbb{Q}$, is infinite dimensional of uncountable dimension.

We close this section by illustrating one of many differences between finite-dimensional linear spaces and infinite-dimensional ones.

Proposition 2.4 *The cardinality of any linearly independent set A in a linear space V is a lower bound for the dimension of V. Moreover, if V is finite dimensional, then any linearly independent set $A \subseteq V$ whose cardinality is equal to the dimension of V is a basis.*

Proof The dimension of V is the cardinality of any basis B, and, as B is in particular a spanning set, it follows from Lemma 2.1 that there is an injection $A \to B$, and so the cardinalities satisfy $|A| \leq |B|$ (refer to Sect. 1.2.11 of the Preliminaries, if needed). The claim now follows since the latter is the dimension of V.

Now, if V is finite dimensional, say of dimension n, and $A = \{x_1, \ldots, x_n\}$ is a set of n linearly independent vectors, then to show A is a basis we just need to prove that it is a spanning set. But if it were not, and $y \in V$ is any vector not in its span, then the set $\{x_1, \ldots, x_n, y\}$ is linearly independent and contains $n + 1$ vectors. But then, by the first part of the proposition, it would follow that $n + 1 \leq n$, an absurdity. $\square$

Remark 2.6 The finite dimensionality assumption is crucial. For instance, in the linear space c_{00} of sequences that are eventually 0 (Example 2.4), consider the vectors $\{e_k\}_{k \geq 1}$ where $e_k = (0, \ldots, 0, 1, 0, \ldots)$, with 1 in the k-th position. These vectors are easily seen to be linearly independent and spanning, thus they form a basis of countably many vectors. The dimension of c_{00} is thus infinitely countable. The set $\{e_2, e_3, \ldots\}$ is clearly also linearly independent, has the same cardinality as the dimension of c_{00}, but it is not spanning, and thus not a basis.

Exercises

Exercise 2.14 Let V be a linear space and $S, S' \subseteq V$ two arbitrary sets of vectors with $S \subseteq S'$. Prove that if S' is linearly independent, then so is S, and prove that if S is a spanning set, then so is S'.

Exercise 2.15 Let V be a finite-dimensional linear space and S a spanning set. Prove that S can be sifted to give a basis, that is, show that there exists a subset $S' \subseteq S$ such that S' is a basis for V. Now do the same without the finite dimensionality assumption.

Exercise 2.16 Consider $\mathbb{R}$ as a linear space over $\mathbb{Q}$ and let α be a transcendental number. Prove that the set $\{1, \alpha, \alpha^2, \alpha^3, \ldots\}$ is linearly independent. Prove that it is not a spanning set by showing that its span is a countable set. If α is only known to be irrational, is the set still necessarily linearly independent?

Exercise 2.17 Consider $\mathbb{C}$ as a linear space over $\mathbb{Q}$ and let z be the complex number $e^{2\pi i/n}$, $n \geq 2$. Show that $\{1, z, z^2, z^3, \ldots\}$ is linearly dependent by writing 0 as a suitable linear combination. What is the minimum number of vectors required to do so?

Exercise 2.18 Show that in the linear space $C(\mathbb{R}, \mathbb{R})$ of all continuous functions $f \colon \mathbb{R} \to \mathbb{R}$ the set $\{1, \sin, \cos\}$ is linearly independent while $\{1, \sin^2, \cos^2\}$ is linearly dependent.

Exercise 2.19 Consider $\mathbb{R}$ as a linear space over $\mathbb{Q}$. Prove that the dimension of $\mathbb{R}$ over $\mathbb{Q}$ is $|\mathbb{R}|$, the cardinality of the real numbers.

Exercise 2.20 Let V be a linear space and $S \subseteq V$ a set of vectors. For a scalar $\alpha \in K$ let us write

$$\alpha S = \{\alpha x \mid x \in S\}.$$

Assuming that $\alpha \neq 0$ is fixed, prove that S is linearly independent (respectively spanning, a basis) if, and only if, αS is linearly independent (respectively spanning, a basis).

Exercise 2.21 Let V be a linear space and $\{x_k\}_{k\in\mathbb{N}}$ countably many vectors in V. For all $m \in \mathbb{N}$, let

$$y_m = \sum_{k=1}^{m} x_k.$$

Prove that $\{x_k\}_{k\in\mathbb{N}}$ is linearly independent (respectively spanning, a basis) if, and only if, $\{y_k\}_{k\in\mathbb{N}}$ is linearly independent (respectively spanning, a basis).

Exercise 2.22 Let B be an arbitrary set and K a field. Consider the linear spaces K^B and $(K^B)_0$. For every $b_0 \in B$ let $x_{b_0} : B \to K$ be the function

$$x_{b_0}(b) = \begin{cases} 1 & \text{if } b = b_0, \\ 0 & \text{otherwise.} \end{cases}$$

Prove that the set $\{x_{b_0}\}_{b_0 \in B}$ is linearly independent in K^B. Is it a basis for K^B? Is it a basis for $(K^B)_0$?

Exercise 2.23 Consider the space c_0 of sequences of complex (or real, if you like) numbers that converge to 0. Let $A_n \in c_0$ be the vector with $1/n$ in the n-th position and zeros elsewhere. Let $s = (1, 1/2, 1/3, \dots)$ be the harmonic sequence. Show that $\{A_n \mid n \geq 1\} \cup \{s\}$ is linearly independent. How many linearly independent vectors can you find in c_0?

Exercise 2.24 For a set B and a field K recall the set $(K^B)_0$ of almost everywhere 0 functions $f : B \to K$. Suppose now that V is a linear space over K and that $B \subseteq V$. Define the function $\psi : (K^B)_0 \to V$ by

$$\psi(f) = \sum_{b \in B} f(b)b.$$

Verify that ψ is well defined and show that

1. B is spanning if, and only if, ψ is surjective.
2. B is linearly independent if, and only if, ψ is injective.
3. B is a basis if, and only if, ψ is bijective.

Exercise 2.25 Let V be a linear space over $\mathbb{C}$, $\dim_\mathbb{C}(V)$ its dimension, and $\dim_\mathbb{R}(V)$ its dimension when considered, via restriction of scalars, as a linear space over $\mathbb{R}$. Prove that if $\dim_\mathbb{R}(V)$ is finite, then $\dim_\mathbb{R}(V) = 2\dim_\mathbb{C}(V)$. What happens in the infinite-dimensional case?

Exercise 2.26 Suppose $F \subseteq K$ are fields and that W is a linear space over K. By restriction of scalars W yields a linear space V over F. Fix a set S of vectors. Show that if S is linearly independent in W, then S is linearly independent in V. Show that if S is spanning in V, then S is spanning in W.

Exercise 2.27 Let V be a linear space, $S \subseteq V$ a spanning set, and $x \in S$. Show that if x is a linear combination of vectors from $S \setminus \{x\}$, then $S \setminus \{x\}$ is a spanning set.

Exercise 2.28 By convention, the empty sum, i.e., the sum of zero vectors, in a linear space is the 0 vector. Show that $\emptyset$ is linearly independent in any linear space V and that it is spanning only when $|V| = 1$.

Exercise 2.29 For a countable set B write $(\mathbb{Q}^B)_0$ as a countable union of countable sets. Can the same be done for $\mathbb{Q}^B$?

2.3 Linear Operators

The definition and properties of the structure preserving functions between linear spaces form the topic of this section. The content is discussed and exemplified in the context of the linear spaces introduced above.

Definition 2.4 Let V and W be linear spaces over the same field K. A function $T : V \to W$ is *additive* if

$$T(x + y) = T(x) + T(y)$$

for all vectors $x, y \in V$, and it is *homogenous* if

$$T(\alpha x) = \alpha T(x)$$

for all vectors $x \in V$ and all scalars $\alpha \in K$. If T is both additive and homogeneous, then it is called a *linear operator*, a condition equivalent to the equality

$$T(\alpha_1 x_1 + \alpha_2 x_2) = \alpha_1 T(x_1) + \alpha_2 T(x_2)$$

for all $x_1, x_2 \in V$ and $\alpha_1, \alpha_2 \in K$.

Remark 2.7 Synonymous terms for linear operator are *linear transformation* and *linear homomorphism*. Throughout the book we will adopt the convention that any reference to a linear operator $T : V \to W$ immediately implies, implicitly at times, that V and W are linear spaces over the same field K.

It is a straightforward proof by induction that a linear operator T preserves any linear combination, that is

$$T \left(\sum_{k=1}^{m} \alpha_k x_k \right) = \sum_{k=1}^{m} \alpha_k T(x_k)$$

for all scalars $\alpha_1, \ldots, \alpha_m \in K$ and vectors $x_1, \ldots, x_m \in V$. In fact, this property can equivalently be taken as the definition of linear operator. The next result shows that the linearity requirement of a linear operator forces any linear operator to also respect the zero vector, additive inverses, and subtraction.

Proposition 2.5 *For a linear operator* $T : V \to W$:

1. $T(0) = 0$.
2. $T(-x) = -T(x)$ *for all* $x \in V$.
3. $T(x - y) = T(x) - T(y)$ *for all* $x, y \in V$.

Proof 1. Notice that $T(0) + T(0) = T(0 + 0) = T(0)$ and use cancelation.
2. $T(x) + T(-x) = T(x + (-x)) = T(0) = 0$.
3. $T(x - y) = T(x + (-y)) = T(x) + T(-y) = T(x) - T(y)$.

$\square$

2.3.1 Examples of Linear Operators

Trivial, but important, examples of linear operators are the following. For any linear space V, the identity function id $: V \to V$ is always a linear operator. On the other extreme, given linear spaces V and W over the same field, the function $T_0 : V \to W$ given by $T_0(x) = 0$ is also immediately seen to be a linear operator.

In the presence of a basis for the domain V, linear operators $T : V \to W$ are easily characterized by the following result, readily yielding an endless supply of examples.

Lemma 2.2 *Let V and W be linear spaces over the same field K and B a basis for V. Then any function $F : B \to W$ extends uniquely to a linear operator $T_F : V \to W$.*

Proof For a given function $F : B \to W$, suppose that $T_F : V \to W$ is a linear operator extending F. Then, given any vector $x \in V$, write

$$x = \sum_{b \in B} \alpha_b \cdot b$$

as a (finite) linear combination of basis elements, and then it follows that

$$T_F(x) = T_F \left(\sum_{b \in B} \alpha_b \cdot b \right) = \sum_{b \in B} \alpha_b \cdot T_F(b) = \sum_{b \in B} \alpha_b \cdot F(b).$$

Noticing that the computation above expresses $T_F(x)$ in terms of vectors of the form $F(b)$ we conclude that an extension T_F, if it exists, is unique. If we now take the equality above as a definition, then we obtain a function $T_F : V \to W$ (relying on the uniqueness of the linear combination expressing x in order to assure that T_F is well defined). Verifying that this T_F is indeed a linear operator and that it extends F is immediate.   □

Remark 2.8 This linear operator $T_F : V \to W$ obtained from the function $F : B \to W$ is said to be obtained by *linearly extending* F to all of V.

Often enough the technique of the last result is inadequate, either because a basis is not readily available or because a more direct formula for the linear operator is obtainable. The following examples illustrate these possibilities.

Example 2.17 The derivation operation d/dt, taking a function f to its derivative df/dt, satisfies the properties

$$\frac{d}{dt}(f+g) = \frac{d}{dt}(f) + \frac{d}{dt}(g)$$

and

$$\frac{d}{dt}(\alpha f) = \alpha \frac{d}{dt}(f),$$

namely, it is additive and homogenous, and thus, if we choose the domain and codomain correctly, we expect it to be a linear operator. To turn this observation into a precise statement, recall the linear spaces $C^k([a, b], \mathbb{R})$ from Example 2.7 of all functions $x : [a, b] \to \mathbb{R}$ with a continuous k-th derivative. Then, for every $k \geq 1$, the observation above can be stated by saying that $d/dt : C^k([a, b], \mathbb{R}) \to C^{k-1}([a, b], \mathbb{R})$ is a linear operator. Similarly, $d/dt : C^\infty([a, b], \mathbb{R}) \to C^\infty([a, b], \mathbb{R})$ is a linear operator on the linear space of infinitely differentiable functions.

Example 2.18 The integral operator $\int_a^b dt\, f(t)$ is also well known to be additive and homogenous, and, since every continuous function on a closed interval is integrable, we obtain that for every interval $I = [a, b]$ the function

$$f \mapsto \int_a^b dt\, f(t)$$

is a linear operator from the linear space $C(I, \mathbb{R})$ of Example 2.7 to $\mathbb{R}$ as a linear space over itself.

Example 2.19 Spaces of infinite sequences, such as $\mathbb{R}^\infty$ from Example 2.3 or the space c_0 from Example 2.4 admit the following operators (we use $\mathbb{R}^\infty$ just to fix one possibility). The *shift operator* $S : \mathbb{R}^\infty \to \mathbb{R}^\infty$, given, for $x = (x_1, x_2, \dots, x_k \dots)$, by $S(x) = (x_2, x_3, \dots, x_k, \dots)$, is easily seen to be a linear operator. It is equally easy to see that the function $T : \mathbb{R}^\infty \to \mathbb{R}^\infty$, given for $x = (x_1, x_2, \dots, x_k \dots)$ as above by $T(x) = (0, x_1, x_2, \ldots, x_k, \ldots)$ is a linear operator. Note that $S \circ T$ is the identity while $T \circ S$ is not, a phenomenon that is known to be impossible in finite-dimensional linear spaces. These operators are the *creation* and *annihilation* operators $a^\dagger$ and a, widely used in Quantum Mechanics.

We see thus that even when a basis can be given explicitly (and certainly when it cannot) it may be much simpler to define a linear operator directly rather than by linear extension on a basis.

2.3.2 Algebra of Operators

With suitable domain and codomain, linear operators can be composed or added. We investigate the algebraic laws governing these operations.

Proposition 2.6 *The composition* $S \circ T : U \to W$ *of any two linear operators* $U \xrightarrow{\ T\ } V \xrightarrow{\ S\ } W$ *(where in particular all linear spaces are over the same field) is a linear operator.*

Proof The additivity of $S \circ T$ follows from the definition of composition and the computation

$$S(T(x + y)) = S(T(x) + T(y)) = S(T(x)) + S(T(y)),$$

valid for all vectors $x, y \in U$, utilizing the additivity of T and S. The homogeneity of the composition follows similarly as the reader can verify. □

Next, any two functions $T, S : V \to W$ between linear spaces over the field K can be added together to form the function $T + S : V \to W$ given by

$$(T + S)(x) = T(x) + S(x)$$

for all $x \in V$. Further, given a scalar $\alpha \in K$, the function $\alpha T : V \to W$ is given by

$$(\alpha T)(x) = \alpha(T(x)).$$

Proposition 2.7 *For all linear operators* $T, S : V \to W$ *and scalars* $\alpha \in K$, *both* $T + S$ *and* αT *are linear operators.*

Proof For all vectors $x, y \in V$ and scalars $\beta, \gamma \in K$:

$$\begin{aligned}
(T + S)(\beta x + \gamma y) &= T(\beta x + \gamma y) + S(\beta x + \gamma y) \\
&= T(\beta x) + T(\gamma y) + S(\beta x) + S(\gamma y) \\
&= \beta T(x) + \gamma T(y) + \beta S(x) + \gamma S(y) \\
&= \beta(T(x) + S(x)) + \gamma(T(y) + S(y)) \\
&= \beta(T + S)(x) + \gamma(T + S)(y).
\end{aligned}$$

The proof for αT is similar and left for the reader. □

The result above shows that the set of all linear operators $T : V \to W$ has naturally defined notions of addition and scalar multiplication. It is a pleasant fact that with these operations one obtains a linear space.

Definition 2.5 Given linear spaces V and W over the same field K we denote by $\mathrm{Hom}(V, W)$ the set of all linear operators $T : V \to W$.

Theorem 2.3 *For linear spaces V and W over the same field K the set* $\mathrm{Hom}(V, W)$, *when endowed with the operations of addition and scalar multiplication as above, is a linear space over K.*

Proof The proof is a straightforward verification of the linear space axioms, so we only give the details for a few of the axioms. For instance, given linear operators $T_1, T_2 : V \to W$, to show that

$$T_1 + T_2 = T_2 + T_1$$

we note that

$$(T_1 + T_2)(x) = T_1(x) + T_2(x) = T_2(x) + T_1(x) = (T_2 + T_1)(x)$$

where the commutativity of vector addition in W was used.

To show the existence of an additively neutral element, recall that $T_0 : V \to W$ given by $T_0(x) = 0$, is always a linear operator, and thus $T_0 \in \mathrm{Hom}(V, W)$. Seeing that T_0 is additively neutral is simply the computation

$$(T + T_0)(x) = T(x) + T_0(x) = T(x) + 0 = T(x).$$

We leave the verification of the other axioms to the reader. □

2.3.3 Isomorphism

The linear spaces $\mathbb{R}^{n+1}$ and P_n, while consisting of very different elements, are, in some sense, essentially identical. It is obvious that one may think of a sequence of $n + 1$ real numbers as the coefficients of a polynomial. Vice versa, one may identify a polynomial function with its list of coefficients and thus obtain $n + 1$ real numbers. Thus, one may rename the elements in one space to obtain the elements of the other, and, moreover, the linear structure under this renaming is respected. This situation is made precise, and generalized, by the concept of isomorphism.

Definition 2.6 A linear operator $T : V \to W$ that, as a function, is a bijection is called a *linear isomorphism* or (simply an *isomorphism* if the linear context is clear). Stated differently, T is an isomorphism precisely when it is *invertible*. When T :

$V \to W$ is an isomorphism the spaces V and W are said to be *isomorphic*, denoted by $V \cong W$.

The following result establishes some expected behavior of isomorphisms.

Proposition 2.8 *Suppose that U, V, and W are linear spaces over the same field K and consider the linear operators $T : U \to V$ and $S : V \to W$. The following then hold.*

1. *The identity function $id : V \to V$ is an isomorphism.*
2. *If T and S are isomorphisms, then so is $S \circ T$.*
3. *If T is an isomorphism, then so is the inverse function T^{-1}.*

Proof 1. We already noted that the identity function is always a linear operator. It is clearly bijective and thus an isomorphism.

2. We have seen that the composition $S \circ T$ of linear operators is a linear operator. The composition of bijective functions is again bijective, and so if T and S are isomorphisms, then so is $S \circ T$.

3. Since $T : U \to V$ is an isomorphism, the inverse function $T^{-1} : V \to U$ exists, and is clearly a bijection. To conclude the proof it remains to be shown that T^{-1} is a linear operator. Indeed, T^{-1} is additive since

$$
\begin{aligned}
T^{-1}(x + y) &= T^{-1}(T(T^{-1}(x)) + T(T^{-1}(y))) \\
&= T^{-1}(T(T^{-1}(x) + T^{-1}(y)) \\
&= T^{-1}(x) + T^{-1}(y)
\end{aligned}
$$

for all vectors $x, y \in V$. Further, T^{-1} is homogenous since

$$
T^{-1}(\alpha x) = T^{-1}(\alpha T(T^{-1}(x))) = T^{-1}(T(\alpha T^{-1}(x))) = \alpha T^{-1}(x)
$$

for all vectors $x \in V$ and scalars $\alpha \in K$.

$\square$

Remark 2.9 The third property deserves further attention. A linear operator $T : U \to V$ is a function between the sets underlying the linear spaces that is required to conserve the linear structure on the sets. Inducing temporary amnesia and forgetting the linear structure altogether focuses just on the functional aspect. The property then says that if T, merely as a function, is invertible, then, upon recovering from the amnesic episode, the fact that T conserves the algebraic structure forces T^{-1} to do the same. Analogous such conservation forcing may very well fail in other situations. For instance, the function $f : [0, 2\pi) \to C$ given by $f(\theta) = e^{i\theta}$ is a continuous bijection onto the unit circle C, but its inverse is not continuous. So, continuity of f does not force the continuity of f^{-1}, unless some further topological restrictions are placed on the spaces involved.

Corollary 2.1 *For all linear spaces U, V, and W over the same field K:*

1. $V \cong V$.
2. If $U \cong V$, then $V \cong U$.
3. If $U \cong V$ and $V \cong W$, then $U \cong W$.

Isomorphic linear spaces are essentially identical, except (possibly) for the names of the elements in them and they thus possess exactly the same linear properties. This is a somewhat vague statement but it is almost always immediate how to turn it into a precise statement. For instance, the dimension of a linear space is a linear property and thus any two isomorphic linear spaces have the same dimension, as we now show.

Proposition 2.9 *If* $V \cong W$, *then the linear spaces have the same dimension.*

Proof By assumption there exists a linear isomorphism $T : V \to W$. The dimension of a linear space is the cardinality of any of its bases, and thus the result will be established by exhibiting a bijection between a basis of V and a basis of W. Let $B \subseteq V$ be a basis and consider its image $T(B) \subseteq W$. Clearly, $T|_B$, the restriction of T to B, is a bijection between B and $f(B)$, so all that remains to be done is show that $f(B)$ is a basis for W, namely that $f(B)$ is a spanning set of linearly independent vectors. Let $y \in W$ be an arbitrary vector. Since T is onto, we may write $y = T(x)$ for some $x \in V$. As B is a basis, we express x as a (finite!) linear combination of basis elements:

$$x = \sum_{b \in B} \alpha_b \cdot b.$$

But then

$$y = T(x) = \sum_{b \in B} \alpha_b \cdot T(b)$$

and thus y is in the span of $T(B)$. As y was arbitrary we conclude that $T(B)$ spans all of W. The proof that $T(B)$ is also linearly independent follows a similar pattern and is left for the reader. $\square$

The converse to this result is also true, as the following result implies.

Theorem 2.4 *If* V *is a linear space and* B *is a basis for it, then* $V \cong (K^B)_0$.

Proof In Proposition 2.3 we established that the function $T : (K^B)_0 \to V$ given by

$$T(f) = \sum_{b \in B} f(b) \cdot b$$

is a bijection. We show now that it is in fact an isomorphism, thus completing the proof. We need to verify that $T(\alpha_1 f + \alpha_2 g) = \alpha_1 T(f) + \alpha_2 T(g)$, which amounts to showing that

$$\sum_{b \in B} (\alpha_1 f + \alpha_2 g)(b) \cdot b = \alpha_1 \sum_{b \in B} f(b) \cdot b + \alpha_2 \sum_{b \in B} g(b) \cdot b,$$

which follows by an immediate computation.   □

Remark 2.10 Every linear space V over K is thus isomorphic to a linear space of the form $(K^B)_0$. The latter space is clearly only dependent on the cardinality of the set B, i.e., if X is any set with the same cardinality as B, then $(K^B)_0 \cong (K^X)_0$. In other words, the dimension of a linear space determines it up to isomorphism. It should be noted however that generally there is no canonical choice for an isomorphism between V and $(K^X)_0$.

In the finite-dimensional case we obtain the following corollary.

Corollary 2.2 *Let V be a finite-dimensional linear space over the field K. Any choice of a basis $x_1, \ldots, x_n$ for V gives rise to an isomorphism $V \rightarrow K^n$. In this way we may identify every vector in V, by means of coordinates, with an n-tuple of scalars.*

In a more general context, in particular in the infinite-dimensional case, where bases are rarely explicitly available, we have the following results.

Corollary 2.3 *If V and W are linear spaces over the same field K and they have the same dimension, then they are isomorphic.*

Proof By Remark 2.10, if we let X be a set of cardinality equal to the common dimensions of V and W, then we obtain an isomorphisms $T_1 : (K^X)_0 \rightarrow V$ and an isomorphism $T_2 : (K^X)_0 \rightarrow W$. It follows that $T_2 \circ T_1^{-1} : V \rightarrow W$ is an isomorphism, showing that $V \cong W$, as claimed.   □

Combining Proposition 2.9 and Corollary 2.3, the discussion above is summarized as follows.

Theorem 2.5 *Two linear spaces over the same field are isomorphic if, and only if, they have the same dimension.*

Remark 2.11 This result is an example of what is known as a rigidity phenomenon. Rigidity refers to a situation where two structures are essentially the same given that they have essentially the same substructure of some kind, typically of a much coarser nature than the original structures. In this case, the rigidity of linear spaces over a fixed field K is that the dimension suffices to determine the linear space up to isomorphism.

Example 2.20 Recall the linear space P_n of polynomial functions from Example 2.6. We can show that $P_n \cong \mathbb{R}^{n+1}$ by constructing an isomorphism between the two spaces. Indeed, referring here to a typical element in $\mathbb{R}^{n+1}$ by $(a_0, \ldots, a_n)$, let

$$T : \mathbb{R}^{n+1} \rightarrow P_n$$

be given by

$$T(a_0, \ldots, a_n) = a_n t^n + \ldots + a_1 t + a_0.$$

It is a trivial matter to verify that T is a linear operator, clearly bijective, and thus an isomorphism. Similarly, the reader is invited to show that $c_{00} \cong P$. By the discussion above, the existence of an isomorphism (but not any explicit isomorphism) could have been deduced by noting that both $\mathbb{R}^{n+1}$ and P_n, as linear spaces over $\mathbb{R}$, have dimension $n + 1$. Similarly, both c_{00} and P have countably infinite dimension and are thus isomorphic.

Remark 2.12 Theorem 2.5 tells us that the dimension of a linear space essentially is all that one needs to know about a linear space in order to study it (since after all, isomorphic linear spaces are essentially the same). It is thus tempting to, once and for all, choose a single representative from each isomorphism class of linear spaces. This certainly suffices for the study of all linear spaces, and seems more economical than having all of this redundancy in the form of multiple isomorphic specimens of linear spaces. However, the example above clearly shows the danger in this approach. While $\mathbb{R}^{n+1}$ and P_n are isomorphic, they have different qualities (at least to us humans). For instance, it is very natural to consider the integral as an operator on polynomials, but not so much on elements in $\mathbb{R}^{n+1}$. The richness of a plethora of different linear spaces, even if they are isomorphic, is a blessing that should not be given up for the illusion of a more economical treatment.

Exercises

Exercise 2.30 Consider the linear space c of all convergent sequences of complex numbers (or real numbers, with the obvious adjustments). Prove that the assignment

$$\{x_m\}_{m \geq 1} \mapsto \lim_{m \to \infty} x_m$$

is a linear operator from c to $\mathbb{C}$.

Exercise 2.31 Provide the necessary definitions to make the assertion that $f \mapsto \lim_{x \to \infty} f(x)$ is a linear operator into a precise statement. Proceed similarly for the assignments $f \mapsto \frac{\partial f}{\partial t}$ and $f \mapsto \frac{d^2 f}{dt^2}$.

Exercise 2.32 Verify that the composition $S \circ T$ of linear operators is homogenous.

Exercise 2.33 Consider the linear space c of convergent complex sequences. Show, for fixed n, that evaluation at n, namely $x \mapsto x_n$ where x_n is the n-th term in the sequence, is a linear operator $\text{ev}_n : c \to \mathbb{C}$. Is every element in $\text{Hom}(c, \mathbb{C})$ a linear combination of evaluation operators?

Exercise 2.34 Finish the proof of Theorem 2.3 showing that $\text{Hom}(V, W)$ is a linear space.

Exercise 2.35 Let V and W be finite-dimensional linear spaces over a field K of dimensions n and m with $B_1 = \{v_1, \ldots, v_n\}$ and $B_2 = \{w_1, \ldots, w_m\}$ bases for V and W, respectively. For any vector $x \in V$ write

$$[x]_{B_1} = (a_1, \ldots, a_n) \in K^n$$

where $a_1, \ldots, a_n$ are the unique scalars such that

$$x = \sum_{k=1}^{n} a_k v_k.$$

The tuple $[x]_{B_1}$ is called the vector of coordinates of x in the basis B_1. The vector $[y]_{B_2}$ is similarly defined for all $y \in W$.

1. Given a linear operator $T : V \to W$ prove the existence of a unique matrix $[T]_{B_1, B_2} \in M_{n,m}(K)$ such that

$$[T(x)]_{B_2} = [T]_{B_1, B_2} \cdot [x]_{B_1}$$

   where on the right-hand side the $\cdot$ stands for the ordinary product of a matrix by a vector. The matrix $[T]_{B_1, B_2}$ is called the *representative* matrix of the linear operator T in the given bases.

2. Prove that the assignment $T \mapsto [T]_{B_1, B_2}$, mapping a linear operator to the matrix representing it in the two given bases, is a linear isomorphism

$$\psi_{B_1, B_2} : \mathrm{Hom}(V, W) \to M_{n.m}(K).$$

3. Suppose now that U is a third linear space with its chosen basis B_3 and that $S : W \to U$ is a linear operator. Prove that $[S \circ T]_{B_1, B_3} = [S]_{B_2, B_3} \cdot [T]_{B_1, B_2}$, where the $\cdot$ on the right-hand side is the ordinary product of matrices.

Note how this exercise illustrates that matrix multiplication is defined the way it is in order to correspond to operator composition.

Exercise 2.36 Prove that $\mathbb{R}^n$ and $\mathbb{C}^n$, $n \geq 1$, when considered as linear spaces over $\mathbb{R}$, are not isomorphic.

Exercise 2.37 Prove that $\mathbb{R}^\infty$ and $\mathbb{C}^\infty$, as linear spaces over $\mathbb{R}$, are isomorphic.

Exercise 2.38 A linear operator $T : V \to V$ is *nilpotent* if there exists an $m \geq 1$ such that T^m, the m-fold composition of T with itself, is the zero operator $0 : V \to V$. Prove that if $T : V \to V$ is nilpotent, then

$$(\mathrm{id}_V - T)^{-1}$$

exists and thus $\mathrm{id}_V - T$ is an isomorphism. What does this exercise have to do with the sum

$$(1 - t)^{-1} = \sum_{k=0}^{\infty} t^k$$

of a geometric series for $|t| < 1$?

Exercise 2.39 Let $T : V \to V$ be an operator with the property that for all $x \in V$ there exists an $m \geq 1$ such that $T^m(x) = 0$. Prove that if V is finite-dimensional, then T is nilpotent (see previous exercise), but if V is infinite-dimensional then T need not be nilpotent.

Exercise 2.40 Prove that for every set $\mathcal{B}$ there exists a linear space with $\mathcal{B}$ as a basis (and consequently for any cardinality κ there exists a linear space whose dimension is κ).

Exercise 2.41 Let $f : \mathbb{R} \to \mathbb{R}$ be a continuous function with $f(x + y) = f(x) + f(y)$ for all $x, y \in \mathbb{R}$. Prove that there exists some $a \in \mathbb{R}$ such that $f(x) = ax$ for all $x \in \mathbb{R}$.

Exercise 2.42 Consider $\mathbb{R}$ as a linear space over itself. Show that the function $T : \mathbb{R} \to \mathrm{Hom}(\mathbb{R}, \mathbb{R})$ given by $T(a)(x) = ax$ is a linear isomorphism, so $\mathbb{R} \cong \mathrm{Hom}(\mathbb{R}, \mathbb{R})$.

Exercise 2.43 Use a Hamel basis for $\mathbb{R}$ as a linear space over $\mathbb{Q}$ to construct a function $f : \mathbb{R} \to \mathbb{R}$ satisfying $f(x + y) = f(x) + f(y)$ for all $x, y \in \mathbb{R}$ that is not of the form $f(x) = ax$ for any $a \in \mathbb{R}$.

Exercise 2.44 Prove the linear analogue of the Cantor-Schröder-Bernstein Theorem: If two linear spaces V and W admit injective linear operators $T : V \to W$ and $S : W \to V$, then $V \cong W$.

Exercise 2.45 Let V be a linear space over a field K and suppose a basis B of V is given. We view K as a linear space over itself. Define, for each $b \in B$, the vector $T_b \in \mathrm{Hom}(V, K)$ so that $T_b(x)$ is the coefficient of b when expressing x as a linear combination of the basis vectors. Show that the set $\{T_b \mid b \in B\}$ is linearly independent and, further, that it is a basis if, and only if, V is finite dimensional.

Exercise 2.46 Prove that for all $m, n \geq 1$ the sets $\mathbb{R}^n$ and $\mathbb{R}^m$ have the same cardinality but that $\mathbb{R}^n \cong \mathbb{R}^m$ as linear spaces over $\mathbb{R}$ if, and only if, $m = n$.

Exercise 2.47 Find all linear isomorphisms $\mathbb{R} \to \mathbb{R}$ when $\mathbb{R}$ is a linear space over itself. Repeat the exercise for $\mathbb{C}$ as a linear space over itself as well as over $\mathbb{R}$.

2.4 Subspaces, Products, and Quotients

Subspaces arise naturally as portions of a given linear space that inherit a linear space structure from the ambient linear space. We show that any linear operator gives rise to two subspaces, its kernel and its image, and we show how to combine two linear spaces to form their product space. We also show how to eliminate a subspace so as to obtain a quotient space. The connection to complementary spaces is made explicit.

2.4.1 Subspaces

Given a subset $A \subseteq V$ of a linear space V the operations of addition and scalar multiplication, when restricted to vectors in A always yield elements in the ambient space V. We are interested in the case where the results of these operations always yield elements in A.

Definition 2.7 A subset $A \subseteq V$ is said to be *closed under addition* if

$$x + y \in A$$

for all vectors $x, y \in A$. Similarly, A is said to be *closed under scalar multiplication* if

$$\alpha x \in A$$

for all vectors $x \in A$ and scalars $\alpha \in K$. The set A is called a *linear subspace* (or simply *subspace* if the linear context is evident) of V if $A \neq \emptyset$ and A is closed under addition and scalar multiplication. A subspace A is called a *proper subspace* if it is properly contained in the ambient space V.

Remark 2.13 If A is a linear subspace of V, then A, with vector addition and scalar multiplication induced from V, is a linear space. Indeed, the verification of each of the axioms follows the same pattern. For instance,

$$x + y = y + x$$

holds for all $x, y \in A$ since the same equality holds for all vectors x and y in the ambient space V. Moreover, notice that the zero vector 0 is always an element in A and is the zero vector of A. Indeed, $0 = 0 \cdot x$ holds for all $x \in A$, and since A is required to be non-empty, at least one $x \in A$ exists.

The subset $\{0\}$ consisting of just the zero vector is always a subspace called the *trivial subspace* of V. Another immediate example of a subspace of V is V itself, i.e., a non-proper subspace. An immediate property of subspaces is their transitive property, that is if $A \subseteq B \subseteq V$, then if B is a subspace of V and A is a subspace of B, then A is a subspace of V.

Example 2.21 Consider the linear space $\mathbb{R}^2$ and its subset A consisting only of the vectors of the form $(x, 0) \in \mathbb{R}^2$. Clearly, A is non-empty, closed under addition, and closed under scalar multiplication, and hence is a subspace of $\mathbb{R}^2$. More generally, a line $l = \{tx \mid t \in \mathbb{R}\}$, with $x \in \mathbb{R}^2$ a non-zero vector, is a linear subspace of $\mathbb{R}^2$. These linear subspaces, together with the subspaces $\{0\}$ and $\mathbb{R}^2$, exhaust all of the linear subspaces of $\mathbb{R}^2$. Similarly, linear subspaces of $\mathbb{R}^n$, for larger values of n, correspond to the origin, lines through the origin, planes through the origin, and hyper-planes through the origin. Notice the following subtle point. The field $\mathbb{R}$, when viewed as a linear space over itself, is obviously isomorphic to $\mathbb{R}^1$, that is to $\mathbb{R}^n$ where $n = 1$.

However, even though the linear space $\mathbb{R} \cong \mathbb{R}^1$ may naturally be identified with either the **X**-axis or the **Y**-axis in $\mathbb{R}^2$, formally speaking, $\mathbb{R}$ itself (or $\mathbb{R}^1$) is not a subspace of $\mathbb{R}^2$. In fact, it is not even a subset of $\mathbb{R}^2$. More generally, $\mathbb{R}^n$ is a subspace of $\mathbb{R}^m$ if, and only if, $n = m$, in which case the two spaces coincide. When $n \leq m$ the space $\mathbb{R}^n$ may be identified in infinitely many ways with various subspaces of $\mathbb{R}^m$ but no such identification is canonical. An analogous discussion can be given for linear spaces over $\mathbb{C}$.

Example 2.22 Consider the linear space $\mathbb{C}^\infty$ from Example 2.3 and the linear spaces c of convergent sequences, c_0 of sequences that converge to 0, and c_{00} of sequences that are eventually 0 (see Example 2.4). Clearly,

$$c_{00} \subset c_0 \subset c \subset \mathbb{C}^\infty$$

and in fact c_{00} is a linear subspace of c_0, which in turn is a linear subspace of c, which is a linear subspace of $\mathbb{C}^\infty$. The verification is immediate.

Other subspaces of $\mathbb{C}^\infty$ include, for each $n \geq 1$, the set of all vectors of the form $(x_1, \ldots, x_n, 0, 0, \ldots)$. This space is clearly isomorphic to $\mathbb{C}^n$, as are infinitely many other subspaces of $\mathbb{C}^\infty$, as the reader is asked to verify. In any event, $\mathbb{C}^n$ is not itself a subspace of $\mathbb{C}^\infty$.

Example 2.23 Consider the linear space P_n (Example 2.6) consisting of polynomial functions with real coefficients of degree not exceeding n, and P, the linear space of all polynomials with real coefficients. For all $n \leq m$, it holds that $P_n \subseteq P_m \subset P$, and it follows easily that P_n is a subspace of P_m, which in turn is a subspace of P. We thus obtain the tower of proper subspace inclusions

$$P_0 \subset P_1 \subset P_2 \subset \cdots \subset P_n \subset \cdots \subset P.$$

Example 2.24 With reference to Example 2.7, the subset of $C([a, b], \mathbb{R})$ consisting of the continuous functions $x : [a, b] \to \mathbb{R}$ which vanish at a and b, that is such that $x(a) = x(b) = 0$, constitutes a linear subspace. On the contrary, the subset of those continuous functions with $x(a) = x(b) = 1$ does not constitute a linear subspace (as it fails to contain the zero vector).

The following family of linear subspaces of $\mathbb{C}^\infty$ is of significant importance.

Example 2.25 (*The ℓ_p Spaces*) Consider the linear space $\mathbb{C}^\infty$ from Example 2.3 and recall that a sequence $x = (x_1, \ldots, x_k, \ldots) \in \mathbb{C}^\infty$ is said to be *bounded* if there exists some $M \in \mathbb{R}$ such that

$$|x_k| < M$$

for all $k \geq 1$. Let $\ell_\infty \subseteq \mathbb{C}^\infty$ be the set of all bounded sequences. Clearly, the zero vector, namely the constantly zero sequence, is in ℓ_∞, the sum of two bounded sequences is again bounded, and thus in ℓ_∞, and if x is bounded by M, then $\alpha x = (\alpha x_1, \ldots, \alpha x_k, \ldots)$ is bounded by $|\alpha| M$, and thus is in ℓ_∞. Hence ℓ_∞ is a linear subspace of $\mathbb{C}^\infty$.

Now, let $p \geq 1$ and consider the set ℓ_p of all sequences of complex numbers that are *absolutely p-power summable*, that is, all sequences $x \in \mathbb{C}^\infty$ such that

$$\sum_{k=1}^\infty |x_k|^p < \infty.$$

Clearly, the constantly zero sequence belongs to ℓ_p, and ℓ_p is evidently closed under scalar multiplication. To show that ℓ_p is a subspace of $\mathbb{C}^\infty$ it remains to be shown that ℓ_p is closed under addition, a task we leave to the reader.

We thus obtain a one-parameter family $\{\ell_p\}_{1 \leq p \leq \infty}$ of linear subspaces of $\mathbb{C}^\infty$. It can be shown that if $1 \leq p < q \leq \infty$, then $\ell_p \subset \ell_q$ (consider a suitable variation of the harmonic series), and so ℓ_p is a proper linear subspace of ℓ_q.

Example 2.26 Recall from Example 2.8 that $\mathbb{R}$ may be viewed as a linear space over $\mathbb{R}$, and that $\mathbb{C}$ can be viewed as a linear space over $\mathbb{C}$ or as a linear space over $\mathbb{R}$. The set inclusion $\mathbb{R} \subseteq \mathbb{C}$ is of course always valid, but $\mathbb{R}$ is a linear subspace of $\mathbb{C}$ only when both are considered to be linear spaces over $\mathbb{R}$ (or over $\mathbb{Q}$). A subspace relation between two linear spaces can only hold if both linear spaces are over the same field, and so, for instance, $\mathbb{C}^n$ as a linear space over $\mathbb{R}$ is not a linear subspace of $\mathbb{C}^n$ as a linear space over $\mathbb{C}$.

We address the relationship between subspaces and dimension. In the finite case there are no surprises. The infinite case, as usual, is more subtle.

Theorem 2.6 *For a linear space V of dimension n and a subspace $U \subseteq V$ of dimension m, the following hold.*

1. $m \leq n$.
2. *If V is finite dimensional and $m = n$, then $U = V$.*

Proof 1. The proof is given in full generality, thus n and m may be infinite cardinals. Let B_V be a basis for V and let B_U be a basis for U. By assumption, the cardinality of B_V is n while the cardinality of B_U is m. Now, by definition of basis, B_U is a linearly independent set of vectors in U, and hence is also linearly independent in V, while B_V is a spanning set in V. By Lemma 2.1 there exists an injection $B_U \to B_V$, and thus $n \leq m$.

2. Assume now that $m = n$ and that V is finite dimensional, namely that $n = m$ is a natural number. We may thus find a basis $\{x_1, \dots, x_n\}$ of U. In particular these vectors are linearly independent in U, and thus also in V. By Proposition 2.4, it follows that the set $\{x_1, \dots, x_n\}$ is a basis of V, so its span is V. But the span is also U, and the claim follows.

$\square$

Remark 2.14 Much as in Remark 2.6, the finite dimensionality assumption cannot be avoided. Indeed, the linear space c_{00} from Example 2.4 is, as we saw, of countably infinite dimension but it does contain proper subspaces of the same dimension. For instance, it is easy to verify that the set c_{000} of all sequences in c_{00} whose first term is 0, is a proper subspace of c_{00} even though it is isomorphic to c_{00}, and thus has the same dimension as the ambient space.

2.4.2 Kernels and Images

With any linear operator $T : V \rightarrow W$ we associate a subspace of the domain, called the kernel of T, and a subspace of the codomain, called the image of T. The kernel suffices to detect injectivity of T.

Definition 2.8 Let $T : V \rightarrow W$ be a linear operator. The set

$$\mathrm{Ker}(T) = \{x \in V \mid T(x) = 0\}$$

is called the *kernel* of T.

Theorem 2.7 *The kernel of a linear operator $T : V \rightarrow W$ is a linear subspace of V.*

Proof In Proposition 2.5 we already noted that $T(0) = 0$ and thus $0 \in \mathrm{Ker}(T)$. All we need to do now is show that the kernel is closed under vector addition and scalar products. Indeed, if $x, y \in \mathrm{Ker}(T)$, then

$$T(x + y) = T(x) + T(y) = 0 + 0 = 0$$

and thus $x + y \in \mathrm{Ker}(T)$. Similarly, for all $\alpha \in K$, if $x \in \mathrm{Ker}(T)$, then

$$T(\alpha x) = \alpha T(x) = \alpha \cdot 0 = 0$$

so $\alpha x \in \mathrm{Ker}(T)$ and the proof is complete. □

Theorem 2.8 *A linear operator $T : V \rightarrow W$ is injective if, and only if, its kernel is trivial, i.e., $\mathrm{Ker}(T) = \{0\}$.*

Proof If T is injective, then since we already know that $T(0) = 0$, it follows that $T(x) = 0$ implies $x = 0$, and thus $\mathrm{Ker}(T) = \{0\}$. Conversely, suppose that the kernel of T is trivial and that

$$T(x) = T(y)$$

for some $x, y \in V$. Then

$$T(x - y) = T(x) - T(y) = 0$$

and so

$$x - y \in \text{Ker}(T).$$

But then $x - y = 0$, and so $x = y$, showing that T is injective.   □

Another subspace naturally obtained from a linear operator is its image.

Definition 2.9 Let $T : V \to W$ be a linear operator. The set $\text{Im}(T) = \{T(x) \mid x \in V\}$, namely the set-theoretic image of T, is called the *image* of T.

Proposition 2.10 *The image of a linear operator $T : V \to W$ is a linear subspace of W.*

Proof Obviously, $T(V)$ is not empty. Further, if $x, y \in \text{Im}(T)$, then $x = T(x')$ and $y = T(y')$ for some vectors $x', y' \in V$. It then follows that

$$x + y = T(x') + T(y') = T(x' + y') \in \text{Im}(T).$$

Similarly, $\alpha x \in \text{Im}(T)$ for all $x \in \text{Im}(T)$ and all scalars $\alpha \in K$.   □

The following result, which the reader may recognize as the rank-nullity theorem, has important consequences for finite-dimensional linear spaces.

Lemma 2.3 *The equality*

$$\dim(V) = \dim(\text{Ker}(T)) + \dim(\text{Im}(T))$$

holds for all linear operators $T : V \to W$ between finite-dimensional linear spaces.

Proof We present the general strategy, inviting the reader to fill-in the details. Choose a basis $\{x_1, \ldots, x_k\}$ for $\text{Ker}(T)$. Then augment this basis by (if needed) adding vectors to it, so as to obtain a basis $\{x_1, \ldots, x_k, y_1, \ldots, y_m\}$ of V and show that $\{T(y_1), \ldots, T(y_m)\}$ is a basis for $\text{Im}(T)$.   □

Remark 2.15 The proof above can be adapted to obtain a similar result for infinite dimensional linear spaces, provided one handles cardinal arithmetic with care, of course. However, such a result is not of great importance and we thus avoid its details.

Corollary 2.4 *For a linear operator $T : V \to W$ between finite-dimensional linear spaces of equal dimension injectivity, surjectivity, and bijectivity coincide.*

Proof We have seen that T is injective if, and only if, its kernel is trivial, namely precisely when $\dim(\text{Ker}(T)) = 0$. It follows then that T is injective if, and only if, $\dim(V) = \dim(\text{Im}(T))$. But since $\dim(V) = \dim(W)$, it follows by Theorem 2.6, that $\dim(V) = \dim(\text{Im}(T))$ holds if, and only if, $\text{Im}(T) = W$, in other words, precisely when T is surjective. This completes the verification of the non-trivial arguments in the proof.   □

2.4.3 Products and Quotients

We present the product construction for two linear spaces V and W over the same field K. We endow the set $V \times W$ with an addition operation and a scalar product that turn it into a linear space, as follows. Given $(x_1, x_2), (y_1, y_2) \in V \times W$ and $\alpha \in K$ the addition operation is given by

$$(x_1, x_2) + (y_1, y_2) = (x_1 + y_1, x_2 + y_2)$$

and the scalar product operation is given by

$$\alpha(x_1, x_2) = (\alpha x_1, \alpha x_2).$$

Theorem 2.9 *For linear spaces V and W over the same field K, the set $V \times W$, when endowed with addition and scalar multiplication as above, is a linear space over the field K.*

Proof The verification of the linear space axioms is straightforward. For instance, the additively neutral element is $(0, 0)$ since

$$(x_1, x_2) + (0, 0) = (x_1 + 0, x_2 + 0) = (x_1, x_2)$$

for all $(x_1, x_2) \in V \times W$. The rest of the verification is for the reader. □

Obviously, one can similarly define the product $V_1 \times \cdots \times V_n$ of any finite number of linear spaces, and even the product of any collection of linear spaces. It is also evident, upon inspection of the linear structure of the linear space $\mathbb{R}^n$, that $\mathbb{R}^n$ is the n-fold product of $\mathbb{R}$ with itself, where $\mathbb{R}$ is viewed as a linear space over itself. Similarly, $\mathbb{C}^n$ is the n-fold product of $\mathbb{C}$ with itself.

We now turn to consider the quotient construction of a linear space by a linear subspace. This construction is, in some sense, the reversal of the product construction. We refer the reader to Sect. 1.2.15 of the Preliminaries for basic facts about equivalence relations.

Definition 2.10 Let V be a linear space and U a subspace of it. Two elements $x_1, x_2 \in V$ are said to be *equivalent modulo U* if

$$x_1 - x_2 \in U$$

that is, if

$$x_1 = x_2 + u$$

for some $u \in U$. This relation is denoted by $x_1 = x_2 (\text{mod } U)$, by $x_1 \equiv_U x_2$, or simply by $x_1 \equiv x_2$ if U is evident.

We show that $\equiv$ is an equivalence relation on V. Indeed, for all $x, y, z \in V$

$$x - x = 0 \in U$$

and thus

$$x \equiv x$$

so that $\equiv$ is reflexive. Further,

$$x \equiv y \implies x - y \in U \implies y - x = (-1) \cdot (x - y) \in U \implies y \equiv x$$

and thus $\equiv$ is symmetric. Finally,

$$x - y, y - z \in U \implies x - z = (x - y) + (y - z) \in U \implies x \equiv z$$

and so $\equiv$ is transitive.

Recall that the equivalence class of any vector $x \in V$ is the set

$$[x] = \{y \in V \mid x \equiv y\}.$$

In other words, it is the set of all vectors of the form $x + u$, where $u \in U$. For that reason the equivalence class $[x]$ is also denoted by

$$x + U.$$

It follows from the general theory of equivalence relations that $\{x + U \mid x \in V\}$ is a partition of V.

Example 2.27 As a simple example, consider the usual 2-dimensional plane $\mathbb{R}^2$ and its subspace $\mathbf{Y}$ consisting of the ordinates. Let the element $\mathbf{r} \in \mathbb{R}^2$ be the vector $O - P$ from the origin to the point P in the plane. The equivalence class $[\mathbf{r}]_{\mathbf{Y}}$, formed by the vectors of $\mathbb{R}^2$ equivalent mod $\mathbf{Y}$ to $\mathbf{r}$, is clearly given by all vectors $O - P_i$ with P_i lying on the parallel line to the $\mathbf{Y}$-axis passing through the point P; all these vectors are equivalent *mod* $\mathbf{Y}$ to each other.

The fact that the entire 2-dimensional space is partitioned by all such equivalence classes is simply the fact that $\mathbb{R}^2$ is the disjoint union of all the lines parallel to the $\mathbf{Y}$ axis, which are precisely the translates $x + \mathbf{Y}$ of $\mathbf{Y}$.

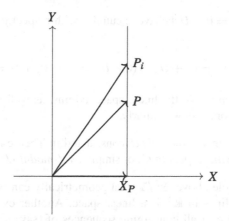

We turn to investigate the quotient set $V/U = \{x + U \mid x \in V\}$ of equivalence classes modulo U. It is natural to introduce an addition and a scalar product on V/U by the formulas

$$(y + U) + (z + U) = (y + z) + U \quad \text{and} \quad \alpha(y + U) = (\alpha y) + U,$$

but we must first verify that these operations are well defined, namely that they do not depend on the chosen representatives. So, for all $x, y, x', y' \in V$ we need to show that if

$$x + U = x' + U \quad \text{and} \quad y + U = y' + U,$$

then

$$(x + y) + U = (x' + y') + U.$$

But the former equalities imply that

$$x' - x \in U \text{ and } y' - y \in U$$

while the latter will be implied by showing that $(x' + y') - (x + y) \in U$. And indeed, since U is closed under addition,

$$(x' + y') - (x + y) = (x' - x) + (y' - y) \in U.$$

A similar argument shows that the scalar product above is well defined and thus the quotient set V/U of equivalence classes is naturally endowed with an addition operation and a scalar product operation.

Theorem 2.10 *Let V be a linear space and $U \subseteq V$ a subspace. The quotient set V/U of equivalence classes modulo U, with the operations of addition and scalar multiplication as given above, is a linear space whose zero vector is U. Moreover, the canonical projection $\pi : V \to V/U$ given by $\pi(x) = x + U$ is a surjective linear operator whose kernel is U.*

Proof To see that $U = 0 + U$ behaves neutrally with respect to addition note that
for all $x + U \in V/U$

$$(x + U) + U = (x + U) + (0 + U) = (x + 0) + U = x + U.$$

The rest of the verification of the linear space axioms, as well as the claims about
the canonical projection, follow similarly.   □

Definition 2.11 The linear space V/U constructed in Theorem 2.10 is called the
quotient space of V with respect to U, or simply as *V modulo U*.

Continuing the example above, $\mathbb{R}^2/\mathbf{Y}$, that geometrically can be thought of as the
space of all vertical lines in $\mathbb{R}^2$, is a linear space. Another example is c/c_{00}, the
quotient of the space c of all converging sequences of (say) real numbers, by the
subspace of all eventually 0 sequences. The quotient may be thought of as the space of
all sequences modulo finite changes. In analysis, the limit of a sequence is insensitive
to finite changes in the sequence, and so it is precisely this quotient that is of interest
already in elementary analysis.

2.4.4   Complementary Subspaces

Contemplating the quotient space associated to Example 2.27, one realizes that $\mathbb{R}^2/\mathbf{Y}$
is isomorphic to $\mathbf{X}$, and that under this isomorphism the canonical projection π :
$\mathbb{R}^2 \to \mathbb{R}^2/\mathbf{Y}$ is nothing but the projection of $\mathbb{R}^2$ onto the $\mathbf{X}$ axis. This observation
generalizes to any quotient space construction.

Definition 2.12 Two subspaces U and W of a given linear space V are said to be
complementary if $U \cap W = \{0\}$ and $U + W = V$.

The second condition states that any vector $x \in V$ can be decomposed as a sum
$x = u + w$ with $u \in U$ and $w \in W$. The first condition forces the uniqueness of this
decomposition.

Proposition 2.11 *If U and W are complementary subspaces of V, then any $x \in V$
can be expressed uniquely as*

$$x = u + w$$

with $u \in U$ and $w \in W$.

The proof is left as an exercise. This unique decomposition is said to express V as a
direct sum of its subspaces U and W, a situation denoted by $V = U \oplus W$.

 The following example shows that a given subspace $U \subseteq V$ may admit more than
one complementary subspace.

Example 2.28 (*Continuing Example 2.27*) In $\mathbb{R}^2$

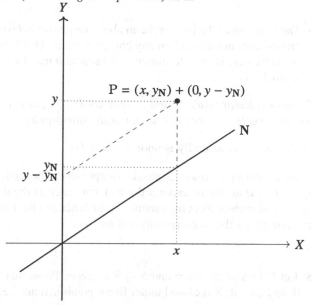

any straight line **N** passing through the origin and not coincident with the subspace **Y** is a complementary subspace. The unique decomposition of a vector pointing at P as a sum of a vector from **N** and a vector from **Y** is formed by the dashed lines forming a parallelogram with the given subspaces.

We may now present the main theorem relating complementary subspaces and the quotient construction.

Theorem 2.11 *Given a linear space V and two complementary subspaces U and W, the quotient space V/U is naturally isomorphic to W (the meaning of 'naturally' is discussed below).*

Proof Recall from Theorem 2.10 that the canonical projection $\pi : V \to V/U$, given by $\pi(x) = x + U$, is a linear operator. The result will be established by showing that the restriction of π to $W \subseteq V$ is an isomorphism between W and V/U. All we have to do is show that the restriction is a bijection. For injectivity we use Theorem 2.8 and show the kernel of the restriction is trivial. For that, suppose that $\pi(w) = 0$ for some $w \in W$, that is (recall that the zero vector in V/U is U)

$$w + U = U$$

and so $w \in U$. But since

$$U \cap W = \{0\}$$

it follows that $w = 0$, as needed. For surjectivity, let $y + U$ be an arbitrary element in V/U. Writing $y = u + w$, with $u \in U$ and $w \in W$, we get that

$$y + U = (u + w) + U = (u + U) + (w + U) = w + U = \pi(w).$$ □

Remark 2.16 The meaning of the isomorphism above being *natural* (or *canonical*) is that its construction does not depend on any choice of basis. This is an important observation since, as we saw, in infinite-dimensional spaces it may be impossible to actually describe any basis.

Corollary 2.5 *Given a linear space V and a subspace $U \subseteq V$, any two subspaces W_1 and W_2 that are complementary to U are naturally isomorphic.*

Proof Each of W_1 and W_2 is naturally isomorphic to V/U.   □

In the case of complementary spaces the stated isomorphism is independent of a basis but it is dependent on the ability to express the ambient space as the direct sum of the given subspace and each of its complements. If the latter can be done explicitly, then an explicit formula for the isomorphism will emerge.

Exercises

Exercise 2.48 Let V be a linear space and $S \subseteq V$ a subset. Prove that S is a linear subspace of V if, and only if, S is closed under linear combinations, i.e., if

$$\sum_{k=1}^{m} \alpha_k s_k \in S$$

for all $\alpha_1, \ldots, \alpha_m \in K$ and $s_1, \ldots, s_m \in S$, and $m \geq 0$. Here we adopt the convention that the empty sum, i.e., the case $m = 0$, is equal to 0.

Exercise 2.49 Is the collection of all sequences with infinitely many terms equal to 0 a subspace of c?

Exercise 2.50 Let V be a linear space and $S \subseteq V$ a subspace. Show that $\dim(S) = 0$ if, and only if, $S = \{0\}$. Assuming that V is finite dimensional prove that $\dim(S) = \dim(V)$ if, and only if, $S = V$.

Exercise 2.51 Let V be a linear space satisfying the property that the only subspace $S \subseteq V$ with $\dim(S) = \dim(V)$ is $S = V$. Prove that V is finite dimensional.

Exercise 2.52 Prove Lemma 2.3: If $T : V \to W$ is a linear operator between finite-dimensional linear spaces, then

$$\dim(V) = \dim(\mathrm{Ker}(T)) + \dim(\mathrm{Im}(T)).$$

Exercise 2.53 Prove Proposition 2.11.

Exercise 2.54 Show that $c = \mathrm{const} \oplus c_0$ where c is the space of convergent sequences and *const* its subspace of constant sequences.

Exercise 2.55 Prove that if $S \subseteq V$ is a subspace of a linear space admitting a unique complementary subspace, then either $S = \{0\}$ or $S = V$.

Exercise 2.56 Suppose that V is a linear space that can be written as $V = U \oplus W$ with $U \cong V \cong W$. Assuming V is finite dimensional prove that $U = V = W = \{0\}$. Show the conclusion fails without the finiteness assumption.

Exercise 2.57 Suppose a linear space can be written as $V = U \oplus W$ and that B_U, B_W are bases for U and W, respectively. Prove that $B_U \cup B_W$ is a basis for V. In the finite-dimensional case this justifies the formula $\dim(U \oplus W) = \dim(U) + \dim(W)$. Argue that the formula remains meaningful in the infinite-dimensional case.

Exercise 2.58 Let V be a linear space and let $\{S_i\}_{i \in I}$ be a family of subspaces of V. Prove that the intersection

$$S = \bigcap_{i \in I} S_i$$

is a linear subspace of V. On the other hand show that if $S_1, S_2 \subseteq V$ are two linear subspaces of V, then $S_1 \cup S_2$ is a linear subspace of V if, and only if, either $S_1 \subseteq S_2$ or $S_2 \subseteq S_1$.

Exercise 2.59 (Refer to Example 2.25) Prove that ℓ_p, for $1 \le p \le \infty$, is a linear space and that if $1 \le p < q \le \infty$, then $\ell_p \subset \ell_q$.

Exercise 2.60 Given linear operators $U \xrightarrow{T} V \xrightarrow{S} W$ show that $\text{Ker}(T) \subseteq \text{Ket}(S \circ T)$ and $\text{Im}(S \circ T) \subseteq \text{Im}(S)$. Show the inclusions can be proper.

Exercise 2.61 Let V and W be two linear spaces over the same field K. Prove that $V \times W \cong W \times V$. Can equality ever hold?

Exercise 2.62 Let V be a linear space and $U \subseteq V$ a linear subspace. Prove that the operation

$$\alpha(x + U) = \alpha x + U,$$

for all vectors $x \in V$ and scalars $\alpha \in K$, is well defined. That is, show that it is independent of the choice of representative for the equivalence class $x + U$.

Exercise 2.63 Let V be a linear space. Prove that $V/V \cong 0$, where 0 denotes any linear space with a single element, necessarily its zero vector. Show also that $V/\{0\} \cong V$, where $\{0\}$ is the trivial subspace of V. Is it possible that any of these isomorphisms is an actual equality?

Exercise 2.64 Let V be a linear space and $U \subseteq V$ a linear subspace. Prove that the canonical projection $\pi : V \to V/U$ given by

$$\pi(x) = x + U$$

is a surjective linear operator.

Exercise 2.65 Let $T: V \to W$ be a surjective linear operator. Prove that $V/\text{Ker(T)}$ is isomorphic to W.

Exercise 2.66 Let $T: V \to W$ be a linear operator. Show that for all $w \in \text{Im}(T)$ the cardinalities of $\text{Ker}(T)$ and $T^{-1}(w)$ are equal.

Exercise 2.67 Let V be a linear space over K and $U \subseteq V$ a subspace. Consider the diagram

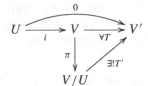

where $i : U \to V$ is the inclusion function, $\pi : V \to V/U$ is the canonical projection to the quotient space, and V' is an arbitrary linear space over K. The rest of the diagram deciphers as follows: prove that for all linear operators $T : V \to V'$ with the property that $T(x) = 0$ for all $x \in U$, there exists a unique linear operator $T' : V/U \to V'$ such that $T' \circ \pi = T$. (This exercise is establishing the *universal property* of the quotient construction.)

Exercise 2.68 Consider the space $C = C([-\pi, \pi], \mathbb{R})$ of functions and c of convergent sequences. Let $C_e \subseteq V$ be the set of even functions, C_o the set of odd functions, $c_e \subseteq c$ consist of sequences with zeros in the even places, and c_o consist of the sequences with zeros in the odd places. Prove that all of these are subspaces and that $C = C_e \oplus C_o$ as well as $c = c_e \oplus c_o$.

Exercise 2.69 Let V and W be two linear spaces over the same field K and consider the product $V \times W$. Show that

1. $U = \{(x, 0) \mid x \in V\}$ is a linear subspace of $V \times W$ that is isomorphic to V. We denote U more suggestively by $V \times \{0\}$.
2. Prove that $(V \times W)/(V \times \{0\}) \cong W$.

Exercise 2.70 Consider the spaces $c_{00} \subset c_0 \subset c$ of sequences of (say) real numbers and the linear operator $\lim: c \to \mathbb{R}$. Show that

1. $\text{Ker}(\lim) = c_0$.
2. $c/c_0 \cong \mathbb{R}$.
3. The function $\lim$ factorises through the canonical projection $\pi : c \to c/c_{00}$, i.e., there exists a linear operator $T : c/c_{00} \to \mathbb{R}$ with $\lim = T \circ \pi$.
4. Is $c/c_{00} \cong \mathbb{R}$?

Exercise 2.71 The previous exercises is a formalisation of the fact that the limit operation on a sequence is insensitive to finitely many changes in the sequence. The definite integral of a function is similarly insensitive to finitely many changes in the function. Provide the necessary ambient definitions to formalise this fact in the spirit of the previous exercise.

Exercise 2.72 Consider linear spaces U, V and the diagram

$$U \xleftarrow{\pi_U} U \times V \xrightarrow{\pi_V} Y$$

in which $\pi_U(x, y) = x$ and $\pi_V(x, y) = y$ are the projection functions. The diagram reads as: for all linear spaces W and linear operators T_U, T_V there exists a unique linear operator S satisfying $\pi_U \circ S = T_U$ and $\pi_V \circ S = T_V$. Prove this is so.

Exercise 2.73 Consider linear spaces U, V and the diagram

$$U \xrightarrow{\iota_U} U \times V \xleftarrow{\iota_V} V$$

in which $\iota_U(x) = (x, 0)$ and $\iota_V(y) = (0, y)$. The diagram reads as: for all linear spaces W and linear operators T_U, T_V there exists a unique linear operator S satisfying $S \circ \iota_U = T_U$ and $S \circ \iota_V = T_V$. Prove this is so. Note the formal duality between this diagram and the one in the previous exercise. It is instructive to compare and contrast this pair of exercises with Exercises 1.12 and 1.13.

2.5 Inner Product Spaces and Normed Spaces

So far in our treatment of linear spaces two geometric aspects are missing: angles between vectors and the length of a vector. This situation is unavoidable since in general linear spaces it need not be possible to coherently define any of these notions. With extra structure these notions become available.

2.5.1 Inner Product Spaces

The reader is most likely familiar with the *standard inner product* on $\mathbb{R}^n$, i.e.,

$$\langle x, y \rangle = \sum_{k=1}^{n} x_k y_k.$$

The definition of an inner product space is a generalization of certain properties the standard inner product has. We will not pause here to motivate this definition any further, except for two comments. The first is that the results below show that an inner product allows one (at least in the real case) to speak of angles between vectors, thus obtaining retrospective justification for the axioms. The second comment pertains to the non-arbitrary nature of the standard inner product; the reader is invited to discover it from elementary geometry, i.e., the law of cosines.

Definition 2.13 (*Inner Product Space*) Let V be a linear space over K where K is either $\mathbb{R}$ or $\mathbb{C}$. Assume that a function $\langle -, - \rangle : V \times V \to K$ is given. The function $\langle -, - \rangle$ is said to be an *inner product* if for all $x, y, z \in V$ and $\alpha, \beta \in K$ the following conditions hold.

Conjugate symmetry or *Hermitian symmetry* $\langle y, x \rangle = \overline{\langle x, y \rangle}$

Linearity in the second argument: $\langle x, \alpha y + \beta z \rangle = \alpha \langle x, y \rangle + \beta \langle x, z \rangle$

Positivity: $\langle x, x \rangle \geq 0$

Definiteness: $\langle x, x \rangle = 0 \implies x = 0$

where $\overline{\langle x, y \rangle}$ is the complex conjugate of $\langle x, y \rangle$. Positivity and definiteness are often juxtaposed to saying that $\langle -, - \rangle$ is *positive definite*. Using the inner product the *norm* of a vector $x \in V$ is $\|x\| = \sqrt{\langle x, x \rangle}$.

The linear space V together with the function $\langle -, - \rangle$ is then called an *inner product space*. To emphasize that $K = \mathbb{R}$ we use the term *real inner product space* while the case $K = \mathbb{C}$ is termed *complex inner product space*. The scalar $\langle x, y \rangle$ is the *inner product* of the given vectors (in that order!).

Remark 2.17 Notice that when $K = \mathbb{R}$ conjugate symmetry reduces to *symmetry*, that is

$$\langle y, x \rangle = \langle x, y \rangle,$$

and the inner product is also linear in its first argument, namely,

$$\langle \alpha x + \beta y, z \rangle = \alpha \langle x, z \rangle + \beta \langle y, z \rangle.$$

In the complex case one generally has

$$\langle \alpha x + \beta y, z \rangle = \overline{\alpha} \langle x, z \rangle + \overline{\beta} \langle y, z \rangle,$$

as follows easily from conjugate symmetry and linearity in the second argument. In both real and complex inner product spaces the inner product is additive in each argument. We mention as well that in mathematical circles it is common to write (x, y) instead of $\langle y, x \rangle$, that is to define an inner product as being linear in the first argument rather than the second. Obviously, the difference is only cosmetic.

Example 2.29 Consider the linear space $\mathbb{R}^n$ and the definition

$$x \cdot y = (x_1, \ldots, x_n) \cdot (y_1, \ldots, y_n) = \sum_{k=1}^{n} x_k y_k.$$

With this definition it is easily seen that $\mathbb{R}^n$ becomes a real inner product space. Similarly, considering $\mathbb{C}^n$, defining

$$x \cdot y = (x_1, \ldots, x_n) \cdot (y_1, \ldots, y_n) = \sum_{k=1}^{n} \overline{x_k} y_k$$

endows $\mathbb{C}^n$ with the structure of a complex inner product space. Notice that for $x, y \in \mathbb{R}^\infty$ the 'obvious definition'

$$x \cdot y = (x_1, \ldots) \cdot (y_1, \ldots) = \sum_{k=1}^{\infty} x_k y_k$$

fails to endow $\mathbb{R}^\infty$ with the structure of an inner product space, simply because the sum may fail to converge, so this is not even a function.

Example 2.30 Consider the space $C(I, \mathbb{R})$ of continuous real valued functions on the interval $I = [a, b]$. The familiar properties of the integral show at once that defining, for all $x, y \in C(I, \mathbb{R})$,

$$\langle x, y \rangle = \int_a^b dt\, x(t) y(t)$$

endows $C(I, \mathbb{R})$ with the structure of a real inner product space. With the proper conjugation in the integral above, the space $C(I, \mathbb{C})$ of continuous complex valued functions becomes a complex inner product space.

Before we present any further examples let us explore some elementary, and not so elementary, general results.

Proposition 2.12 *For all vectors x, y in an inner product space V (either real or complex)*

$$\|x + y\|^2 \le \|x\|^2 + 2|\langle x, y \rangle| + \|y\|^2.$$

Proof Expanding $\|x + y\|^2 = \langle x + y, x + y \rangle$ we see that

$$\langle x + y, x + y \rangle = \langle x, x + y \rangle + \langle y, x + y \rangle = \langle x, x \rangle + \langle x, y \rangle + \langle y, x \rangle + \langle y, y \rangle$$

and thus the claim will follow by showing that

$$\langle x, y \rangle + \langle y, x \rangle \le 2 \cdot |\langle x, y \rangle|,$$

which is immediate if V is a real inner product space. In the complex case suppose that $\langle x, y \rangle = u + iv$. Then

$$\langle x, y \rangle + \langle y, x \rangle = \langle x, y \rangle + \overline{\langle x, y \rangle} = 2 \cdot u$$

and since

$$|u| = \sqrt{u^2} \le \sqrt{u^2 + v^2} = |u + iv| = |\langle x, y \rangle|$$

the result follows.   □

The geometric interpretation of the norm $\|x\|$ is that it is the length of the vector x. As for the geometric meaning of the inner product, notice that any non-zero vector $x \in V$ can be *normalized* by defining $\hat{x} = x/\|x\|$, a vector of *unit* length. For any vector x of unit length and an arbitrary vector y, the scalar $\langle x, y \rangle$ is interpreted as the *component* of the vector y in the direction x. Consequently, we make the following definition.

Definition 2.14 Let V be an inner product space. Two vectors $x, y \in V$ are said to be *orthogonal* or *perpendicular* if $\langle x, y \rangle = 0$.

Theorem 2.12 (Pythagoras' Theorem) *The equality*

$$\|x + y\|^2 = \|x\|^2 + \|y\|^2$$

holds for all orthogonal vectors x and y in an inner product space.

Proof $\|x + y\|^2 = \langle x + y, x + y \rangle = \|x\|^2 + \langle x, y \rangle + \langle y, x \rangle + \|y\|^2 = \|x\|^2 + \|y\|^2.$ □

We see that the vanishing of the inner product $\langle x, y \rangle$ is the algebraic manifestation of the angle formed by the vectors being a right angle. The relationship between angles and the inner product goes deeper.

2.5.2 The Cauchy-Schwarz Inequality

The following inequality is among the most important inequalities in mathematics. It has numerous uses, some of which we will see immediately and some later on.

Theorem 2.13 (Cauchy-Schwarz Inequality) *The inequality*

$$|\langle x, y \rangle| \leq \|x\| \|y\|$$

holds for all vectors x, y in an inner product space V. Equality holds if, and only if, x and y are co-linear, namely one is a scalar multiple of the other.

Proof For simplicity let us assume V is a real inner product space (a slight adaptation is needed for the complex case). The inequality is trivial when $x = 0$ and so we proceed assuming that $x \neq 0$. Let $\hat{x} = x/\|x\|$ be the normalization of x, so in particular $\|\hat{x}\| = 1$. Dividing the desired inequality by $\|x\|$ we see that we need to establish the inequality

$$|\langle \hat{x}, y \rangle| \leq \|y\|$$

(and so the Cauchy-Schwarz Inequality acquires the interpretation that the absolute value of a component of y in a given unit direction sets a lower bound on the length of y). Rewriting y to identify its component in the $\hat{x}$ direction we obtain

$$y = y_{\hat{x}} + (y - y_{\hat{x}})$$

where $y_{\hat{x}} = \langle \hat{x}, y \rangle \hat{x}$. Since

$$\langle \hat{x}, y - y_{\hat{x}} \rangle = \langle \hat{x}, y \rangle - \langle \hat{x}, y_{\hat{x}} \rangle = \langle \hat{x}, y \rangle - \langle \hat{x}, y \rangle \langle \hat{x}, \hat{x} \rangle = 0$$

it follows that $y_{\hat{x}}$ and $y - y_{\hat{x}}$ are orthogonal. Pythagoras' Theorem now yields

$$\|y\|^2 = \|y_{\hat{x}}\|^2 + \|y - y_{\hat{x}}\|^2 \geq \|y_{\hat{x}}\|^2 = \langle \hat{x}, y \rangle^2.$$

Extracting square roots completes the proof (and the clause regarding equality is left as an exercise). $\qquad\square$

Since we now know that in a real inner product space

$$-1 \leq \frac{\langle x, y \rangle}{\|x\| \|y\|} \leq 1$$

for all non-zero vectors x and y, it follows that

$$\theta = \arccos \frac{\langle x, y \rangle}{\|x\| \|y\|}$$

is defined. Stated differently,

$$\langle x, y \rangle = \cos \theta \, \|x\| \|y\|$$

and so we define θ to be the angle between the vectors x and y. With that done we can write

$$\|x - y\|^2 = \|x\|^2 + \|y\|^2 - 2 \cos(\theta) \|x\| \|y\|$$

simply by expanding the left-hand side. This is the law of cosines.

2.5.3 Normed Spaces

In an inner product space every vector can be assigned a norm. A normed space is an abstraction of some key properties of this norm. This is useful since, as we will see, many linear spaces fail to admit an inner product but do admit the structure of a normed space. In such spaces one may still speak of the length of a vector, but not (necessarily) of angles between vectors.

Definition 2.15 (*Normed Space*) A linear space V with a function $x \mapsto \|x\|$, that associates with every vector $x \in V$ a real number $\|x\|$, called the *norm* of x, is said to be a *normed space* if for all vectors $x, y \in V$ and $\alpha \in K$

Positivity: $\|x\| > 0$ provided $x \neq 0$

Homogeneity: $\|\alpha x\| = |\alpha| \|x\|$

Triangle inequality: $\|x + y\| \leq \|x\| + \|y\|.$

The following result is an immediate consequence of the axioms.

Proposition 2.13 *Let V be a normed linear space. For all vectors $x, y \in V$:*

1. $\|x\| = 0$ *if, and only if, $x = 0$.*
2. $\|-x\| = \|x\|.$
3. $\|x - y\| \leq \|x\| + \|y\|.$
4. $|\|x\| - \|y\|| \leq \|x - y\|.$

Proof 1. Using homogeneity, $\|0\| = \|0 \cdot 0\| = 0 \cdot \|0\| = 0$, and together with positivity the claim follows.

2. Using homogeneity, $\|-x\| = \|(-1) \cdot x\| = 1 \cdot \|x\| = \|x\|.$
3. Using (2) and the triangle inequality, $\|x - y\| = \|x + (-y)\| \leq \|x\| + \|y\|.$
4. We need to show that $-\|x - y\| \leq \|x\| - \|y\| \leq \|x - y\|$. Using (3) and the triangle inequality, $\|x\| = \|(x + y) - y\| \leq \|x + y\| + \|y\|$ and similarly $\|y\| \leq \|x\| + \|x - y\|$, as needed.

$\square$

The definition of normed space was motivated by certain intuitive properties that the lengths of vectors, as modeled by the norm in an inner product space, satisfy. The next result shows that the inner product formalism indeed gives rise to a normed space, as expected.

Lemma 2.4 *An inner product space with its associated norm is a normed space.*

Proof Recall that the norm in an inner product space is given by $\|x\| = \sqrt{\langle x, x \rangle}$, and so positivity of the norm is immediate. The computation

$$\|\alpha x\|^2 = \langle \alpha x, \alpha x \rangle = \alpha \overline{\alpha} \langle x, x \rangle = |\alpha|^2 \|x\|^2$$

yields homogeneity. As for the triangle inequality, by Proposition 2.12

$$\|x + y\|^2 = \langle x + y, x + y \rangle \leq \|x\|^2 + 2|\langle x, y \rangle| + \|y\|^2$$

and by applying the Cauchy-Schwarz inequality we obtain that

$$\|x + y\|^2 \leq \|x\|^2 + 2\|x\|\|y\| + \|y\|^2 = (\|x\| + \|y\|)^2$$

from which the result follows by extracting square roots. $\square$

We saw that an inner product space determines a norm by means of $\|x\| = \sqrt{\langle x, x \rangle}$. In that case the inner product can be expressed in terms of the norm through the *polarization identity*

real case: $\qquad \langle x, y \rangle = \frac{1}{4} \left(\|x + y\|^2 - \|x - y\|^2 \right)$

complex case: $\quad \langle x, y \rangle = \frac{1}{4} \left(\|x + y\|^2 - \|x - y\|^2 + i\|x + iy\|^2 - i\|x - iy\|^2 \right)$

as a straightforward computation reveals. A natural question is whether every normed space arises in this way from an inner product space. The polarization identity shows that if so, then the inner product is unique. However, not every norm is the norm associated to an inner product. Recall, from elementary geometry, that for a parallelogram the sum of the squares of the sides equals the sum of the squares of the diagonals. Since only lengths are involved in the statement it can be interpreted in any normed space, but it only holds when the norm is that associated with an inner product.

Proposition 2.14 *In an inner product space V the* parallelogram identity

$$2\|x\|^2 + 2\|y\|^2 = \|x + y\|^2 + \|x - y\|^2$$

holds for all $x, y \in V$.

Proof A straightforward computation. $\qquad\qquad\qquad\qquad\qquad\qquad\qquad\qquad$ □

Remarkably, the converse is also true.

Theorem 2.14 *If V is a normed space in which the parallelogram identity holds for all vectors, then V admits an inner product that induces the given norm.*

Proof We focus only on the real case. The polarization identity dictates the first step: if such an inner product exists it must be given by

$$\langle x, y \rangle = \frac{1}{4} \left(\|x + y\|^2 - \|x - y\|^2 \right)$$

for all $x, y \in V$. We need to verify that $\langle -, - \rangle$ is indeed an inner product. Symmetry of $\langle -, - \rangle$ follows immediately from the fact that $\|y - x\| = \|x - y\|$:

$$\langle y, x \rangle = \frac{1}{4} \left(\|y + x\|^2 - \|y - x\|^2 \right) = \frac{1}{4} \left(\|x + y\|^2 - \|x - y\|^2 \right) = \langle x, y \rangle.$$

Since

$$\langle x, x \rangle = \frac{1}{4} \left(\|2x\|^2 \right) = \|x\|^2$$

we see that $\langle -, - \rangle$ is positive definite and that its associated norm is the one given. So far we made no use of the parallelogram identity and so, expectedly, it must appear when establishing the remaining properties, namely additivity and homogeneity of $\langle -, - \rangle$ in the second argument. Unfortunately, the proof is lengthy and rather technical but otherwise uninformative and not entertaining, and we thus omit it.   □

Let us consider an illustrative and important family of normed spaces where we can see the above theorem in action. All of the norms will be defined on the same underlying linear space, namely $\mathbb{R}^2$. For all $p > 0$ let

$$\|(x_1, x_2)\|_p = \left(|x_1|^p + |x_2|^p\right)^{1/p},$$

defined for all vectors $(x_1, x_2) \in \mathbb{R}^2$. We also set $\|(x_1, x_2)\|_\infty = \max\{|x_1|, |x_2|\}$. In the next section we consider a more general family of normed spaces and from the results there it follows that for each $p \geq 1$ the definition of $\| - \|_p$ endows $\mathbb{R}^2$ with an inner product. For now, we present some geometric aspects by visualizing the set of vectors forming the unit circle, namely

$$\{(x_1, x_2) \in \mathbb{R}^2 \mid \|(x_1, x_2)\|_p = 1\}$$

for various values of p, shown in Fig. 2.1.

Firstly, when $p = 2$ the norm is the one associated to the standard inner product in $\mathbb{R}^2$. Its geometry is the familiar Euclidean geometry of the plane. The fact that $\| - \|_2$ satisfies the parallelogram identity is thus the algebraic incarnation of the same identity from elementary geometry. For $p \geq 1$ each $\| - \|_p$ is still a norm but only for $p = 2$ is it induced by an inner product. This is so since the parallelogram identity fails. For instance, $2\left(\|x\|_1^2 + \|y\|_1^2\right) = \|x + y\|_1^2 + \|x - y\|_1^2$ fails for the vectors $x = (1, 0)$, $y = (0, 1)$ by direct computation; the diagonals are too long. A common feature of these unit circles for $p \geq 1$ is convexity: the straight line connecting any two points on each such circle is contained within the circle. It is also apparent that each unit circle is symmetric about the origin and bounded along each direction. These properties are characteristic of unit balls in normed spaces.

Lemma 2.5 *Let V be a normed space and $B = \{x \in V \mid \|x\| \leq 1\}$ the unit ball in it. Then B is convex, namely $tx + (1 - t)y \in B$ for all $x, y \in B$ and $0 \leq t \leq 1$, B is symmetric about the origin, namely $-x \in B$ whenever $x \in B$, and the intersection of B with any straight line through the origin is a finite segment.*

Proof Given $x, y \in B$ and $0 \leq t \leq 1$ the triangle inequality yields

$$\|tx + (1 - t)y\| \leq \|tx\| + \|(1 - t)y\| = t\|x\| + (1 - t)\|y\| = 1$$

so $tx + (1 - t)y \in B$. The symmetry claim and the intersection with a line follow at once from homogeneity.   □

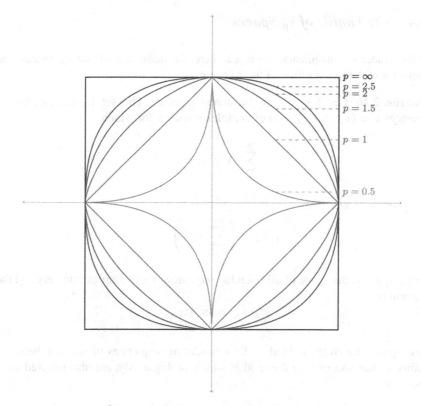

$p = \infty$
$p = 2.5$
$p = 2$
$p = 1.5$
$p = 1$
$p = 0.5$

Fig. 2.1 Unit circles in $\mathbb{R}^2$ for different values of p

We can now conclude that $\| - \|p$ for $p < 1$ is not a norm. Together with a topological property, namely B being closed, the three properties in the lemma are also sufficient; given such a closed set B there exists a unique norm on the ambient space whose unit ball is precisely B. Closed sets will be discussed in the chapter on topological spaces. In $\mathbb{R}^2$ it is the usual notion, i.e., a subset containing all of its limit points with respect to the Euclidean metric.

We close the section, and the chapter, by introducing two very important families of normed spaces. The ℓ_p spaces, introduced already in Example 2.25, are spaces of sequences and each carries its own norm, the ℓ_p norm. The second family of spaces, the pre-L_p spaces, allude to the family of L_p spaces and the L_p norm. These spaces are studied in more detail in Chap. 5.

2.5.4   The Family of ℓ_p Spaces

For the reader's convenience we repeat here the definition of the ℓ_p spaces and augment it with the definition of the ℓ_p norm.

Definition 2.16 Let $1 \leq p < \infty$ be a real number. The set ℓ_p consists of all sequences $x = (x_1, \ldots, x_k, \ldots)$ of complex numbers for which

$$\sum_{k=1}^{\infty} |x_k|^p < \infty.$$

The ℓ_p-*norm* of x is

$$\|x\|_p = \left(\sum_{k=1}^{\infty} |x_k|^p \right)^{1/p}.$$

The ℓ_∞ space is the space of all bounded sequences x of complex numbers and the ℓ_∞ norm is

$$\|x\|_\infty = \sup_k |x_k|.$$

These spaces are over the field $\mathbb{C}$. By considering sequences of real numbers one obtains similar spaces over the field $\mathbb{R}$ which, ambiguously, are also referred to as ℓ_p spaces.

It is immediate that $\|x\|_p = 0$ if, and only if, $x = 0$. To further investigate the structure of ℓ_p, we need the following classical result.

Proposition 2.15 (Young's Inequality) *The inequality*

$$ab \leq \frac{a^p}{p} + \frac{b^q}{q}$$

holds for all positive real numbers p and q satisfying $1/p + 1/q = 1$ and non-negative real numbers a and b, with equality if, and only if, $a^p = b^q$.

Proof Recall the weighted arithmetic-geometric means inequality

$$\sqrt[w]{x_1^{w_1} x_2^{w_2}} \leq \frac{w_1 x_1 + w_2 x_2}{w}$$

which holds for all non-negative real numbers x_1, x_2, and all positive weights w_1, w_2, with $w = w_1 + w_2$. Applying this inequality to the numbers $x_1 = a^p$ and $x_2 = b^q$, with weights $w_1 = 1/p$ and $w_2 = 1/q$, yields the result. $\square$

The next inequality is used often when manipulating elements of the spaces ℓ_p. It is used below to establish the triangle inequality when proving that each ℓ_p space, with its ℓ_p norm, is a normed space. It is convenient to first introduce the following notation. For any two sequences $x = (x_1, \ldots, x_k, \ldots)$ and $y = (y_1, \ldots, y_k, \ldots)$, let

$$xy = (x_1 y_1, \ldots, x_k y_k, \ldots),$$

namely, the vector of component-wise products of the given vectors.

Theorem 2.15 (Hölder's Inequality) *Given positive real numbers p and q for which $1/p + 1/q = 1$, the inequality*

$$\|xy\|_1 \le \|x\|_p \cdot \|y\|_q$$

holds for all $x \in \ell_p$ and $y \in \ell_q$.

Proof The case where either $\|x\|_p = 0$ or $\|y\|_q = 0$ is trivial and so we may assume this is not so. By component-wise division of x and y, respectively, by $\|x\|_p$ and $\|y\|_q$, we may assume that $\|x\|_p = \|y\|_q = 1$, namely that

$$\sum_{k=1}^{\infty} |x_k|^p = 1 = \sum_{k=1}^{\infty} |y_k|^q,$$

and we need to show that $\|xy\|_1 \le 1$, or in other words that

$$\sum_{k=1}^{\infty} |x_k y_k| \le 1.$$

By Young's Inequality we have that

$$|x_k y_k| = |x_k||y_k| \le \frac{|x_k|^p}{p} + \frac{|y_k|^q}{q}$$

and therefore

$$\sum_{k=1}^{\infty} |x_k y_k| \le \frac{1}{p} + \frac{1}{q} = 1,$$

as required. $\square$

Theorem 2.16 (Minkowski's Inequality) *Let $1 \le p < \infty$. The inequality*

$$\|x + y\|_p \le \|x\|_p + \|y\|_p$$

holds for all $x, y \in \ell_p$.

Proof Notice that

$$\|x+y\|_p^p = \sum_{k=1}^{\infty} |x_k+y_k|^p = \sum_{k=1}^{\infty} |x_k||x_k+y_k|^{p-1} + \sum_{k=1}^{\infty} |y_k||x_k+y_k|^{p-1}.$$

Applying Hölder's inequality with the given p and the associated $q = p/(p-1)$, we obtain

$$\sum_{k=1}^{\infty} |x_k||x_k+y_k|^{p-1} \le \left(\sum_{k=1}^{\infty} |x_k|^p\right)^{\frac{1}{p}} \left(\sum_{k=1}^{\infty} |x_k+y_k|^p\right)^{1-\frac{1}{p}} = \|x\|_p \frac{\|x+y\|_p^p}{\|x+y\|_p}$$

and a similar inequality for the second summand. It follows that

$$\|x+y\|_p^p \le (\|x\|_p + \|y\|_p)\frac{\|x+y\|_p^p}{\|x+y\|_p}$$

and, by simplifying, the claimed inequality is established. $\square$

Theorem 2.17 *For all* $1 \le p \le \infty$ *the space* ℓ_p *is a normed linear space.*

Proof The case $p = \infty$ is left for the reader, so assume that $p < \infty$. The non-negativity of $\|x\|_p$ is clear from the definition of the norm and it was already noted that it is immediate that $\|x\| = 0$ implies $x = 0$. Homogeneity, is also immediate since

$$\|\alpha x\|_p^p = \sum_{k=1}^{\infty} |\alpha x_k|^p = |\alpha|^p \|x\|_p^p$$

and finally the triangle inequality

$$\|x+y\|_p \le \|x\|_p + \|y\|_p$$

is precisely Minkowski's Inequality. $\square$

2.5.5 The Family of Pre-L_p Spaces

With the large family $\{\ell_p\}_{1\le p\le\infty}$ of normed spaces at hand one can put the general theory to use for sequences. Noticing that $\ell_p \subset \ell_q$ for all $1 \le p < q \le \infty$, and with sequences in ℓ_p converging faster to 0 than sequences in ℓ_q, given a bounded sequence one would typically try to identify a suitable p such that the sequence is found in ℓ_p, and proceed to apply the general theory to the problem at hand.

As long as sequences are concerned the ℓ_p spaces are adequate, but quite often when modelling a physical problem mathematically one obtains a function rather than a sequence. It is thus desirable to consider normed spaces of functions. Obtaining the correct analogue of the ℓ_p spaces, namely the L_p spaces, requires some more sophisticated machinery than what we currently have. At this point we present what we call the pre-L_p spaces.

Consider the linear space $C([a, b], \mathbb{R})$ of all continuous functions $x : [a, b] \to \mathbb{R}$ (we note that with the necessary modifications in the coming definitions and proofs, one may alter the domain, as well as consider complex-valued functions). The pre-L_p space K_p, for $1 \leq p < \infty$, is the space $C([a, b], \mathbb{R})$ with the L_p norm given by

$$\|x\|_p = \left(\int_a^b dt \, |x(t)|^p \right)^{1/p}$$

and we immediately note that $\|x\| = 0$ implies $x = 0$ (due to the continuity of x). Lastly, K_∞ the space $C([a, b], \mathbb{R})$ with the L_∞ norm given by

$$\|x\|_\infty = \sup_{a \leq t \leq b} \{|x(t)|\}.$$

The rest of the section is devoted to showing that K_p with the L_p norm is a normed space for all $1 \leq p \leq \infty$. In fact, the proof is formally identical to the case of the ℓ_p spaces. Indeed, the only ingredient one needs is the following version of Hölder's Inequality.

Theorem 2.18 (Hölder's Inequality) *Given positive real numbers p and q such that $1/p + 1/q = 1$, the inequality*

$$\|xy\|_1 \leq \|x\|_p \cdot \|y\|_q$$

holds for all $x \in K_p$ and $y \in K_q$, where xy is the function $(xy)(t) = x(t)y(t)$.

Proof An inspection of the proof of Hölder's Inequality for sequences reveals that it was obtained by a point-wise application of Young's Inequality as well as properties of summation that are well known to hold for integration as well. The same argument can thus be used to establish Hölder's Inequality for functions. □

The details of the proof of the following result are formally identical to the proof of Theorem 2.17

Theorem 2.19 *For all $1 \leq p \leq \infty$ the pre-L_p space K_p is a normed linear space.*

Remark 2.18 Note again that the domain of the functions x in a pre-L_p space can be altered (sometimes with necessary extra care) and that the codomain can be replaced by $\mathbb{C}$ (with obvious adaptations to the definition of the L_p norm). All such spaces are still called pre-L_p spaces and denoted, ambiguously, by K_p. The L_p spaces will be introduced in Chap. 5.

Exercises

Exercise 2.74 Prove the generalized Pythagoras' Theorem: in an inner product space, given m pairwise orthogonal vectors $x_1, \ldots, x_m$, the equality

$$\|x_1 + \cdots + x_m\|^2 = \|x_1\|^2 + \cdots + \|x_m\|^2$$

holds.

Exercise 2.75 Show that in an inner product space $\|x + y\|^2 = \|x\|^2 + \|y\|^2 + 2\mathrm{Re}(\langle x, y \rangle)$.

Exercise 2.76 The converse of Pythagoras' Theorem in an inner product space states that if $\|x + y\|^2 = \|x\|^2 + \|y\|^2$, then $\langle x, y \rangle = 0$. Prove it true in the real case and show that it fails in the complex case.

Exercise 2.77 In the spirit of the proof of the Cauchy-Schwarz Inequality in the real case prove that equality holds if, and only if, the two vectors are co-linear. Proceed as follows:

1. Dispose of the case where either vector is 0.
2. Reduce the general case to the case where both vectors are normalized.
3. Assuming $\langle x, y \rangle = 1$, compute $\|x - y\|^2$ and conclude that $x = y$.
4. Handle the case $\langle x, y \rangle \neq 1$.

Can you tackle the complex case?

Exercise 2.78 Let $V = C^1([-\pi, \pi], \mathbb{R})$ be the linear space of continuously differentiable functions. Show that

$$\langle f, g \rangle = \int_{-\pi}^{\pi} dt \ f(t)g(t) + f'(t)g'(t)$$

is an inner product in V. Use it to show that for any continuously differentiable function $f : [-\pi, \pi] \to \mathbb{R}$

$$\left| \int_{-\pi}^{\pi} dt \ f(t) \cos t - f'(t) \sin t \right| \leq \sqrt{2\pi} \int_{-\pi}^{\pi} dt \ |f(t)^2| + |f'(t)|^2.$$

Exercise 2.79 Let V be an inner product space. Show that $x \mapsto \langle x, - \rangle$ defines a function $V \to \mathrm{Hom}(V, K)$. Prove that in the finite-dimensional case this function is surjective. Hint: an orthonormal basis is useful.

Exercise 2.80 Prove that ℓ_2 when endowed with the operation

$$\langle x, y \rangle = \sum_{k=1}^{\infty} \overline{x_k} y_k$$

is an inner product space.

Exercise 2.81 Prove that ℓ_∞ with the ℓ_∞ norm is a normed space and that K_∞ with the L_∞ norm is a normed space.

Exercise 2.82 Let V be an inner product space. Prove the parallelogram identity

$$2(\|x\|^2 + \|y\|^2) = \|x + y\|^2 + \|x - y\|^2$$

for all $x, y \in V$.

Exercise 2.83 Consider the linear space $\mathbb{R}^2$ and a closed subset $B \subseteq V$ that is closed, convex, symmetric, and segmented as in Lemma 2.5. Define the *Minkowski functional* $\| - \|_B : \mathbb{R}^2 \to [0, \infty)$ by

$$\|x\|_B = \inf \left\{ \alpha > 0 \mid \frac{1}{\alpha} x \in B \right\}.$$

Prove that $\| - \|_B$ turns V into a normed space whose unit ball is B.

Exercise 2.84 Prove that for $p \geq 1$, $p \neq 2$ the normed space ℓ_p is not an inner product space.

Exercise 2.85 Prove that for $0 < p < 1$ the function $\| - \|_p$ on $\mathbb{R}^2$ is not a norm.

Exercise 2.86 Does the equality

$$\ell_1 = \bigcap_{p > 1} \ell_p$$

hold?

Exercise 2.87 For $x \in \ell_p$, with $1 \leq p < \infty$, does the equality

$$\lim_{q \to \infty} \|x\|_q = \|x\|_\infty$$

hold?

Exercise 2.88 Prove that K_2 is an inner product space.

Exercise 2.89 Let $1 \leq p \leq \infty$ and consider for any $x \in K_p$ the sequence of samples $s(x) = (x(1), x(2), \ldots, x(k), \ldots)$. Prove that s is a linear operator $s : K_p \to \ell_p$. How do the norms $\|x\|_p$ and $\|s(x)\|_p$ compare?

Exercise 2.90 Prove Hölder's Inequality for functions and prove that K_p with the L_p norm is a normed space.

Further Reading

Introductory level texts on linear algebra typically treat linear spaces with a strong emphasis on techniques of matrices. When bases are not explicitly available, as in the infinite-dimensional case, this approach must be replaced by a coordinate-free approach, as was done in this chapter. For an introductory level text treating linear algebra in a coordinate-free fashion see Chaps. 10–12 of [2]. A more advanced text with a coordinate-free approach, as well as an algebraically much deeper approach to linear algebra, making the connection between linear spaces and modules, is [3]. Another text that bridges the gap between matrix-centric linear algebra and Hilbert space theory is [1]. The reader seeking to enhance her intuition and ability with the Cauchy-Schwarz inequality is advised to consider [4].

References

1. Brown, W.C.: A Second Course in Linear Algebra, p. xii+264. A Wiley-Interscience Publication, Wiley, New York (1988)
2. Dummit, D.S., Foote, R.S.: Abstract Algebra, 3rd edn., p. xii+932. Wiley, Hoboken, NJ (2004)
3. Roman, S.: Advanced Linear Algebra. Graduate Texts in Mathematics, vol. 135, 3rd edn., p. xviii+522. Springer, New York (2008)
4. Steele, J.M.: The Cauchy-Schwarz Master Class. MAA Problem Books Series, p. x+306. Mathematical Association of America, Washington, DC; Cambridge University Press, Cambridge (2004)

Chapter 3
Topological Spaces

Topology is often described as rubber-sheet geometry, namely the study of geometric properties that are insensitive to stretching and shrinking, without tearing or gluing. Famously, for a topologist, there is no difference between a cup and a donut. The impact of topology on modern mathematics is difficult to quantify but virtually impossible to exaggerate. Without a doubt its place as one of the pillars of mathematics is secured.

The definition of a topological space is an abstraction of part of the structure of the real line, specifically that part that allows one to speak of continuity and convergence. The choice for the topological notions introduced in this chapter is dictated by two needs. One need is to familiarize the reader with those topological concepts that are most relevant in the context of this book, the other is to acclimatize the reader to topology, its main results, and its techniques. These two needs pull in slightly different directions and this chapter represents a compromise. We provide the reader with sufficient motivation for the concepts while keeping things firmly grounded in analysis.

Section 3.1 presents the definition of a topology and the accompanying notion of continuity, followed by a detailed construction of the Euclidean topology on the real numbers $\mathbb{R}$. The associated notion of continuity is shown to be equivalent to the usual one, motivating the definitions. Section 3.2 is a basic study of convergence in the topological setting, its relation to cluster points, and its behaviour under the first countability axiom. Section 3.3 is concerned with techniques for constructing topologies, in particular by forcing a collection of sets to be open sets and by forcing a collection of functions to be continuous. Coproducts, products, and quotients are then given as particular examples. Section 3.4 discusses the Hausdorff separation property and topological connectivity and Sect. 3.5 is a study of compactness, developing enough material to prove the Heine-Borel Theorem.

© Springer Nature Switzerland AG 2021
C. Alabiso and I. Weiss, *A Primer on Hilbert Space Theory*, UNITEXT for Physics,
https://doi.org/10.1007/978-3-030-67417-5_3

3.1 Topology—Definition and Elementary Results

The definition of a topology is quite simple; a certain collection of subsets of a set satisfying easily stated axioms. However, without further motivation, the axioms may appear arbitrary and not too palatable. To assist the reader digest the notion, the definitions below are immediately succeeded by a detailed motivating discussion taking place in the familiar setting of the real numbers $\mathbb{R}$. The motivation we give is of a topology as the result of stripping away redundancies in the definition of continuity. We then explore various examples of topological spaces.

3.1.1 Definition and Motivation

We present the definition of a topological space followed by the definition of a continuous function between topological spaces. The abstract definitions are then exemplified in the context of the real numbers, establishing the Euclidean topology on $\mathbb{R}$, and, through the process, justifying the abstract definitions.

Definition 3.1 A *topology* τ_X on an arbitrary set X is a collection of subsets of X such that the following conditions hold.

1. Both the empty set $\emptyset$ and the entire set X are members of τ_X.
2. τ_X is stable under arbitrary unions. That is, if $\{U_i\}_{i \in I}$ is a family of elements in τ_X, then
$$\bigcup_{i \in I} U_i \in \tau_X.$$

3. τ_X is stable under finite intersections. That is, if $U_1, \ldots, U_m$ are members of τ_X, then
$$U_1 \cap \cdots \cap U_m \in \tau_X.$$

The pair (X, τ_X) is called a *topological space* and the members of τ_X are referred to as *open* sets. We often speak of the topological space X without explicitly mentioning τ_X. We write (X, τ) if there is no need to syntactically remind ourselves that the collection τ consists of the open sets of X.

Remark 3.1 Clearly, stability under finite intersections is equivalent to the condition that $U_1 \cap U_2 \in \tau_X$ for any *two* open sets $U_1, U_2 \in \tau_X$.

For the following definition, and as a general advise for safely taking the first few steps into the realm of topology, the reader may wish to review the basic properties of inverse images (e.g., Sect. 1.2.9 of the Preliminaries).

Definition 3.2 Let (X, τ_X) and (Y, τ_Y) be topological spaces. A function $f : X \to Y$ is said to be *continuous* if $f^{-1}(U)$ is an open set in X whenever U is an open set in Y. In other words, f is continuous if

$$U \in \tau_Y \implies f^{-1}(U) \in \tau_X.$$

The following results establish a particular topology on the set $\mathbb{R}$ of real numbers, and investigate the notion of continuity with respect to that topology. The aim is two-fold; to exemplify the definitions in a concrete and familiar setting, and to motivate the definitions.

Recall that a function $f : \mathbb{R} \to \mathbb{R}$ is said to be continuous if, intuitively, small changes in the value of x cause small changes in the value of $f(x)$. Formally, f is continuous if for every point $x_0 \in \mathbb{R}$ and $\varepsilon > 0$ one can find a $\delta > 0$ such that if $|x - x_0| < \delta$, then $|f(x) - f(x_0)| < \varepsilon$. The absolute value function is used here in the role of a measurement device for distances. However, the notion of continuity is quite insensitive to numerous changes one can perform on the absolute value function. Continuity is a local phenomenon, and so if one truncates the absolute value function and defines

$$|x|_t = \begin{cases} |x| & \text{if } |x| < 1 \\ 1 & \text{otherwise,} \end{cases}$$

then continuity with respect to $|x|$ and with respect to $|x|_t$ coincide. Scaling effects, such as defining $|x|_2 = 2|x|$, also have no effect on continuity in the precise sense that continuity with respect to $|x|$ and with respect to $|x|_2$ coincide.

The decision to use the absolute value function in formulating the definition of continuity (or, more generally, that of limit) is thus revealed to be an arbitrary choice among many topologically equivalent alternatives. It is a natural question whether one can distill a definition of continuity devoid of any arbitrary and irrelevant structure and consequently furnish a better understanding of the concept of continuity. The affirmative answer is given as follows. Call a subset $U \subseteq \mathbb{R}$ *open* if for every $x \in U$ there exists an $\varepsilon > 0$ such that $B_\varepsilon(x) \subseteq U$, where we define

$$B_\varepsilon(x) = \{y \in \mathbb{R} \mid |x - y| < \varepsilon\} = (x - \varepsilon, x + \varepsilon),$$

the *open interval* of radius ε about x. With the aim of utilizing these sets to define a topology on $\mathbb{R}$, we first prove that any open interval is an open set.

Proposition 3.1 *The open interval* $B_\varepsilon(x) = (x - \varepsilon, x + \varepsilon)$, *for all* $x \in \mathbb{R}$ *and* $\varepsilon > 0$, *is an open set.*

Proof Given $y \in B_\varepsilon(x) = (x - \varepsilon, x + \varepsilon)$, we are required to find a $\delta > 0$ such that $B_\delta(y) = (y - \delta, y + \delta) \subseteq B_\varepsilon(x)$. Let

$$\delta = \varepsilon - |x - y|$$

and note that $\delta > 0$. Elementary algebra shows this δ suits our purposes. $\qquad \square$

We can now present our first example of a topology; an important enough example that we state it as a theorem.

Theorem 3.1 *The collection τ of all open sets $U \subseteq \mathbb{R}$ is a topology on $\mathbb{R}$.*

Proof The empty set vacuously satisfies the condition for being an open set, simply as it has no points at all. The set $\mathbb{R}$ itself is clearly an open set, since if $x \in \mathbb{R}$, then $x \in B_1(x) \subseteq \mathbb{R}$. Thus the first condition in the definition of a topology is satisfied. To show stability under arbitrary unions, suppose that $\{U_i\}_{i \in I}$ is a family of open sets in $\mathbb{R}$ and we need to show that

$$U = \bigcup_{i \in I} U_i$$

is open. Indeed, if $x \in U$, then $x \in U_i$ for some $i \in I$. But since U_i is open, there exists a $\delta > 0$ with

$$B_\delta(x) \subseteq U_i \subseteq U,$$

as required. It remains to show for any two open sets $U_1, U_2 \subseteq \mathbb{R}$ that $U_1 \cap U_2$ is open. Indeed, if $x \in U_1 \cap U_2$, then there are $\delta_1 > 0$ and $\delta_2 > 0$ with

$$B_{\delta_1}(x) \subseteq U_1$$
$$B_{\delta_2}(x) \subseteq U_2.$$

Setting $\delta = \min\{\delta_1, \delta_2\}$ (and noting that $\delta > 0$), one easily sees that

$$B_\delta(x) \subseteq B_{\delta_1}(x) \cap B_{\delta_2}(x) \subseteq U_1 \cap U_2,$$

as required, and thus completing the proof.   □

Definition 3.3 The collection of all open sets $U \subseteq \mathbb{R}$ as just defined is called the *Euclidean topology* on $\mathbb{R}$. It is also commonly referred to as the *standard* topology or the *ordinary* topology.

The open sets turn out to contain enough information about the space $\mathbb{R}$ in order to completely characterize continuity, as follows.

Theorem 3.2 *A function $f : \mathbb{R} \to \mathbb{R}$ is continuous in the usual meaning if, and only if, $f^{-1}(U)$ is an open set for every open set $U \subseteq \mathbb{R}$.*

Proof Suppose f is continuous and let U be an open set. Our aim is to show that $f^{-1}(U)$ is open. Let $x_0 \in f^{-1}(U)$, that is $y_0 = f(x_0) \in U$, and our aim is to find a $\delta > 0$ such that
$$B_\delta(x_0) \subseteq f^{-1}(U).$$

Since U is open, there exists an $\varepsilon > 0$ such that

$$B_\varepsilon(y_0) = (y_0 - \varepsilon, y_0 + \varepsilon) \subseteq U.$$

By continuity of f, there exists a $\delta > 0$ such that $|f(x) - y_0| < \varepsilon$, provided that $|x - x_0| < \delta$. But that means that for every $x \in B_\delta(x_0)$

$$f(x) \in B_\varepsilon(y_0) \subseteq U.$$

In other words, $x \in f^{-1}(U)$ for all x in the set $B_\delta(x_0)$, and so

$$B_\delta(x_0) \subseteq f^{-1}(U),$$

which is what we set out to obtain.

In the other direction, suppose that $f^{-1}(U)$ is an open set whenever U is an open set. Given a point $x_0 \in \mathbb{R}$ and an $\varepsilon > 0$, let $y_0 = f(x_0)$ and consider the set $U = B_\varepsilon(y_0)$, and notice that U is itself an open set. By assumption, the set $f^{-1}(U)$ is an open set, and it contains x_0. There is thus a $\delta > 0$ such that $B_\delta(x_0) \subseteq f^{-1}(U)$. Thus, if $|x - x_0| < \delta$, then $x \in B_\delta(x_0)$, and therefore $f(x) \in B_\varepsilon(y_0)$, showing that $|f(x) - f(x_0)| < \varepsilon$. In other words, f is continuous at x_0. Since x_0 was arbitrary, f is continuous. □

Recalling Definition 3.2 the result above can be restated as follows.

Theorem 3.3 *A function $f : \mathbb{R} \to \mathbb{R}$ is continuous in the usual sense if, and only if, it is continuous with respect to the Euclidean topology on $\mathbb{R}$.*

This result justifies and motivates Definition 3.2 and, a-priori, the definition of a topology. Let us pause to reflect on the situation right now. The familiar notion of continuity for functions $f : \mathbb{R} \to \mathbb{R}$ was seen to be faithfully captured by the collection of open sets $U \subseteq \mathbb{R}$. The collection of open sets was defined in terms of $B_\varepsilon(x) = \{y \in \mathbb{R} \mid |x - y| < \varepsilon\}$, thus still directly using the absolute value function. It would thus seem that we had accomplished nothing except for hiding the absolute value function in some fancy notation. However, something quite substantial was achieved. Consider the scaled absolute value function $|x|_2 = 2 \cdot |x|$, and let us define

$$B_{2,\varepsilon}(x) = \{y \in \mathbb{R} \mid |x - y|_2 < \varepsilon\} = \left(x - \frac{\varepsilon}{2}, x + \frac{\varepsilon}{2}\right)$$

for all $x \in \mathbb{R}$ and $\varepsilon > 0$. Let us further say that a set $U \subseteq \mathbb{R}$ is 2-open if for all $x \in U$ there exists an $\varepsilon > 0$ such that $B_{2,\varepsilon}(x) \subseteq U$. In other words, we repeat the definition of open sets, replacing the absolute value function by a scaled version of it.

Proposition 3.2 *The collection of open sets and the collection of 2-open sets are the same.*

Proof Notice that

$$B_{\varepsilon/4}(x) \subseteq B_{2,\varepsilon}(x) \subseteq B_\varepsilon(x)$$

holds for all $x \in \mathbb{R}$ and $\varepsilon > 0$. Thus, if U is an open set and $x \in U$, then

$$B_\varepsilon(x) \subseteq U$$

for some $\varepsilon > 0$, but then

$$B_{2,\varepsilon}(x) \subseteq B_\varepsilon(x) \subseteq U$$

and so U is also 2-open. Conversely, if U is 2-open and $x \in U$, then

$$B_{2,\varepsilon}(x) \subseteq U$$

for some $\varepsilon > 0$, but then

$$B_{\varepsilon/4}(x) \subseteq B_{2,\varepsilon}(x) \subseteq U$$

and so U is also open. $\square$

Similarly, one may show that the collection of open sets is insensitive to other changes in the absolute value function, such as scaling by other factors or various truncations. It is not a coincidence that those changes to the absolute value function that are immaterial to the notion of continuity are also immaterial to the concept of open set. Indeed, Theorem 3.2 characterizes continuity in terms of the open sets alone.

Thus, by concentrating on the open sets one is freed from the irrelevant (for the purposes of continuity) particularities of the absolute value function. The definition of a topology is thus an abstraction of a particular formulation of continuity that avoids directly mentioning any particular concept of $\mathbb{R}$ that is irrelevant to the notion of continuity. As with any abstraction, once it is made, a plethora of new situations emerges, situations that superficially have little to do with the geometric notion of continuity. Introducing a topology on a set immediately allows one to import a significant amount of geometric intuition, machinery, and analogy into an a-priori non-geometric context.

3.1.2 More Examples

We present more topological spaces, illustrating the versatility of the formalism. We remark that there are far more interesting examples of topological spaces than can be recounted in this book. The choice of the examples given below is dictated by the focus of the book; the examples are chosen to elucidate the forthcoming topological concepts, while avoiding delving into the intricacies of the more obscure examples of topological spaces, thus allowing the reader to ease into the topological framework while remaining firmly grounded in analysis.

Example 3.1 (*Discrete And Indiscrete Topologies*) Any set X supports two extreme topologies on it. One is the collection $\{\emptyset, X\}$, called the *indiscrete topology* on X and the other is the collection of all subsets of X, called the *discrete topology* on X. It is immediate to verify that these are indeed topologies, and that they are distinct topologies except when $|X| \leq 1$.

Example 3.2 (*The Sierpiński Space*) Let $\mathbb{S} = \{0, 1\}$ be a set with two elements. The collection

$$\{\emptyset, \mathbb{S}, \{1\}\}$$

is readily seen to be a topology on $\mathbb{S}$ giving rise to what is known as the *Sierpiński* space.

Example 3.3 Let $X = \{a, b, c, d, e\}$, be a nonempty set consisting of five elements. Of the collections

$$\tau = \{X, \emptyset, \{a\}, \{c, d\}, \{a, c, d\}, \{b, c, d, e\}\}$$
$$G = \{X, \emptyset, \{a\}, \{c, d\}, \{a, c, d\}, \{b, c, d\}\}$$
$$H = \{X, \emptyset, \{a\}, \{c, d\}, \{a, c, d\}, \{a, b, d, e\}\}$$

τ is a topology on X as the axioms are easily confirmed by inspection. G is not a topology on X, since $\{a\} \cup \{b, c, d\} = \{a, b, c, d\} \notin G$. The collection H is also not a topology on X, since $\{c, d\} \cap \{a, b, d, e\} = \{d\} \notin H$.

Example 3.4 (*The Cofinite and Cocountable Topologies*) Let X be a set and τ the family of all subsets U of X whose complement $X \setminus U$ is finite, and if needed, manually include $\emptyset$ in τ. We show that τ is a topology on X. First, $\emptyset \in \tau$ by definition, and since $X \setminus X = \emptyset$ is finite, it follows that $X \in \tau$, and so the first condition in the definition of topology is satisfied. Stability under arbitrary unions is immediate, so finally consider two sets U_1, U_2 with finite complements in X. Since

$$X \setminus (U_1 \cap U_2) = (X \setminus U_1) \cup (X \setminus U_2)$$

and since both $(X \setminus U_1)$ and $(X \setminus U_2)$ are finite, it follows that $U_1 \cap U_2$ has finite complement too. This topology is called the *cofinite topology* on X. Notice that if X is finite, then the cofinite topology coincides with the discrete topology on X (Example 3.1).

Similarly, call a subset $U \subseteq X$ *cocountable* if its complement $X \setminus U$ is a countable set. The collection of all cocountable subsets of X (with $\emptyset$ added in manually) also forms a topology on X called the *cocountable topology*. The proof is similar to the one given above, with the necessary care when handling countable sets (see Sect. 1.2.13 of the Preliminaries).

3.1.3 Elementary Observations

We explore immediate properties of topological spaces and of continuous functions, and we discuss the notion of homeomorphism. We start off with a useful criterion for detecting open sets in an arbitrary topological space.

Proposition 3.3 *A subset V of a topological space X is open if, and only if, for all $x \in V$ there exists an open set U_x such that*

$$x \in U_x \subseteq V.$$

Proof If V is open, then clearly for every $x \in V$ one may choose $U_x = V$. In the other direction, the stated condition implies that

$$V = \bigcup_{x \in V} U_x$$

and so V is expressed as the union of open sets, and is thus itself open. $\square$

Proposition 3.4 *The composition $g \circ f : X \to Z$ of any two continuous functions $X \xrightarrow{\ f\ } Y \xrightarrow{\ g\ } Z$ between topological spaces is itself continuous.*

Proof Given an open set U in Z, we need to verify that $(g \circ f)^{-1}(U)$ is open in X. The inverse image function satisfies

$$(g \circ f)^{-1}(U) = f^{-1}(g^{-1}(U))$$

and since g is continuous it follows that $g^{-1}(U)$ is open in Y, and since f, too, is continuous it follows that $f^{-1}(g^{-1}(U))$ is open in X. $\square$

Proposition 3.5 *Given two topologies τ_1 and τ_2 on the same set X, the identity function id : $(X, \tau_1) \to (X, \tau_2)$ is continuous if, and only if, $\tau_2 \subseteq \tau_1$.*

Proof Notice that for the identity function id : $X \to X$ and any subset $U \subseteq X$ (open or not) one has $\mathrm{id}^{-1}(U) = U$. Thus, the condition for continuity is precisely that $U \in \tau_1$ for all $U \in \tau_2$, in other words, that $\tau_2 \subseteq \tau_1$. $\square$

Definition 3.4 Given two topologies τ_1 and τ_2 on the same set X, if $\tau_1 \subseteq \tau_2$, then τ_1 is said to be *coarser* than τ_2 while τ_2 is said to be *finer* than τ_1.

Notice that the discrete topology on X is the finest topology among all topologies on X while the indiscrete topology is the coarsest one. In general, different topologies on the same set may be incomparable, i.e., one need not be contained in the other.

Definition 3.5 A function $f : X \to Y$ between topological spaces is said to be a *homeomorphism* if f is bijective and both $f : X \to Y$ and $f^{-1} : Y \to X$ are continuous functions. If there exists a homeomorphism between X and Y, then X and Y are said to be *homeomorphic* spaces.

Remark 3.2 One should not confuse between the terms homeomorphism and homomorphism. The term homomorphism in algebra is typically used to indicate a structure preserving function, with isomorphism used for the invertible structure preserving functions. In topology, the structure preserving functions are precisely the continuous functions, and the invertible structure preserving functions are called homeomorphisms.

Note that $f : X \to Y$ is a homeomorphism if, and only if, for all $U \subseteq Y$

$$f^{-1}(U) \in \tau_X \iff U \in \tau_Y.$$

Homeomorphic spaces thus have essentially the same collections of open sets, up to a renaming of the elements. Intuitively, two spaces are homeomorphic if one can be continuously changed, without gluing or tearing, to obtain the other. While homeomorphic spaces must have the same cardinality (since a homeomorphism is in particular a bijection), the converse quite often fails since both the bijection and its inverse must be continuous. For instance, the circumference of a circle and a line segment (say in $\mathbb{R}^2$) have the same cardinality, but they are not homeomorphic. This is intuitively quite clear since (it would appear that) the only way to create the hole that the circle encloses is to glue the two ends of the line segment, but that changes the topology (i.e., glueing is a continuous operation but its inverse, tearing, is not). However, turning such intuitive arguments into formal ones is usually not straightforward.

Generally speaking, it can be quite tricky to prove that two topological spaces are not homeomorphic. A common technique in establishing such a result is the following one. A property P about topological spaces is said to be a *topological invariant* if for all homeomorphic topological spaces X and Y, X has property P precisely when Y does. To show that two spaces are not homeomorphic it thus suffices to find a topological invariant that only one of the two spaces possesses. The topological concepts introduced in the rest of this chapter often provide one with a suitable topological invariant.

3.1.4 Closed Sets

With every subset S of a set X one may associate the complementary set $S^c = X \setminus S$, giving rise to a bijective operation $\mathcal{P}(X) \to \mathcal{P}(X)$ on the set $\mathcal{P}(X)$ of all subsets of X. It is thus clear that any concept given in terms of subsets of X gives rise, by taking complements, to what must essentially be an equivalent concept. This general observation holds for the definition of a topology, and we now devote some time for the relevant details.

We first observe that the Sierpiński space $\mathbb{S} = \{0, 1\}$ from Example 3.2 can be used to identify the open sets in an arbitrary topological space.

Proposition 3.6 *There exists a bijective correspondence between the open sets of a topological space X and continuous functions $f : X \to \mathbb{S}$.*

Proof Recall (e.g., Sect. 1.2.10 of the Preliminaries) that with every subset $S \subseteq X$ one may associate the indicator function

$$f_S(x) = \begin{cases} 1 & \text{if } x \in S \\ 0 & \text{if } x \notin S \end{cases}$$

and that this correspondence is a bijection between all subsets of X and all functions $X \to \mathbb{S}$. For $f_S : X \to \mathbb{S}$ to be continuous, the inverse image of every open set in $\mathbb{S}$, namely the sets $\emptyset$, $\mathbb{S}$, and $\{1\}$, must be open in X. The set $f_S^{-1}(\emptyset) = \emptyset$ is always open in X and similarly so is $f_S^{-1}(\mathbb{S}) = X$. So, the only further condition imposed by continuity is that $f_S^{-1}(\{1\})$ must be open too, and since the latter set is precisely S we see that continuous functions $f : X \to \mathbb{S}$ correspond bijectively to the open subsets of X. $\square$

Clearly, every function $f : X \to \mathbb{S}$ is equally well completely determined by the inverse image of 0 instead of the inverse image of 1. We thus see that there is also a correspondence between continuous functions $f : X \to \mathbb{S}$ and the *complements of the open sets* in X, giving rise to the following definition.

Definition 3.6 A subset F of a topological space X is said to be *closed* provided its complement $X \setminus F$ is open.

The preceding discussion implies that a topology can be specified by the collection of closed sets. The details are given in the following result.

Theorem 3.4 *Let X be an arbitrary set. A given collection $\mathcal{F}$ of subsets of X is the collection of closed sets for some topology on X if, and only if, the following conditions hold.*

1. *Both the empty set $\emptyset$ and the entire set X are members of $\mathcal{F}$.*
2. *$\mathcal{F}$ is stable under finite unions. That is, if $F_1, \ldots, F_m$ are members of $\mathcal{F}$, then*

$$F_1 \cup \cdots \cup F_m \in \mathcal{F}.$$

3. *$\mathcal{F}$ is stable under arbitrary intersections. That is, if $\{F_i\}_{i \in I}$ are members of $\mathcal{F}$, then*

$$\bigcap_{i \in I} F_i \in \mathcal{F}.$$

Moreover, a collection $\mathcal{F}$ satisfying the conditions above is the collection of closed sets for a unique topology on X.

Proof Applying De Morgan's Laws (see Sect. 1.2.5 of the Preliminaries), the details are straightforward and are left to the reader. We only stress out here that the collection

$$\{X \setminus F \mid F \in \mathcal{F}\}.$$

is the unique topology determined by $\mathcal{F}$. $\square$

Example 3.5 If a set X is given the discrete topology, then every subset of it is closed, while if X is given the indiscrete topology, then the only closed subsets of X are $\emptyset$ and X itself. In the Sierpiński space $\mathbb{S} = \{0, 1\}$, the closed sets are: $\emptyset$, $\mathbb{S}$, and $\{0\}$.

Example 3.6 Consider $\mathbb{R}$ with the Euclidean topology (Definition 3.3). It is easy to verify that for $a, b \in \mathbb{R}$ with $a < b$, the open interval (a, b) is an open set and the closed interval $[a, b]$ is a closed set. The open rays (a, ∞) and $(-\infty, a)$ are also open sets, and the closed rays $[a, \infty)$ and $(-\infty, a]$ are closed sets. Every singleton set $\{a\}$ is a closed set. Consequently, every finite set $F \subseteq \mathbb{R}$ is closed.

Remark 3.3 It should be emphasized that in general an arbitrary union of closed sets need not be closed nor does an arbitrary intersection of open sets need be open. For instance, consider $\mathbb{R}$ with the Euclidean topology (Definition 3.3). Since every singleton set $\{x\}$ is closed, every subset $X \subseteq \mathbb{R}$ is the union of closed sets, i.e.,

$$X = \bigcup_{x \in X} \{x\}$$

but, of course, not every subset of $\mathbb{R}$ is closed. Similarly, for any $x \in \mathbb{R}$, one has

$$\{x\} = \bigcap_{\varepsilon > 0} (x - \varepsilon, x + \varepsilon)$$

showing that $\{x\}$ is an intersection of open sets, but is itself not open.

Remark 3.4 It is important to realize that a set is not a door. A set need not be open nor closed, while it may be both open and closed. For instance, consider the topology τ from Example 3.3. One easily verifies that $\{a, c\}$ is not closed nor open and that $\{a\}$ is both closed and open. Sets that are both closed and open are said to be *clopen*. Notice that the empty set $\emptyset$ and X itself are clopen in any topological space. If these are the only clopen sets in X, then X is said to be *connected*, a property we explore further in Sect. 3.4.

The dual nature of open and closed sets is seen in the following result. Recall that a function $f : X \to Y$ between topological spaces is continuous if $f^{-1}(U)$ is open in X for every open set U in Y.

Theorem 3.5 *A function $f : X \to Y$ is continuous if, and only if, $f^{-1}(F)$ is a closed set in X for every closed set F in Y.*

The proof is, as it should be, a tautology that the reader is urged to clarify for herself.

The duality between open and closed sets is quite helpful since, while dual to each other, open and closed sets have different topological qualities and thus it may be more natural to establish a result in terms of closed sets rather than open ones, or vice versa. The reader should be warned though that not all concepts that may appear dual are in fact dual.

Definition 3.7 A function $f : X \to Y$ between topological spaces is

- an *open mapping* if $f(U)$ is open in Y for every open set U in X.
- a *closed mapping* if $f(F)$ is closed in Y for every closed set F in X.

The reader is invited to find examples of open mappings that are not closed as well as closed ones that are not open.

3.1.5 Bases and Subbases

Generally speaking, a topology is a very large collection of subsets. For instance, the Euclidean topology on $\mathbb{R}$ has uncountably many open sets. It is thus often desirable to obtain smaller (or at least more manageable) collections that give one access to the entire topology. This is what bases and subbases are designed to achieve.

Definition 3.8 A collection $\mathcal{B}$ of open sets in a topological space X is a *basis* (or a *base*) for the topology τ_X if for every open set $U \subseteq X$ and $x \in U$ there exists an element $B \in \mathcal{B}$ such that

$$x \in B \subseteq U.$$

The elements of $\mathcal{B}$ are called *basis elements*.

Example 3.7 For every topological space (X, τ_X) the collection $\mathcal{B} = \tau_X$ is, quite trivially, a basis. More interestingly, the collection $\{B_\varepsilon(x) \mid x \in \mathbb{R}, \ \varepsilon > 0\}$ is a basis for the Euclidean topology on $\mathbb{R}$. Indeed, if $U \subseteq \mathbb{R}$ is open, then, by definition, for each $x \in U$ there is an $\varepsilon > 0$ such that

$$x \in B_\varepsilon(x) \subseteq U.$$

If X is an arbitrary set endowed with the indiscrete topology, then the collection $\{X\}$ is a basis: there are at most two open sets in the indiscrete topology, namely $\emptyset$ and X, so if U is open and $x \in U$, then necessarily $U = X$ and clearly

$$x \in X \subseteq X.$$

If X is endowed with the discrete topology, then the collection $\{\{x\} \mid x \in X\}$ of all singleton subsets of X is a basis. Indeed, in the discrete topology every $U \subseteq X$ is open, and if $x \in U$, then clearly

$$x \in \{x\} \subseteq U.$$

Definition 3.9 A collection $\mathcal{A}$ of open sets in a topological space X is a *subbasis* for the topology if the collection $\{A_1 \cap \cdots \cap A_m \mid A_1, \ldots, A_m \in \mathcal{A}, \ m \geq 0\}$ of all finite intersections of elements of $\mathcal{A}$ is a basis for the topology.

Remark 3.5 By convention, the empty intersection (i.e., the case $n = 0$ above) is interpreted to be the set X itself.

Example 3.8 Consider the set of real numbers $\mathbb{R}$ with the Euclidean topology. The collection

$$\{(a, \infty) \mid a \in \mathbb{R}\} \cup \{(-\infty, b) \mid b \in \mathbb{R}\}$$

of all open rays is a subbasis. Indeed, the open rays are open sets and since

$$B_\varepsilon(a) = (a - \varepsilon, \infty) \cap (-\infty, a + \varepsilon)$$

we see that the collection of all finite intersections of the open rays contains the basis we presented above. It follows that the collection is a basis since, in general, if $\mathcal{B}' \supseteq \mathcal{B}$ are collections of open sets and $\mathcal{B}$ is a basis, then so is $\mathcal{B}'$.

We now show that bases can be used directly to detect continuity of functions.

Proposition 3.7 *Let $f : X \to Y$ be a function between topological spaces and let $\mathcal{B}$ be a basis for the topology on Y. Then f is continuous if, and only if, $f^{-1}(B)$ is open in X for all basis elements $B \in \mathcal{B}$.*

Proof If f is continuous, then $f^{-1}(U)$ is open for all open sets $U \subseteq Y$. Since any $B \in \mathcal{B}$ is in particular an open set in Y it follows that $f^{-1}(B)$ is open in X. In the other direction, assume that $f^{-1}(B)$ is open for all $B \in \mathcal{B}$ and let U be an arbitrary open set in Y. It follows at once from the definition of basis that we may express U as a union of basis elements

$$U = \bigcup_{x \in U} B_x.$$

It then holds that

$$f^{-1}(U) = \bigcup_{x \in U} f^{-1}(B_x)$$

and since each $f^{-1}(B_x)$ is assumed open, the set $f^{-1}(U)$ is thus expressed as a union of open sets, and so is open. Thus $f^{-1}(U)$ is open for all open sets U in Y, namely f is continuous. $\qquad\square$

We close this section by introducing a local variant of the notion of basis.

Definition 3.10 Let X be a topological space and $x \in X$. A collection $\mathcal{B}$ of open subsets of X is said to be a *local basis* at x if for all $B \in \mathcal{B}$

$$x \in B$$

and if for every open set U in X with $x \in U$ there is an element $B_x \in \mathcal{B}$ (called a *basis element*) with

$$B_x \subseteq U.$$

Every basis $\mathcal{B}$ for the topology on X clearly gives rise to the local basis

$$\mathcal{B}_x = \{B \in \mathcal{B} \mid x \in B\}$$

at x. It is also the case that if for every $x \in X$ one has a local basis $\mathcal{B}_x$ at x, then the collection

$$\mathcal{B} = \bigcup_{x \in X} \mathcal{B}_x$$

is a basis for the topology on X. We leave the verification of these simple claims to the reader.

Exercises

Exercise 3.1 How many topologies are there on a set with two elements? Of the topological spaces you just found, which are homeomorphic?

Exercise 3.2 Characterize all continuous functions $f : \mathbb{S} \to \mathbb{R}$ from the Sierpiński space to $\mathbb{R}$ with the Euclidean topology.

Exercise 3.3 Consider the set $\{0, 1\}$ with the discrete topology. Given a topological space X and a subset $S \subseteq X$ show that the indicator function $\chi : X \to \{0, 1\}$ is continuous if, and only if, S is clopen.

Exercise 3.4 Prove Theorem 3.5.

Exercise 3.5 Let X and Y be sets. When Y is equipped with the indiscrete topology and X with an arbitrary topology, show that every function $f : X \to Y$ is continuous.

Exercise 3.6 Show that a function $f : X \to Y$ between topological spaces is continuous if, and only if, $f^{-1}(F)$ is closed in X for all closed $F \subseteq Y$.

Exercise 3.7 Show that open function, closed function, and continuous function are distinct notions.

Exercise 3.8 Verify that the only finite open subset of $\mathbb{R}$ with the Euclidean topology is the empty set.

Exercise 3.9 For $\alpha \geq 0$ consider the truncated absolute value function given by $|x|_\alpha = |x|$ if $|x| \leq \alpha$ and $|x| = 1$ otherwise. Declare a subset $U \subseteq \mathbb{R}$ to be α-open if for all $x \in S$ there is $\varepsilon > 0$ with $\{y \in \mathbb{R} \mid |x - y|_\alpha < \varepsilon\} \subseteq U$. Prove that for all $\alpha \geq 0$ this results in a topology and that when $\alpha > 0$ it is the Euclidean topology. What happens when $\alpha = 0$?

Exercise 3.10 Prove the following version of the Cantor-Schröder-Bernstein Theorem: if $f : X \to Y$ and $g : Y \to X$ are injective continuous functions between finite topological spaces, then X and Y are homeomorphic.

Exercise 3.11 Give an example of a topological space with infinitely many open sets in which any two non-empty open sets have non-empty intersection.

Exercise 3.12 Let X and Y be sets. When X is equipped with the discrete topology and Y with an arbitrary topology, show that every function $f : X \to Y$ is continuous.

Exercise 3.13 Show that any constant function $f : X \to Y$ between topological spaces is continuous.

Exercise 3.14 Show that any continuous function $f : X \to Y$, when Y is given the discrete topology and X is given the indiscrete topology, is a constant function.

Exercise 3.15 Given topological spaces X, Y, Z show that X is homeomorphic to X and that if X is homeomorphic to Y which in turn is homeomorphic to Z, then X is homeomorphic to Z.

Exercise 3.16 By considering the discrete and indiscrete topologies on a given set demonstrate that a bijective continuous function need not be a homeomorphism.

Exercise 3.17 Prove Theorem 3.4.

Exercise 3.18 We have seen that with respect to the Euclidean topology a function $f : \mathbb{R} \to \mathbb{R}$ is continuous if, and only if, it satisfies the familiar $\varepsilon - \delta$ continuity condition. Give $\varepsilon - \delta$ formulations for f being open and for f being closed.

Exercise 3.19 Verify that the collection of all sets $B_\varepsilon(x)$, where $x \in \mathbb{Q}$ and $\varepsilon > 0$ is rational, is a basis for the Euclidean topology on $\mathbb{R}$. What is its cardinality?

Exercise 3.20 Let $X = \mathbb{R}$ be endowed with the Euclidean topology and $Y = \mathbb{R}$ with the cofinite topology. Is the identity function id : $X \to Y$ continuous? Is the identity function id : $Y \to X$ continuous?

Exercise 3.21 A function $f : X \to Y$ between topological spaces is *locally constant* if for every $x \in X$ there is an open set $U \subseteq X$ with $x \in U$ and so that the restriction of f to U is a constant function. Show that if Y is a discrete space, then the continuity of f implies it is locally constant.

3.2 Subspaces, Point-Set Relationships, and Countability Axioms

This section introduces geometric notions regarding various relationships between a point and a set in a topological space. We also consider countability axioms and study the consequences they entail for some of these notions.

3.2.1 Subspaces and Point-Set Relationships

Any subset of a topological space inherits a topology and thus one may speak of subspaces, which we discuss first. Next, various qualitative aspects of a given point and a subset of a topological space are introduced and some basic results are established.

Definition 3.11 (*Subspace Topology*) Let (X, τ_X) be a topological space and $Y \subseteq X$ a subset. The collection

$$\tau_Y = \{U \cap Y \mid U \in \tau_X\}$$

is a topology on Y called the *subspace* topology. The topological space (Y, τ_Y) is called a *topological subspace* of (X, τ_X), or simply a *subspace* if the topological context is clear.

Proposition 3.8 *The subspace topology on a subset Y of a topological space X is the smallest topology on Y making the inclusion function $i : Y \to X$ continuous.*

Proof Noticing that

$$i^{-1}(U) = Y \cap U$$

for all $U \subseteq X$ one sees that the inclusion function $i : Y \to X$ is continuous precisely if $Y \cap U$ is open in Y whenever $U \subseteq X$ is open in X. This condition clearly holds for the subspace topology τ_Y on Y. Further, if τ is any topology on Y, then the condition that $i : Y \to X$ is continuous with respect to the topologies τ on Y and τ_X on X amounts to the condition

$$U \in \tau_X \implies U \cap Y \in \tau$$

and thus $\tau_Y \subseteq \tau$. This establishes the the minimality claim of the subspace topology τ_Y.   □

Given a subset and a point in a topological space the open sets may be used to identify the qualitative relative location of the point in relation to the set. This is the subject of the following definition.

Definition 3.12 Let $S \subseteq X$ be a subset of a topological space and $x \in X$ a point.

1. x is an *interior* point of S if there exists an open set U such that $x \in U \subseteq S$.
2. x is an *exterior* point of S if there exists an open set U such that $x \in U \subseteq X \setminus S$.
3. x is a *boundary* point of S if every open set U containing x intersects S as well as $X \setminus S$ non-trivially.
4. x is a *cluster* point or an *accumulation* point of S if every open set U containing x intersects $S \setminus \{x\}$ non-trivially.
5. x is an *isolated* point of S if $\{x\}$ is an open set.

Example 3.9 Let us consider the space $\mathbb{R}$ with its Euclidean topology. It is easily seen that for $S = [a, b]$, a closed interval, as well as $S = (a, b)$, an open interval:

1. a point x is interior precisely when $a < x < b$
2. a point x is exterior precisely when either $x < a$ or $x > b$
3. a point x is a boundary point precisely when $x = a$ or $x = b$
4. a point x is a cluster point precisely when $a \leq x \leq b$.
5. S has no isolated points, but the space $X = [0, 1] \cup \{2\}$, as a subspace of $\mathbb{R}$ with the Euclidean topology, has the point $x = 2$ as its only isolated point.

Further, in $\mathbb{R}$ with the Euclidean topology, it is easy to see that every real number x is a cluster point of the subset $\mathbb{Q}$ of rational numbers. In general, an isolated point x is never a cluster point of any set S, simply since for the open set $U = \{x\}$ one always has $U \cap (S \setminus \{x\}) = \emptyset$.

Definition 3.13 Let X be a topological space and $S \subseteq X$ a subset.

1. The *interior* of S is the set $\text{int}(S) = \{x \in X \mid x \text{ is an interior point of } S\}$.
2. The *exterior* of S is the set $\text{ext}(S) = \{x \in X \mid x \text{ is an exterior point of } S\}$.
3. The *boundary* of S is the set $\partial(S) = \{x \in X \mid x \text{ is a boundary point of } S\}$.
4. The *derived set* of S is the set $S' = \{x \in X \mid x \text{ is a cluster point of } S\}$.
5. The *closure* of S is the set $\overline{S} = S \cup S'$.

Proposition 3.9 *Let X be a topological space and $S \subseteq X$ a subset. It then holds that*

1. *The interior $\text{int}(S)$ is the union of all open sets contained in S.*
2. *The exterior $\text{ext}(S)$ is the union of all open sets disjoint from S.*
3. *The closure $\overline{S}$ is the intersection of all closed sets that contain S.*
4. *$\overline{S} = S \cup \partial(S)$.*

Proof 1. Let $\mathcal{U}$ be the set of all open sets $U \subseteq X$ with $U \subseteq S$, and let

$$W = \bigcup_{U \in \mathcal{U}} U$$

so we need to show that $W = \text{int}(S)$. If $x \in W$, then $X \in U$ for some $U \in \mathcal{U}$, and since $x \in U \subseteq S$ we may conclude that $x \in \text{int}(S)$. Conversely, if $x \in \text{int}(S)$, then there exists an open set U such that $x \in U \subseteq S$, and thus $x \in W$.

2. Notice that $\text{ext}(S) = \text{int}(X \setminus S)$, and now apply 1.

3. Let $\mathcal{F}$ be the set of all closed sets $F \subseteq X$ with $S \subseteq F$, and let

$$G = \bigcap_{F \in \mathcal{F}} F$$

so we need to show that $G = \overline{S}$. Let $x \in G$, and thus $x \in F$ for all $F \in \mathcal{F}$, and our aim is to show that $x \in \overline{S}$. If $x \in S$, then $x \in S \cup S' = \overline{S}$ so we may proceed under the assumption that $x \notin S$. We will show that in this case $x \in S'$. Indeed, let $U \subseteq X$ be an open set with $x \in U$ and suppose that $U \cap (S \setminus \{x\}) = \emptyset$, which simplifies to $U \cap S = \emptyset$ since $x \notin S$. But then the closed set $F = X \setminus U$ contains S, and thus $x \in F$, contradicting the fact that $x \in U$. In the other direction, suppose that $x \in \overline{S} = S \cup S'$, and we aim to show that $x \in G$, namely that $x \in F$ for every closed set F with $S \subseteq F$. If $x \in S$, then clearly $x \in F$, and we may thus proceed under the assumption that $x \notin S$, and thus that $x \in S'$. Consider the open set $U = X \setminus F$ and note that $U \cap (S \setminus \{x\}) = U \cap S = \emptyset$. If $x \notin F$, then $x \in U$ and we obtain a contradiction with $x \in S'$. It follows that $x \in F$, as needed.

4. To show that $\overline{S} \subseteq S \cup \partial S$ suppose that $x \in \overline{S}$ but $x \notin S$. Given an open set U with $x \in U$ it follows that $U \cap S \neq \emptyset$. Since $x \notin S$, namely $x \in X \setminus S$, we also have that $x \in U \cap (X \setminus S)$. In other words, any open set containing x contains points from S as well as from $X \setminus S$, thus x is a boundary point of S. In the other direction, suppose that $x \in S \cup \partial S$, and we need to show that $x \in \overline{S} = S \cup S'$.

If $x \in S$, then clearly $x \in \bar{S}$ so assume that $x \notin S$, and thus $x \in \partial S$. But then if U is an open set containing x, then U contains a point from S (and from $X \setminus S$, though this is irrelevant here), but then $U \cap (S \setminus \{x\}) \neq \emptyset$ since $x \notin S$. Thus x is a cluster point, as needed.

$\square$

Corollary 3.1 *Let X be a topological space and $S \subseteq X$ a subset. Then*

1. *The interior int(S) is the largest open set in X that is contained in S. In particular, int(S) is open.*
2. *S is open in X if, and only if, $S = $ int(S). In other words, S is open precisely when all its points are interior.*
3. *The exterior ext(S) is the largest open set in X that is disjoint from S. In particular, ext(S) is open.*
4. *S is closed in X if, and only if, $X \setminus S = $ ext(S).*
5. *The closure $\bar{S}$ is the smallest closed set in X that contains S. In particular, $\bar{S}$ is closed.*
6. *S is closed in X if, and only if, $S' \subseteq S$ (equivalently, if $\bar{S} = S$).*

3.2.2 Sequences and Convergence

We turn to look at sequences in topological spaces, what it means for such sequences to converge, and how this notion relates to the topological point-set relationships presented above.

Definition 3.14 Let $\{x_m\}_{m \geq 1}$ be a sequence of points in a topological space X. A point $x_0 \in X$ is said to be a *limit* of the sequence $\{x_m\}_{m \geq 1}$ if for every open set $U \subseteq X$ with $x_0 \in U$ there exists an $N \in \mathbb{N}$ such that

$$m > N \implies x_m \in U.$$

The sequence is then said to *converge* to x_0 and to be a *convergent sequence*.

Just as continuity of functions can be detected using a basis, so can cluster points and limit points be so detected.

Proposition 3.10 *In a topological space X with a local basis $\mathcal{B}$ at a point x_0:*

1. *x_0 is a cluster point of a subset $S \subseteq X$ if, and only if, $B \cap (S \setminus \{x_0\}) \neq \emptyset$, for all basis elements $B \in \mathcal{B}$.*
2. *A sequence $\{x_m\}_{m \geq 1}$ converges to x_0 if, and only if, for every $B \in \mathcal{B}$ there exists an $N \in \mathbb{N}$ such that*
$$m > N \implies x_m \in B.$$

Proof As the two proofs are very similar we only establish the first claim. If x_0 is a cluster point and $B \in \mathcal{B}$ is a basis element, then, in particular, B is open and $x_0 \in B$, implying $B \cap (S \setminus \{x_0\}) \neq \emptyset$. Conversely, suppose the stated condition holds and let U be an arbitrary open set with $x_0 \in U$. By definition of basis at a point there exists a basis element $B \in \mathcal{B}$ with $x_0 \in B \subseteq U$. By hypothesis then $B \cap (S \setminus \{x_0\}) \neq \emptyset$ and since $U \cap (S \setminus \{x_0\}) \supseteq B \cap (S \setminus \{x_0\})$, the proof is complete. □

Example 3.10 Consider the Euclidean topology on $\mathbb{R}$ and recall that a local basis at x_0 is given by $\{B_\varepsilon(x_0)\}_{\varepsilon > 0}$. Thus, a sequence $\{x_m\}_{m \geq 1}$ in $\mathbb{R}$ converges to $x_0 \in \mathbb{R}$ if, and only if, for every $\varepsilon > 0$ there exists an $N \in \mathbb{N}$ such that

$$m > N \implies x_m \in B_\varepsilon(x_0).$$

In other words: if $m > N$, then $|x_m - x| < \varepsilon$. We thus see that convergence in the topological sense given above is equivalent to convergence of sequences in $\mathbb{R}$ in the usual sense.

While it is reassuring that the notion of convergence can be captured topologically, one should be cautious not to expect the familiar behaviour of limits in $\mathbb{R}$ to remain valid in arbitrary topological spaces, as we illustrate.

Example 3.11 Let X be an arbitrary set endowed with the indiscrete topology. Since the only open sets are $\emptyset$ and X any sequence $\{x_m\}_{m \geq 1}$ in X converges and *any* element $x \in X$ serves as its limit. Consequently, a sequence may converge to more than one point; uniqueness of limits is lost.

Given a subset S of a topological space X we can entertain (at least) two reasonable formalizations for the idea of the set of all points that are infinitesimally close to S. One possibility is given by the closure $\overline{S} = S \cup S'$, where we adjoin to S all of its derived points. A second possibility is obtained by considering the set of all limit points of sequences in S. A-priori, it is not clear how these two possibilities compare. We limit the discussion just to the point of developing enough theory to suit the needs of this book. It should be noted that this is just the tip of the iceberg. The important observation is that, generally, sequences do not suffice to capture cluster points (but either nets or filters do, two concepts we will not discuss). Having said that, we do identify below a broad enough class of topological spaces in which the limit behaviour is more intuitive.

Theorem 3.6 *Let X be a topological space and $S \subseteq X$ a subset. If a non-eventually-constant sequence $\{s_m\}_{m \geq 1}$ in S converges to $x_0 \in X$, then x_0 is a cluster point of the subset S.*

Proof Suppose that U is an open set and $x_0 \in U$. Since $\{s_m\}_{m \geq 1}$ converges to x_0 there exists an $N \in \mathbb{N}$ such that $s_m \in U$ for all $m > N$, and since the sequence is not eventually constant an $m > N$ exists with $s_m \neq x_0$. In particular, $U \cap (S \setminus \{x_0\}) \neq \emptyset$, and thus x_0 is a cluster point of S. □

Theorem 3.7 *If $f : X \to Y$ is a continuous function between topological spaces and $x_m \to x_0$ in X, then $f(x_m) \to f(x_0)$ in Y.*

Proof Given an open set $U \subseteq Y$ with $f(x_0) \in U$ we have that $x_0 \in f^{-1}(U)$ and thus there exists an $N \in \mathbb{N}$ such that

$$m > N \implies x_m \in f^{-1}(U)$$

and thus

$$m > N \implies f(x_m) \in U$$

as required. $\square$

Remark 3.6 The converses of these two theorems are generally false. To see that, it suffices to note that if X is an uncountable set endowed with the cocountable topology, then the only convergent sequences are the eventually constant ones, while every point $x \in X$ is a cluster point of any cofinite set S (i.e., one where where $X \setminus S$ is finite).

3.2.3 Second Countable and First Countable Spaces

We present two axioms out of a class of properties called countability axioms. The conditions we present are quite strong and have pleasant consequences to the behaviour of limit and cluster points. While many topological spaces fail to satisfy either countability axiom, the class of those that do is broad enough to supply one with plenty of topological spaces that are closer to one's intuition than their wilder topological cousins.

Definition 3.15 A topological space X satisfies the *second axiom of countability* or is said to be *second countable* if there is a countable basis for the topology on X.

Thus, second countability means that there exist a countable index set I and a family $\{U_i\}_{i \in I}$ of open sets in X such that any set U can be expressed as

$$U = \bigcup_{j \in J_U} U_j$$

for some $J_U \subseteq I$.

Example 3.12 The space $\mathbb{R}$ with the Euclidean topology is second countable. We exhibit a countable basis for the topology, namely the collection

$$\{B_{1/n}(q)\}_{q \in \mathbb{Q}, n \geq 1}.$$

Firstly, note that this is a countable collection as it is indexed by the set $\mathbb{Q} \times (\mathbb{N} \setminus \{0\})$, (see Sect. 1.2.13 of the Preliminaries). Next, given an open set $U \subseteq \mathbb{R}$ and a point $x \in U$, there exists a $\delta > 0$ such that $B_\delta(x) \subseteq U$, and let $n \in \mathbb{N}$ satisfy $1/n < \delta/2$. Since the rationals are dense in the reals, we may find a rational number q for which

$|x - q| < 1/n$. It now easily follows that $x \in B_{1/n}(q) \subseteq B_\delta(x) \subseteq U$, as required by the definition of basis.

Definition 3.16 A subset $S \subseteq X$ in a topological space is *dense* in X if $\overline{S} = X$. The space X is called *separable* if it contains a countable dense subset.

A familiar example of a separable topological space is $\mathbb{R}$ with the Euclidean topology: the countable set $\mathbb{Q}$ is dense in $\mathbb{R}$. In fact, $\mathbb{R}$ is seen to be separable by the following general result.

Theorem 3.8 *Every second countable space X is separable.*

Proof Let $\mathcal{B}$ be a countable basis for the topology. For each $B \in \mathcal{B}$ choose an arbitrary point $x_B \in B$ (if $B = \emptyset$, simply skip it). The set $S = \{x_B \mid B \in \mathcal{B}\}$ is clearly countable so it remains to show that it is dense. Indeed, fix a point $x \in X$. To show that x is in the closure of S suppose $x \notin S$ and let U be an open set with $x \in U$. We now need to show that $U \cap S \neq \emptyset$. But, by definition, there exists a basis element B with $x \in B \subseteq U$. Then x_b must be found in $U \cap S$, establishing the claim. $\square$

Many spaces of interest are not second countable but they do satisfy a weaker condition known as the first axiom of countability. Such spaces still exhibit general behaviour that in some respect is quite close to what our intuition about convergence and closures in $\mathbb{R}$ dictates.

Definition 3.17 A topological space X satisfies the *first axiom of countability* or is said to be *first countable* if there exists, at every point $x \in X$, a countable local basis.

Evidently a second countable space is also first countable.

Recall that the converses of Theorems 3.6 and 3.7 are not generally true in arbitrary topological spaces. For first countable spaces we have the following pleasing result.

Theorem 3.9 *Let X be a first countable space with, at each point $x_0 \in X$, a countable local basis $\mathcal{B}_{x_0} = \{B_m^{x_0}\}_{m \geq 1}$ at x_0.*

1. *For a subset $S \subseteq X$, a point $x_0 \in X$ is in $\overline{S}$ if, and only if, $s_m \to x_0$ for a sequence $\{s_m\}_{m \geq 1}$ in S.*
2. *A subset $F \subseteq X$ is closed if, and only if, F contains all limit points of sequences in it.*
3. *A function $f : X \to Y$ to any topological space Y is continuous if, and only if, it preserves limits of sequences, i.e.,*

$$x_m \to x_0 \implies f(x_m) \to f(x_0)$$

for all sequences $\{x_m\}_{m \geq 1}$ in X.

Proof 1. By Theorem 3.6 every limit point of a non-eventually constant sequence in S is a cluster point of S, and thus is in $\overline{S}$. The limits of eventually constant sequences are, of course, in S and thus also in $\overline{S}$. Suppose now that $x_0 \in X$ is in $\overline{S}$ and, of course, the interesting case is when $x_0 \notin S$. In that case x_0 is a cluster point of S. Therefore, for every $m \geq 1$ the open set $B_1^{x_0} \cap \cdots \cap B_m^{x_0}$ intersects $S \setminus \{x_0\}$ non-trivially, and let s_m be an arbitrary element in the intersection. Clearly the sequence $\{s_m\}_{m \geq 1}$ is in S and its limit is x_0. Indeed, given any open set $U \subseteq X$ with $x_0 \in U$, there exists a basis element $B_N^{x_0}$ with $x_0 \in B_N^{x_0} \subseteq U$. But then

$$m > N \implies s_m \in B_m^{x_0} \subseteq B_N^{x_0} \subseteq U.$$

2. By Corollary 3.1, F is closed if, and only if, $\overline{F} = F$, and we just proved that the elements of $\overline{F}$ coincide with limit points of sequences in F.
3. By Theorem 3.7, if $f : X \to Y$ is continuous, then it preserves limits of sequences. Suppose thus that $f : X \to Y$ preserves limits of sequences, and our aim is to show that f is continuous. We need to show that $f^{-1}(U)$ is open whenever $U \subseteq Y$ is open, or (equivalently!) that $f^{-1}(F)$ is closed whenever $F \subseteq Y$ is closed. It suffices to show that $f^{-1}(F)$ contains its limit points, so suppose that

$$x_m \to x_0$$

in X with $\{x_m\}_{m \geq 1}$ a sequence in $f^{-1}(F)$. We show that $x_0 \in f^{-1}(F)$: since f preserves limits of sequences we have $f(x_m) \to f(x_0)$, but then $f(x_0)$ is a limit point of a sequence in F and since the latter is closed, Theorem 3.6 implies that $f(x_0) \in F$, showing that $x_0 \in f^{-1}(F)$.

$\square$

Exercises

Exercise 3.22 Prove that the boundary $\partial(S)$ of any subset S of a topological space X is a closed set.

Exercise 3.23 Verify that the subspace topology on a subset S of a topological space X is indeed a topology.

Exercise 3.24 Let (X, τ) be a topological space and $Z \subseteq Y \subseteq X$ subsets. Show that the subspace topology on X as a subset of itself is τ. Show further that the subspace topology on Z as a subspace of X coincides with the subspace topology on Z as a subset of Y, where Y is given its subspace topology as a subset of X.

Exercise 3.25 Prove that $\partial S = \partial(X \setminus S)$ for any subset S of a topological space X.

Exercise 3.26 Prove that a subset S of a topological space X is closed if, and only if, it contains its boundary.

Exercise 3.27 Prove that a set S of a topological space is open if, and only if, it is disjoint from its boundary.

Exercise 3.28 Prove that $S \subseteq T \implies \bar{S} \subseteq \bar{T}$ for all subsets S, T of a topological space X.

Exercise 3.29 Prove that $\bar{\bar{S}} = \bar{S}$ for all subsets S of a topological space X.

Exercise 3.30 Let (X, τ) be a topological space. Define the function $\neg\colon \tau \to \tau$ by $\neg(U) = \mathrm{int}(X \setminus U)$. Prove that $U \subseteq V \implies \neg(V) \subseteq \neg(U)$. In $\mathbb{R}$ with the Euclidean topology find an open subset U with $\neg\neg(U) \neq U$.

Exercise 3.31 Find an example of a topological space X and a non-empty subset $S \subseteq X$ such that
$$\mathrm{int}(S) = \mathrm{ext}(S) = \emptyset.$$

Exercise 3.32 Prove the second assertion of Proposition 3.10.

Exercise 3.33 Let X be an uncountable set endowed with the cocountable topology. Show that the only convergent sequences in X are the eventually constant ones while for every cofinite set $S \subseteq X$ (i.e., $X \setminus S$ is finite) $\overline{S} = X$.

Exercise 3.34 In $\mathbb{R}$ with the Euclidean topology show that $\mathbb{Q}$ is dense. Show that the irrationals $\mathbb{R} \setminus \mathbb{Q}$ are also dense.

Exercise 3.35 Construct a first countable space that is not second countable.

Exercise 3.36 Let $\mathbb{N}^+ = \mathbb{N} \cup \{\infty\}$ where ∞ is a symbol. Declare $U \subseteq \mathbb{N}^+$ to be open if $\infty \in U$ and $\mathbb{N} \setminus U$ is finite. Show that together with the empty set these sets form a topology on $\mathbb{N}^+$. Verify that a local basis for the topology is provided by the family $\uparrow(n) = \{k \in \mathbb{N} \mid k \geq n\} \cup \{\infty\}$.

Exercise 3.37 A sequence $\{s_m\}_{m \geq 1}$ in a topological space X can be identified with a function $s\colon \mathbb{N} \to X$, i.e., $m \mapsto s_m$. Given a point $x_0 \in X$ the function s extends to $s^+\colon \mathbb{N}^+ \to X$ by defining $s(\infty) = x_0$. Continuing the previous exercise, prove that $x_m \to x_0$ in X if, and only if, the extended function $f^+\colon \mathbb{N}^+ \to X$ is continuous.

Exercise 3.38 Prove that a countable first countable space is second countable. That is, if X is a countable set endowed with a first countable topology, then it is in fact a second countable topology.

Exercise 3.39 Prove that if X is a second countable topological space, then $|\tau_X|$, the cardinality of τ_X, is at most $|\mathbb{R}|$, the cardinality of the real numbers. Does the converse hold? Does the same result hold if second countability is replaced by first countability?

3.3 Constructing Topologies

In this section we discuss two ways of constructing topologies. The first situation is where one has an arbitrary set and a collection of subsets of it that one would like to have as the open sets in a topology on X. The second scenario concerns a collection of functions one would like to force to be continuous. This construction is used to obtain the product, coproduct, and quotient constructions for topological spaces.

3.3.1 Generating Topologies

As with the Euclidean topology on $\mathbb{R}$ it is often convenient to have a manageable basis for a topology since one can effectively work with the basis elements rather than the entire topology. Given an arbitrary set X, without any prior notion of topology, and a collection $\mathcal{B}$ of subsets of X it is natural to ask whether there exists a topology τ on X for which $\mathcal{B}$ is a basis, and, if such a topology exists, is it unique. The answers are quite decisive.

Theorem 3.10 (Basis Generating Topology) *A collection $\mathcal{B}$ of subsets of a set X is a basis for a topology τ on X if, and only if, the following conditions are met.*

1. *For every $x \in X$ there exists a $B \in \mathcal{B}$ such that $x \in B$. Equivalently,*

$$X = \bigcup_{B \in \mathcal{B}} B.$$

2. *For all $B_1, B_2 \in \mathcal{B}$, if $x \in B_1 \cap B_2$, then there exists $B_3 \in \mathcal{B}$ such that*

$$x \in B_3 \subseteq B_1 \cap B_2.$$

Moreover, when the conditions are met, the topology τ is unique, namely τ is the smallest topology on X such that every $B \in \mathcal{B}$ is open.

Proof Declare $U \subseteq X$ open if for every $x \in U$ there exists a $B \in \mathcal{B}$ such that

$$x \in B \subseteq U.$$

Equivalently, $U \subseteq X$ is open if U is a union of elements from B. We show that the collection τ of all such open sets forms a topology on X. The empty set $\emptyset$ is vacuously open since it has no points, and the set X itself is open by condition 1. Stability under arbitrary unions is straightforward, so let us establish stability under finite intersections. Let $U_1, U_2 \subseteq X$ be open and we need to show that $U = U_1 \cap U_2$ is open. Indeed, if $x \in U$, then $x \in U_1$ and thus there is a $B_1 \in \mathcal{B}$ such that $x \in B_1 \subseteq U_1$. Similarly, there is a $B_2 \in \mathcal{B}$ with $x \in B_2 \subseteq U_2$. As $x \in B_1 \cap B_2$, condition 2 gives us an element $B_3 \in \mathcal{B}$ with

$$x \in B_3 \subseteq B_1 \cap B_2 \subseteq U_1 \cap U_2 = U,$$

as needed.

The claim that $\mathcal{B}$ is a basis for τ is immediate from the construction. As for the minimality claim about τ, note that if τ' is any topology on X for which all $B \in \mathcal{B}$ are open, then any $U \in \tau$, being a union of elements from B, must also be open in τ'. In other words, $\tau \subseteq \tau'$. So, as claimed, among all topologies where each $B \in \mathcal{B}$ is open, τ is the smallest one.   □

We may now pose the exact same question as above, replacing basis by subbasis. Thus, given an arbitrary collection $\mathcal{B}$ of subsets of X, is there a topology τ on X for which $\mathcal{B}$ is a subbasis, and is this topology unique.

Theorem 3.11 (Subbasis Generating Topology) *A collection $\mathcal{B}$ of subsets of a set X is a subbasis for a topology τ on X if, and only if,*

$$\bigcup_{B \in \mathcal{B}} B = X.$$

Moreover, in that case the topology τ is unique, namely τ is the smallest topology on X such that every $B \in \mathcal{B}$ is open.

Proof One may reduce the proof to that of the previous result by noting that the collection

$$\mathcal{B}' = \{B_1 \cap \cdots \cap B_m \mid B_1, \ldots, B_m \in \mathcal{B}, \, m \geq 0\},$$

consisting of all finite intersections of elements of $\mathcal{B}$, satisfies the conditions of being a basis for a topology. In other words, we declare a subset $U \subseteq X$ to be open if it is an arbitrary union of finite intersections of elements from $\mathcal{B}$. The details are left to the reader. □

These are powerful topology generating tools. Any topology constructed from a collection $\mathcal{B}$ by the above techniques is said to be *generated* by $\mathcal{B}$.

Shifting attention from subsets to functions, the next result states, roughly, that given any collection of functions with a common domain or codomain, a canonical topology exists rendering all the given functions continuous. We remark that, in a sense, topological spaces exist as the servants of continuous functions; we define topological spaces since we are interested in continuous mappings. This result thus allows one to tailor a suitable topology from a given collection of functions.

Proposition 3.11 (Topologies Induced by Functions) *Let X be a set and $\{X_i\}_{i \in I}$ a non-empty collection of topological spaces.*

1. *Given functions $\{f_i : X \to X_i\}_{i \in I}$ there exists a unique smallest topology on X such that every $f_i : X \to X_i$ is continuous.*
2. *Given functions $\{f_i : X_i \to X\}_{i \in I}$ there exists a unique largest topology on X such that every $f_i : X_i \to X$ is continuous.*

Proof 1. Suppose that τ is a topology on X such that each $f_i : X \to X_i$ is continuous. That means that $f_i^{-1}(U) \in \tau$ for each $U \in \tau_{X_i}$. In other words $\tau \supseteq \mathcal{B}$ where

$$\mathcal{B} = \{f_i^{-1}(U) \mid i \in I, U \in \tau_{X_i}\}$$

and so we see that we must define τ to be the topology generated by $\mathcal{B}$. This is feasible since $\mathcal{B}$ is easily seen to satisfy the condition of Theorem 3.11.

2. Suppose that τ is a topology on X such that each $f_i : X_i \to X$ is continuous. That means that τ cannot contain any subset $S \subseteq X$ for which there exists an $i \in I$ with $f_i^{-1}(S) \notin \tau_{X_i}$. So, to obtain the largest possible topology on X, we consider the collection

$$\tau = \{S \subseteq X \mid f_i^{-1}(S) \in \tau_{X_i} \text{ for all } i \in I\}.$$

Showing that τ is a topology would complete the proof. Notice that

$$\tau = \bigcap_{i \in I} \{S \subseteq X \mid f_i^{-1}(S) \in \tau_{X_i}\}$$

and that (the reader is invited to prove) the intersection of a family of topologies on X is again a topology on X. Thus we only need to establish that the collection $\{S \subseteq X \mid f_i^{-1}(S) \in \tau_{X_i}\}$, for a single $i \in I$, is a topology. This follows immediately from well-known properties of the inverse image function, as the reader may verify. $\square$

When a topology τ is constructed as above we say that it is generated by the given collection of functions. We note that it is quite possible that the collection of functions consists of just one single function. We already saw such a situation: the subspace topology on a subset Y of a topological space X is precisely the topology generated by the inclusion function $i : Y \to X$.

3.3.2 Coproducts, Products, and Quotients

The constructions above are utilized in in three particular scenarios.

Definition 3.18 Let X and Y be topological spaces and assume $X \cap Y = \emptyset$. The topology on $X \cup Y$ generated by $\{i_X : X \to X \cup Y, i_Y : Y \to X \cup Y\}$, where i_X and i_Y are the inclusion functions, is called the *coproduct topology* on $X \cup Y$, and the space $X \cup Y$ is called the *coproduct* of X and Y.

Examples are abundant, but some care is required in order to hone the intuition. For instance, in $\mathbb{R}$ with the Euclidean topology, the subspaces $\mathbb{Q}$ and $\mathbb{R} \setminus \mathbb{Q}$ are disjoint and while obviously their union recovers the ambient space $\mathbb{R}$, the coproduct topology on $\mathbb{R}$ is not the Euclidean one.

The situation generalizes to infinitely many spaces. Suppose $\{X_i\}_{i \in I}$ is a non-empty collection of topological spaces that are pair-wise disjoint. The *coproduct* of all the X_i is then the set

$$X = \bigcup_{i \in I} X_i$$

endowed with the topology generated by the family of inclusions $\{X_i \to X\}_{i \in I}$. As an example, a topological space X is discrete if, and only if, it is the coproduct of its singleton subspaces.

Definition 3.19 Let X and Y be two topological spaces, and consider the cartesian product $X \times Y$. The topology generated by the set of projections

$$\{\pi_X : X \times Y \to X, \pi_Y : X \times Y \to Y\}$$

is called the *product topology* on $X \times Y$, and the space $X \times Y$ is called the *cartesian product* or simply the *product* of X and Y.

This situation naturally extends to the infinite case. If $\{X_i\}_{i \in I}$ is a non-empty family of topological spaces, then their cartesian product is the set X, the set-theoretic cartesian product of the X_i, endowed with the topology generated by the family of projections $\{\pi_i : X \to X_i\}_{i \in I}$.

Let us consider the case of a product of countably many spaces $\{X_m\}_{m \geq 1}$. The set-theoretic cartesian product $X = X_1 \times \cdots \times X_m \times \cdots$ consists of all sequences $x = (x_1, \ldots, x_m, \ldots)$ with $x_m \in X_m$ for each $m \geq 1$. The projection $\pi_m : X \to X_m$ is given by $\pi_m(x) = x_m$. Following the definitions above it follows that the product topology on X is the one generated by the sets of the form $U_1 \times \cdots \times U_m \times \cdots$ where each U_m is an open set in X_m and $U_m = X_m$ for all but finitely many m.

Example 3.13 A familiar example is $\mathbb{R}^n$, the n-fold product of the topological space $\mathbb{R}$ with itself, when given the Euclidean topology. For $n \geq 2$ the resulting product topology on $\mathbb{R}^n$ is also called the Euclidean topology. Another example is $\mathbb{R}^\infty$ as a countable product of $\mathbb{R}$ with itself.

The final construction we present is that of the quotient topology. Let X be a set with an equivalence relation on it. Recall (e.g., Sect. 1.2.15 of the Preliminaries) that there is then the associated quotient set $X/\sim = \{[x] \mid x \in X\}$ of all equivalence classes $[x]$, and the canonical projection $\pi : X \to X/\sim$ given by $\pi(x) = [x]$. When this situation is enriched by the presence of a topology on X, it is natural to seek out a topology on $X/\sim$ as well. With the tools we now have the path to take is clear.

Definition 3.20 Let X be a topological space and $\sim$ an equivalence relation on the set X. The topology on $X/\sim$ generated by the projection function $\pi : X \to X/\sim$ is called the *quotient topology* on $X/\sim$, and the space $X/\sim$ is called the *quotient of X modulo $\sim$*.

Example 3.14 Consider $[0, 1]$ as a subspace of $\mathbb{R}$ with its standard topology. Define the equivalence relation $\sim$ on $[0, 1]$ by $x \sim y$ when either $x = y$ or $x, y \in \{0, 1\}$. Informally, $\sim$ glues the two ends of the interval $[0, 1]$, resulting in a circle. More precisely, it can be shown that the quotient space $[0, 1]/\sim$ is homeomorphic to a circle in $\mathbb{R}^2$ with the Euclidean topology.

Exercises

Exercise 3.40 Let X be a set and $\{\tau_i\}_{i \in I}$ a collection of topologies on X. Prove that

$$\bigcap_{i \in I} \tau_i$$

is a topology as well. On the contrary, prove that the union of two topologies on X need not be a topology.

Exercise 3.41 Let X be a set and consider the set $\text{Top}(X)$ of all topologies on X. Ordered by inclusion, show that $\text{Top}(X)$ is a complete lattice.

Exercise 3.42 For a subbasis $\mathcal{B}$ on a set X show that the collection of all finite intersections of elements from $\mathcal{B}$ is a basis.

Exercise 3.43 Given functions $X \xrightarrow{f} Y \xrightarrow{g} Z$ where X, Z are sets and (Y, τ) a topological space show that the collection $\{f^{-1}(U) \mid U \in \tau\}$ of inverse images is a topology on X while the collection of direct images $\{f(U) \mid U \in \tau\}$ need not be a topology on Z.

Exercise 3.44 Note that a singleton set carries a unique topology. Let X be a set and consider, for each $x \in X$, the function $f_x : \{x\} \to X$ given by $f_x(x) = x$. Show that the topology on X generated by $\{f_x : \{x\} \to X\}_{x \in X}$ is the discrete topology.

Exercise 3.45 Let X be a topological space and consider, for each $x \in X$, the function $f_x : X \to X$ given by $f_x(t) = x$. Show that the topology on the set X generated by the family $\{f_x : X \to X\}_{x \in X}$ is the indiscrete topology.

Exercise 3.46 Let X be a set and $\mathcal{B}$ a collection of subsets of X satisfying the conditions of Theorem 3.11. Prove that the intersection of all topologies τ with $\mathcal{B} \subseteq \tau$ is precisely the topology generated by $\mathcal{B}$.

Exercise 3.47 Prove that the collection $\{[a, b) \mid a, b \in \mathbb{R}, \ a < b\}$ is a basis for a topology on $\mathbb{R}$, known as the lower-limit topology. When $\mathbb{R}$ is equipped with this topology it is called the *Sorgenfrey line*, denoted by $\mathbb{R}_l$. Let $\mathbb{R}_E$ denote $\mathbb{R}$ with the Euclidean topology. Compare these two topologies on $\mathbb{R}$ and investigate the meaning of continuity of functions $f : \mathbb{R}_l \to \mathbb{R}_E$.

Exercise 3.48 Let X be a topological space. Prove that X is discrete if, and only if, X is the coproduct of all its singleton subspaces.

Exercise 3.49 Consider $\mathbb{R}$ with the Euclidean topology, the rationals $\mathbb{Q}$ as a subspace, and the irrationals $\mathbb{R} \setminus \mathbb{Q}$ as another subspace. Let R be the coproduct of the two subspaces. Show that R and $\mathbb{R}$ have the same underlying set but their topologies are incomparable: each has an open subset that is not open in the other.

Exercise 3.50 Prove that for all topological spaces X, Y, Z, the spaces $X \times (Y \times Z), X \times Y \times Z$, and $(X \times Y) \times Z$ are homeomorphic. Is there any situation in which these spaces are equal?

Exercise 3.51 Let $\{X_i\}_{i \in I}$ be a collection of topological spaces and assume each X_i consists of just two points. Let X be the product of this collection.

1. If all spaces are indiscrete, show that X is indiscrete.

2. If all spaces are discrete and I is finite, show that X is discrete.
3. If all spaces are discrete and I is infinite, show that X is not discrete.

Exercise 3.52 Let X and Y be topological spaces and consider the relation on the set $X \times Y$ where $(x, y) \sim (x', y')$ precisely when $x = x'$ (obviously an equivalence relation). Prove that when $X \times Y$ is endowed with the product topology, the quotient space $(X \times Y)/\sim$ is homeomorphic to X. Can equality ever hold?

Exercise 3.53 Consider $\mathbb{R}$ with the Euclidean topology. Prove that the collection $\{U_1 \times \cdots \times U_m \times \cdots \mid U_1, \ldots, U_m, \cdots \subseteq \mathbb{R} \text{ are open}\}$ is a basis for a topology on $\mathbb{R}^\infty$. This topology is called the *box* topology. Of the box topology and the product topology on $\mathbb{R}^\infty$, which one is finer?

Exercise 3.54 Let X be a topological space and $\sim$ an equivalence relation on X. Prove that a set $U \subseteq X/\sim$ (which is a collection of equivalence classes, thus of subsets of X) is open in the quotient topology if, and only if,

$$\bigcup_{[x] \in U} [x]$$

is open in X.

Exercise 3.55 Construct a quotient space of the square $[0, 1] \times [0, 1]$ that is homeomorphic to the surface of a torus in $\mathbb{R}^3$.

Exercise 3.56 Consider topological spaces X, Y and the diagram

$$X \xleftarrow{\pi_X} X \times Y \xrightarrow{\pi_Y} Y$$
$$\forall f_X \nwarrow \quad \exists! g \uparrow \quad \nearrow \forall f_Y$$
$$T$$

in which $\pi_X(x, y) = x$ and $\pi_Y(x, y) = y$ are the projection functions. The diagram reads as: for all topological spaces T and continuous functions f_X, f_Y there exists a unique continuous function g satisfying $\pi_X \circ g = f_X$ and $\pi_Y \circ g = f_Y$. Prove this is so.

Exercise 3.57 Consider disjoint topological spaces X, Y and the diagram

$$X \xrightarrow{\iota_X} X \coprod Y \xleftarrow{\iota_Y} Y$$
$$\forall f_X \searrow \quad \exists! g \downarrow \quad \swarrow \forall f_Y$$
$$T$$

in which $X \coprod Y$ is the coproduct, $\iota_X(x) = x$ and $\iota_Y(y) = y$. The diagram reads as: for all topological spaces T and continuous functions f_X, f_Y there exists a unique continuous function g satisfying $g \circ \iota_X = f_X$ and $g \circ \iota_Y = f_Y$. Prove this is so.

Exercise 3.58 Consider $[0, 1]$ as a subspace of $\mathbb{R}$ with the Euclidean topology. Let S^1 be the unit circle in $\mathbb{R}^2$ with respect to the norm induced by the standard inner product. Since $S^1 \subseteq \mathbb{R}^2$ it can be viewed as a subspace once $\mathbb{R}^2$ is given a topology. To obtain such a topology consider $\mathbb{R}^2$ as the product of $\mathbb{R}$ with itself when endowed with the Euclidean topology. Now use the function $t \mapsto e^{2\pi i t}$ to construct a homeomorphism $[0, 1]/\sim \to S^1$ where $\sim$ identifies 1 and 0.

3.4   Separation and Connectedness

Separation properties of a topological space relate to the ability of the open sets to separate distinct points, a point from a set, or two disjoint sets. We only consider one such property, the Hausdorff separation property, which is the strongest of the separation axioms pertaining to the ability to separate distinct points. We then turn our attention to the notion of connectedness; a notion that is somewhat more subtle than one might initially expect.

3.4.1   The Hausdorff Separation Property

If X is endowed with the indiscrete topology, then the open sets (of which there are only two) are, in a sense, blind to the individual points in the space. The open sets are not sufficiently refined to distinguish between different points. This is an extreme situation of course and it goes against one's intuition of how distinct points should (in some sense) behave. The following separation property restores some intuition.

Definition 3.21 A topological space X is said to satisfy the *Hausdorff separation property*, or simply to be *Hausdorff*, if for all distinct points $x, y \in X$ there exist disjoint open sets U and V such that $x \in U$ and $y \in V$. Such open sets are said to separate x and y.

Remark 3.7 The Hausdorff separation property is commonly referred to as T_2, indicating its position in a hierarchy of several separation axioms. We will not delve into this issue here.

Example 3.15 $\mathbb{R}$ with the Euclidean topology is Hausdorff. Indeed, if $x \neq y$ are real numbers, then setting

$$\delta = \frac{|x - y|}{2} > 0$$

one easily sees that

$$B_\delta(x) \cap B_\delta(y) = \emptyset$$

thus exhibiting open sets that separate x and y.

Example 3.16 An example of a non-Hausdorff space is given by the Sierpiński space $\mathbb{S} = \{0, 1\}$. Its two points clearly can not be separated since the topology is given by $\{\emptyset, \{1\}, \{0, 1\}\}$. In fact, a finite topological space X is Hausdorff if, and only if, it is discrete. Indeed any discrete space is clearly Hausdorff since any two distinct points x and y are separated by the open sets $\{x\}$ and $\{y\}$. In the other direction, suppose X is finite and Hausdorff, and choose some $x \in X$. For each $y \in X \setminus \{x\}$ we may find disjoint open sets U_y and V_y such that $x \in U_y$ and $y \in V_y$. Then, the intersection

$$U = \bigcap_{y \in X \setminus \{x\}} U_y$$

is an open set (this is where the finiteness assumption on X is used) and it contains x. Further, since $U \subseteq U_y$, it follows that

$$U \cap V_y \subseteq U_y \cap V_y = \emptyset$$

and thus U is an open set that contains x but does not contain any $y \in X \setminus \{x\}$. In other words, $U = \{x\}$, and thus we proved that $\{x\}$ is an open set. Since x was arbitrary, it follows that X is discrete.

One immediate consequence of the Hausdorff property is the following result.

Theorem 3.12 *A convergent sequence $\{x_m\}_{m \geq 1}$ in a Hausdorff space X converges to a unique limit.*

Proof Suppose that $\{x_m\}_{m \geq 1}$ converges to two different points x and y. There exist, by the Hausdorff property, two disjoint open sets U_x and U_y with $x \in U_x$ and $y \in Y_y$. By Definition 3.14, there exist $N_x, N_y \in \mathbb{N}$ such that

$$m > N_x \implies x_m \in U_x \quad \text{and} \quad m > N_y \implies x_m \in U_y.$$

But then, for $m > \max\{N_x, N_y\}$, we have that $x_m \in U_x \cap U_y$, a contradiction since U_x and U_y are disjoint. $\qquad\square$

3.4.2 Path-Connected and Connected Spaces

The second property we discuss is that of connectedness which, to immediately dispel any misconception, is *not* opposing the Hausdorff separation property (or any other separation property). There are two notions to discuss—path connectivity and connectivity—the former more immediately intuitive than the latter, and thus is the one we present first.

Definition 3.22 A space X is said to be *path-connected* if for all points $x, y \in X$ there exists a continuous function $\gamma : [0, 1] \to X$ with $\gamma(0) = x$ and $\gamma(1) = y$.

Here the topology given to the interval $[0, 1]$ is the subspace topology induced by the Euclidean topology on $\mathbb{R}$. A continuous function $\gamma : [0, 1] \rightarrow X$ is naturally thought of as a *path* in X with initial point $\gamma(0)$ and final destination $\gamma(1)$. Thus, X is path-connected if any two points in the space can be connected by a path in the space.

Example 3.17 The space $\mathbb{R}^n$, the n-fold product of $\mathbb{R}$ with the Euclidean topology, is path-connected for all $n \geq 1$. Indeed, given points $x = (x_1, \ldots, x_n)$ and $y = (y_1, \ldots, y_n)$, it is easily verified that $\gamma : [0, 1] \rightarrow \mathbb{R}^n$ given by

$$\gamma(t) = t(y_1, \ldots, y_n) + (1 - t)(x_1, \ldots, x_n)$$

is a path connecting x and y. The same argument shows that any subset of $\mathbb{R}^n$ of the form

$$[a_1, b_1] \times \cdots \times [a_n, b_n],$$

a product of intervals, is path-connected. In particular, any interval $[a, b]$ in $\mathbb{R}$ is path-connected. More generally, any convex subset of $\mathbb{R}^n$ is path-connected. The reader is invited to verify that another example of a connected space is $S = \{x \in \mathbb{R}^n \mid \|x\| = 1\}$, the unit sphere, viewed as a subspace of $\mathbb{R}^n$ with the Euclidean topology, provided that $n > 1$.

An example of a space that is not path-connected is $\mathbb{Q}$ as a subspace of $\mathbb{R}$ with the Euclidean topology. In fact, no two distinct points $x, y \in \mathbb{Q}$ can be connected by a path. To see that, one can show that any path $\gamma : [0, 1] \rightarrow \mathbb{Q}$ is constant. However, we establish the same result through the more general notion of topological connectedness.

Definition 3.23 A topological space X is said to be *disconnected* if there exist non-empty open sets U and V such that $U \cap V = \emptyset$ and $U \cup V = X$. A space X is said to be *connected* if it is not disconnected.

Theorem 3.13 *Any closed interval $[a, b]$, as a subspace of $\mathbb{R}$ with the Euclidean topology, is connected.*

Proof Without loss of generality let us assume that $[a, b] = [0, 1]$. Suppose that $[0, 1]$ is not connected. Then there exist non-empty open subsets U and V in the subspace topology such that $U \cup V = [a, b]$ and $U \cap V = \emptyset$. Since a singleton set is not open in the subspace topology on $[0, 1]$ it follows that not U nor V is a singleton set. In particular, we may assume that there exist $a \in U$ and $b \in V$ with $0 < a < b < 1$. The set $S = \{x \in U \mid x < b\}$ is non-empty and bounded above, and so admits a supremum s, and notice that $0 < s < 1$.

Now, since $[0, 1] = U \cup V$ we must have that either $s \in U$ or $s \in V$. If $s \in U$ then $s < b$ and, U being open, there exists an $a' \in U$ with $s < a' < b$, contrary to the choice of s. On the other hand, if $s \in V$ then, V being open, there exist an $\varepsilon > 0$ with $(s - \varepsilon, s + \varepsilon) \subseteq V$, again contrary to the choice of s. The assumption of the existence of U and V as above thus leads to a contradiction, and so $[0, 1]$ is connected. $\square$

Remark 3.8 The intuition behind the concept of connectedness is perhaps better explained by referring to the word 'samenhangend', the term used for this property in the Dutch language. The literal meaning of 'samenhangend' is 'hanging together', and it is this topological quality that the definition of connectedness captures. For instance, it is intuitively clear that if a space is path-connected, then it must be 'hanging together'. This is indeed the case.

Lemma 3.1 *If a topological space X is path-connected, then it is connected.*

Proof Suppose that X is disconnected. That means that there exist disjoint non-empty open sets $U, V \subseteq X$ such that

$$U \cup V = X.$$

Choose points $x \in U$ and $y \in V$ and, utilizing the assumption that the space X is path-connected, let $\gamma : [0, 1] \to X$ be a path from x to y. Since γ is continuous, it follows that $\gamma^{-1}(U)$ and $\gamma^{-1}(V)$ are open sets in $[0, 1]$. Further,

$$\gamma^{-1}(U) \cap \gamma^{-1}(V) = \gamma^{-1}(U \cap V) = \gamma^{-1}(\emptyset) = \emptyset$$

and

$$\gamma^{-1}(U) \cup \gamma^{-1}(V) = \gamma^{-1}(U \cup V) = \gamma^{-1}(X) = [0, 1].$$

Lastly, since $x = \gamma(0)$ and $x \in U$ it follows that $0 \in \gamma^{-1}(U)$, and in particular $\gamma^{-1}(U) \neq \emptyset$. Similarly one shows that $\gamma^{-1}(V) \neq \emptyset$.

The conclusion is thus that $\gamma^{-1}(U)$ and $\gamma^{-1}(V)$ separate $[0, 1]$. But $[0, 1]$ is connected and thus so is X. □

Example 3.18 We return to the example of $\mathbb{Q}$ and show that it is disconnected, and thus also not path-connected. Indeed, let α be an arbitrary irrational number and let $U = \{x \in \mathbb{Q} \mid x < \alpha\}$ and $V = \{x \in \mathbb{Q} \mid x > \alpha\}$. Clearly, $U \cap V = \emptyset$ and $U \cup V = \mathbb{Q}$, and thus all that is left to show is that U and V are open. Indeed, given any $x \in U$, setting $\delta = \frac{\alpha - x}{2}$, one easily sees that $(x - \delta, x + \delta) \cap \mathbb{Q} \subseteq U$. But this shows that U is open in the subspace topology on $\mathbb{Q}$, as required. The claim that V, too, is open follows similarly.

Theorem 3.14 *If $f : X \to Y$ is a surjective continuous mapping between topological spaces and X is connected, then so is Y.*

Proof Suppose that Y is disconnected, namely that $Y = U \cup V$ for some non-empty open sets $U, V \subseteq Y$ which further satisfy $U \cap V = \emptyset$. Note that

$$X = f^{-1}(Y) = f^{-1}(U) \cup f^{-1}(V)$$

and that both $f^{-1}(U)$ and $f^{-1}(V)$ are open in X. Further,

$$f^{-1}(U) \cap f^{-1}(V) = f^{-1}(U \cap V) = f^{-1}(\emptyset) = \emptyset.$$

Since X is connected, we may conclude that either $f^{-1}(U)$ or $f^{-1}(V)$ is empty. But, since f is surjective, that is impossible and so Y must be connected. □

Exercises

Exercise 3.59 Show that being Hausdorff is hereditary: if X is Hausdorff, then so is any of its subspaces.

Exercise 3.60 Prove that if X, Y are Hausdorff topological spaces, then so is $X \times Y$. Does the converse hold?

Exercise 3.61 Find a non-Hausdorff topological space X with a non-trivial Hausdorff quotient as well as a Hausdorff topological space X with a non-Hausdorff quotient.

Exercise 3.62 Let X be a topological space and $\mathcal{B}$ a basis for its topology. Show that X is Hausdorff if, and only if, for all distinct points $x, y \in X$ there exist disjoint basis elements $U_x, U_y \in \mathcal{B}$ with $x \in U_x$ and $y \in U_y$.

Exercise 3.63 Call two points x, y in a topological space X *inseparable* if for all open sets $U, x \in U \iff y \in U$. Prove that inseparability is an equivalence relation on X. Is the quotient space $X/\sim$ Hausdorff?

Exercise 3.64 Prove that a topological space X is Hausdorff if, and only if, the diagonal set $\{(x, x) \mid x \in X\}$ is closed in the product space $X \times X$.

Exercise 3.65 In Example 3.18 justify why the chosen δ is positive.

Exercise 3.66 Considering $\mathbb{Q}$ and $[0, 1]$ as subspaces of $\mathbb{R}$ with the Euclidean topology show that any continuous function $f : [0, 1] \to \mathbb{Q}$ is constant.

Exercise 3.67 Prove that each of the Hausdorff separation property, connectedness, and path-connectedness is a topological invariant.

Exercise 3.68 Show that for a topological space X the following conditions are equivalent.

1. X is connected.
2. The only clopen (i.e., closed and open) subsets of X are $\emptyset$ and X.
3. The only continuous functions $X \to \{0, 1\}$ are constant, where the codomain is given the discrete topology.
4. The only continuous functions $X \to Y$ are constant, where Y is an arbitrary discrete space.

Exercise 3.69 Consider $\mathbb{R}$ with the Euclidean topology.

1. Prove that removing a single point from $\mathbb{R}$ results in a disconnected subspace.
2. Show that removing a single point from $\mathbb{R}^2 = \mathbb{R} \times \mathbb{R}$ results in a connected subspace.

3. Conclude that $\mathbb{R}$ and $\mathbb{R}^2$ are not homeomorphic.
4. Generalize to show that $\mathbb{R}$ and $\mathbb{R}^n$, for $n > 1$, are not homeomorphic.
5. Show that removing a straight infinite line from $\mathbb{R}^2$ results in a disconnected subspace.
6. Show that removing a straight infinite line from $\mathbb{R}^3$ results in a connected subspace.
7. Clarify for yourself why one cannot deduce at this point that $\mathbb{R}^2$ and $\mathbb{R}^3$ are not homeomorphic.

It is true that $\mathbb{R}^n$ and $\mathbb{R}^m$ are homeomorphic if, and only if, $m = n$ but the proof is hard.

Exercise 3.70 Let X be a topological space and $S \subseteq X$ a subset. Prove that if S (as a subspace) is connected, then so is its closure $\bar{S}$.

Exercise 3.71 Characterize all of the connected subsets of $\mathbb{R}$ under the Euclidean topology.

Exercise 3.72 Prove the Intermediate Value Theorem. Let $f : X \to \mathbb{R}$ be a continuous function from a topological space X to the topological space $\mathbb{R}$ with the Euclidean topology. If X is connected, then for all $x, z \in X$ and $c \in \mathbb{R}$ with

$$f(x) < c < f(z),$$

there exists a $y \in X$ with
$$f(y) = c.$$

Exercise 3.73 Prove that if X and Y are connected (respectively path-connected) topological spaces, then so is $X \times Y$.

Exercise 3.74 Prove that a quotient space of a connected space is connected.

Exercise 3.75 Show the Sierpiński space is path-connected.

Exercise 3.76 Consider the function $f : (0, 1] \to \mathbb{R}$ given by $f(x) = \sin(1/x)$, and $\mathbb{R}$ equipped with the Euclidean topology.

1. Show that $X = \{(x, f(x)) \mid x \in (0, 1]\}$, the graph of f, as a subspace of $\mathbb{R}^2$ is path-connected (and hence connected).
2. The *topologist's sine curve* is the space $Y = X \cup \{(0, 0)\}$ as a subspace of $\mathbb{R}^2$. Prove that Y is connected.
3. Show that Y is not path-connected.

3.5 Compactness

The reader is probably aware of the importance of closed intervals in $\mathbb{R}$ and, more generally, of closed and bounded subsets of $\mathbb{R}$ or $\mathbb{R}^n$. For instance, the Bolzano-Weierstrass Theorem states that every infinite sequence in a closed interval $[a, b]$

admits a convergent subsequence, or the theorem that a continuous function on a closed interval is uniformly continuous. It turns out that the crucial property allowing for the mentioned results is topological: the sets are compact. The aim of this section is to present the notion of compactness and prove several results leading to the characterization of the closed and bounded subsets of $\mathbb{R}$ (under the Euclidean topology) as precisely the compact sets. The section concludes with the statement of the important result, known as Tychonoff's Theorem, that the product of any family of compact topological spaces is itself compact.

Definition 3.24 Let X be a topological space and $C \subseteq X$ a subset. An *open covering* of C is a collection $\{U_i\}_{i \in I}$ of open sets whose union contains the entire set C, i.e.,

$$C \subseteq \bigcup_{i \in I} U_i.$$

The set C is said to be *compact* if given any open covering $\{U_i\}_{i \in I}$ of C, there exists a *finite subcovering* of C. That is, there exists finitely many sets $U_{i_1}, \ldots, U_{i_m}$ from the given covering such that

$$C \subseteq U_{i_1} \cup \cdots \cap U_{i_m}.$$

Notice that C may coincide with the set X, so we may speak of the space X itself being compact.

Example 3.19 Any finite set C in any topological space is compact. This is clear since any open covering of a finite set must admit a finite subcovering, using at most as many open sets from the covering as there are points in C.

Example 3.20 Any unbounded set $Y \subseteq \mathbb{R}$, when $\mathbb{R}$ is given the Euclidean topology, is not compact. To see that, consider the open covering

$$Y \subseteq \bigcup_{k=1}^{\infty} (-k, k).$$

Clearly, no finite subcovering suffices to cover Y precisely because Y was chosen to be unbounded. In other words, any compact set in $\mathbb{R}$ must be bounded. Similarly, an open interval (a, b) is not compact in $\mathbb{R}$ either, since the open covering

$$(a, b) \subseteq \bigcup_{k=1}^{\infty} \left(a + \frac{1}{k}, \infty \right)$$

admits no finite subcovering of (a, b), as is easy to verify.

Example 3.21 For any set X, if X is endowed with the indiscrete topology, then there are only two open sets (or sometimes less) in existence and so every subset

$C \subseteq X$ is certainly compact. If, however, X is given the discrete topology, then every singleton set $\{x\}$ is open, and thus any subset $C \subseteq X$ admits the open covering

$$C = \bigcup_{x \in C} \{x\}$$

and thus is compact if, and only if, C is finite.

The next example is important enough to be stated as a theorem.

Theorem 3.15 *A closed interval* $[a, b]$ *in* $\mathbb{R}$ *with the Euclidean topology is compact.*

Proof Suppose that an open covering $\{U_i\}_{i \in I}$ of $[a, b]$ is given. Our goal is to find a finite subcovering. Define the set S of all $a \leq s \leq b$ for which a finite subcovering of $[a, s]$ exists. The goal is reduced to showing that $b \in S$. Now, since $[a, a] = \{a\}$ can certainly be covered by just one element from the given covering, it follows that $a \in S$ and thus S is not empty. By definition, S is bounded and thus has a supremum $t \leq b$. We proceed to show that $t \in S$. Take any U_{i_0} from the given covering for which $t \in U_{i_0}$. As U_{i_0} is open, we have

$$t \in (t - \varepsilon, t + \varepsilon) \subseteq U_{i_0}$$

for some $\varepsilon > 0$, and since $t - \varepsilon < t$ it follows that $t - \varepsilon \in S$. There is then a finite subcovering of $[a, t - \varepsilon]$, and by adjoining U_{i_0} to it we obtain a finite subcovering of $[a, t]$, so indeed $t \in S$. Now, t is the supremum of S and so, for all $\delta > 0$ with $t + \delta < b$ no finite subcovering of $[a, t + \delta]$ exists. However, we already have a finite covering of [a,t] by open sets and any such covering also covers $[a, t + \delta]$ for some $\delta > 0$. If it was the case that $t < b$, then such a delta can be chosen to satisfy $t + \delta < b$, leading to a contradiction. The conclusion is that $b = t \in S$, which was the goal.   □

Not all closed subsets of $\mathbb{R}$ are compact. For instance, $[a, \infty)$, being unbounded, is not compact. However, if the ambient space itself is compact, then closed subsets are compact.

Lemma 3.2 *A closed subset* C *of a compact space* X *is compact.*

Proof Given an open covering of C simply note that adding to it the set $X \setminus C$, which is open since C is assumed closed, yields an open covering of X. But X is compact so a finite subcovering of X exists and discarding $X \setminus C$ from it if necessary yields a finite subcovering of C, as required.   □

The following example of a compact subset that is not closed shows that the converse of the lemma need not be true.

Example 3.22 Let X be an arbitrary set and consider the cofinite topology on it. The open sets (other than $\emptyset$) are those whose complements are finite, and thus we may identify the closed sets (other than X itself) as precisely the finite subsets. We now show that any subset $C \subseteq X$ is compact. Suppose an open covering $\{U_i\}_{i \in I}$ of C is given. To extract a finite subcovering consider any non-empty member U_i of the covering, and note that since it is open it misses at most finitely many of the elements of C. Thus only finitely many other open sets from the given covering are required to cover C, and thus a finite subcovering exists. In particular, if X is infinite, then not every compact subset of it is closed.

The next result shows the crucial obstacle in this example is that the ambient space is not Hausdorff.

Proposition 3.12 *A compact subset C of a Hausdorff space X is closed.*

Proof If $C = X$, then C is closed, so we may assume $X \setminus C$ is not empty as we proceed to show that it is open. Let $x_0 \in X \setminus C$ be an arbitrary point. For every $y \in C$, using the Hausdorff assumption on X, there exist disjoint open sets U_y and V_y such that $x_0 \in U_y$ and $y \in V_y$. Clearly, the collection $\{V_y\}_{y \in C}$ is an open covering of C, and since C is compact there exists a finite subcovering $V_{y_1}, \ldots, V_{y_k}$ such that

$$C \subseteq V_{y_1} \cup \cdots \cup V_{y_k}.$$

Let

$$U = U_{y_1} \cap \cdots \cap U_{y_k}$$

and note that U is open, that $x_0 \in U$, and that $U \subseteq X \setminus C$. In other words, x_0 is an interior point of $X \setminus C$ and since x_0 was an arbitrary point, every point of $X \setminus C$ is interior, and thus, by Corollary 3.1, $X \setminus C$ is open.   □

We can now characterize the compact subsets of $\mathbb{R}$.

Theorem 3.16 (Heine-Borel) *With respect to the Euclidean topology on $\mathbb{R}$ a subset $C \subseteq \mathbb{R}$ is compact if, and only if, C is closed and bounded.*

Proof Assume that $C \subseteq \mathbb{R}$ is a compact subset. Since $\mathbb{R}$ is Hausdorff it follows by Proposition 3.12 that C is closed while the fact that C must be bounded was already noted in Example 3.20. In the other direction, if C is bounded, then it is contained in some closed interval $[a, b]$ which, by Theorem 3.15, is a compact set. If C is also closed, then it is a closed subset of the compact set $[a, b]$ and thus, by Lemma 3.2, is itself compact.   □

The next result is a generalization of the Bolzano-Weierstrass Theorem.

Theorem 3.17 *An infinite subset S of a compact topological space X has a cluster point in X.*

Proof Suppose that S does not have any cluster points in X. That means that for every $x \in X$ there exists an open set U_x with $x \in U_x$ and such that $U_x \cap S$ is either empty or the set $\{x\}$. Clearly, $\{U_x\}_{x \in X}$ is an open cover of X and thus admits a finite subcover, say $U_{x_1}, \ldots, U_{x_n}$. Since

$$S \subseteq U_{x_1} \cup \cdots \cup U_{x_n}$$

and as $U_{x_i} \cap S$ contains at most one element, we conclude that $|S| \leq n$, so S is finite. Consequently, an infinite set S has at least one cluster point. □

We end this section be mentioning the following important result.

Theorem 3.18 (Tychonoff Theorem) *The product of any number (finite or infinite) of compact topological spaces is compact.*

The proof of this result is slightly beyond the scope of this book.

Exercises

Exercise 3.77 Consider $\mathbb{R}$ with the Euclidean topology. Show that any covering of a closed interval $[a, b]$ by open sets is also a covering of $[a, b + \delta]$ for some $\delta > 0$ which depends on the given covering.

Exercise 3.78 Show that a subset of a compact space may be compact but may also be non-compact. Similarly, show that a subset of a non-compact space may be compact but may also be non-compact.

Exercise 3.79 Let X be a topological space and $\{S_i\}_{i \in I}$ a family of compact subsets in it. Show that if I is finite, then $\bigcup_{i \in I} S_i$ is compact and that the finiteness assumption on I cannot be dropped.

Exercise 3.80 Let X, Y be disjoint topological spaces. Prove that the coproduct $X \coprod Y$ is compact if, and only if, X and Y are compact.

Exercise 3.81 Prove, without appealing to Tychonoff's Theorem, that the product of any finite number of compact topological spaces is compact.

Exercise 3.82 Prove that if C is a non-empty compact subset of $\mathbb{R}$ with respect to the Euclidean topology, then C admits both a minimum and a maximum.

Exercise 3.83

1. Prove that if $f : X \to Y$ is a continuous function between topological spaces and $C \subseteq X$ is compact, then $f(C)$ is compact in Y.
2. Prove the following Extreme Value Theorem. Let $f : X \to \mathbb{R}$ be a continuous function from a topological space X to the topological space $\mathbb{R}$ with the Euclidean topology. If X is compact, then f attains both a maximum and a minimum, i.e., there exist points $x_m, x_M \in X$ such that

$$f(x_m) \leq f(x) \leq f(x_M)$$

for all $x \in X$.
3. Deduce Weierstrass' Theorem: a function $f : [a, b] \to \mathbb{R}$ attains both a minimum and a maximum.

Exercise 3.84 Prove that compactness is a topological invariant, namely if X and Y are homeomorphic topological spaces and one is compact, then so is the other.

Exercise 3.85 Prove that a quotient space of a compact space is compact.

Exercise 3.86 Construct a topological space X with a non-empty proper subset that is both open and compact.

Exercise 3.87 Show that if X is an uncountable set endowed with the cocountable topology, then X is not compact. What are its compact subsets?

Exercise 3.88 A topological space X satisfies the *finite intersection property* if for any collection $\mathcal{F}$ of closed sets, if

$$F_1 \cap \cdots \cap F_m \neq \emptyset$$

for all $F_1, \ldots, F_m \in \mathcal{F}$, then

$$\bigcap_{F \in \mathcal{F}} F \neq \emptyset.$$

Prove that a topological space X has the finite intersection property if, and only if, X is compact.

Exercise 3.89 Show that the topological analogue of the Cantor-Schröder-Bernstein theorem fails using the subspace inclusion $[-1, 1] \to \mathbb{R}$, the latter with the Euclidean topology. Find a continuous injection $\mathbb{R} \to [-1, 1]$ and prove that $[-1, 1]$ and $\mathbb{R}$ are not homeomorphic.

Exercise 3.90 In general, a continuous bijection $f \colon X \to Y$ need not be a homeomorphism. Prove that if X is compact and Y is Hausdorff, then such an f is a homeomorphism. Hint: It suffices to show that $f(C)$ is closed in Y for all closed sets $C \subseteq X$.

Further Reading

For a perspective on some of the ideas that led to the birth of topology see [5]. A comprehensive introduction to topology, including most of the results in this chapter and far more, can be found in [3, 4]. For an unexpected usage of topology, illustrating its versatility, see Furstenberg's proof of the infinitude of primes ([1]). For a collection of essays on various topological developments, including an essay devoted to topology and physics, see [2].

References

1. Furstenberg, H.: On the infinitude of primes. Am. Math. Monthly **62**, 355 (1955)

2. James, I.M. (ed.): History of Topology, p. x+1056. North-Holland, Amsterdam (1999)
3. McCluskey, A., McMaster, B.: Undergraduate Topology: A Working Textbook. Oxford University Press, Oxford (2014)
4. Munkres, J.R.: Topology: A First Course, p. xvi+413. Prentice-Hall, Inc, Englewood Cliffs, N.J. (1975)
5. Richeson, D.S.: Euler's Gem, The Polyhedron Formula and The Birth of Topology, First paperback printing, p. xiv+317. Princeton University Press, Princeton, NJ (2012)

Chapter 4
Metric Spaces

Metric spaces were introduced by Maurice Fréchet in 1906 in his Ph.D. dissertation devoted to functional analysis. The axioms given by Fréchet (almost identical to the ones we give below) form an abstraction of the notion of distance thus allowing for a unified treatment of numerous particular cases under a single formalism. The ubiquity of metric spaces in Mathematics and Physics is overwhelming but, naturally, this chapter is mainly concerned with examples and results most relevant in the context of this book.

Section 4.1 contains the definition of metric space and explores some examples, primarily of normed spaces as metric spaces. Section 4.2 is concerned with the topology induced by the distance function, the associated concept of convergence, and basic topological observations. Section 4.3 considers two types of structure preservation for mappings between metric spaces: non-expanding mappings and uniformly continuous ones. Section 4.4 is concerned with the concept of completeness in metric spaces, the Banach Fixed Point Theorem and Baire's Theorem, and Cantor's metric completion process by means of Cauchy sequences. Finally, Sect. 4.5 examines compactness for metric spaces and establishes the fact that a metric space is compact precisely when it is complete and totally bounded.

4.1 Metric Spaces—Definition and Examples

Our intuitive understanding of the behaviour of distance, in the plane for instance, is clearly represented in the definition of metric space that we give below. Of particular importance is the simple fact that any normed space has a natural metric structure associated with it (though many metric spaces certainly do not arise in this manner). Consequently, the normed spaces from Chap. 2 are all revisited here through a metric lens.

© Springer Nature Switzerland AG 2021

C. Alabiso and I. Weiss, *A Primer on Hilbert Space Theory*, UNITEXT for Physics, https://doi.org/10.1007/978-3-030-67417-5_4

Before we proceed with the definition and a list of examples we note that for the general study of metric spaces there are technical advantages in allowing infinite distance between points. We thus first adjoin the symbol ∞ as an extended real number and denote by $\mathbb{R}_+ = \{x \in \mathbb{R} \mid x \geq 0\} \cup \{\infty\}$ the set of *extended non-negative real numbers*.

Remark 4.1 We accept that $x < \infty$ for all real x (and we also accept that $\infty \leq \infty$). Addition is extended to include ∞ by setting $x + \infty = \infty + x = \infty$ for all extended real $x \geq 0$.

Definition 4.1 Let X be a set. A function $d : X \times X \to \mathbb{R}_+$ is said to be a *metric* or a *distance function* on X if the properties:

$$\begin{array}{ll}
\text{Distance is symmetric:} & d(x, y) = d(y, x) \\
\text{The triangle inequality:} & d(x, z) \leq d(x, y) + d(y, z) \\
\text{Distance separates points:} & d(x, y) = 0 \iff x = y
\end{array}$$

are satisfied for all $x, y, z \in X$. The pair (X, d) is then called a *metric space* and $d(x, y)$ the *distance* between x and y. Quite often we refer to the metric space X without explicitly mentioning the distance function d. Moreover, when two (or more) metric spaces are considered we often overload the symbol d to stand for the metric function in each space, relying on semantics to resolve any confusion. If needed, we refer to the metric function of X by d_X.

Remark 4.2 There is significant flexibility in the choice of axioms above, giving rise to several structures that may resemble metric spaces to varying degrees. For instance it is possible that $d(x, y) = \infty$, that is we allow the distance between points to be infinity. Some authors do not permit the distance function to attain the value ∞. The difference between these two possibilities is largely cosmetic. It is also possible to replace the separation condition by the demand that $d(x, x) = 0$, giving rise to what are known as *semimetric* spaces, and again, their theory is quite similar to that of metric spaces. However, if one neglects symmetry, giving rise to what are known as *quasimetric* spaces, the resulting theory is in many respects strikingly different.

Observing that the metric space axioms are universally quantified it follows at once that if $S \subseteq X$ is a subset of a metric space X, then the restriction of the distance function $d : X \times X \to \mathbb{R}_+$ to the subset $S \times S$ is a distance function on S.

Definition 4.2 (*Metric Subspace*) For a subset S of a metric space X the pair (S, d_S), where d_S is the restriction of d to $S \times S$, is called a *metric subspace* of X (or simply *subspace* if the metric context is clear).

There are various ways for a function f between metric spaces to interact with the metric structures, namely to preserve the distances to various degrees. Here is the strictest option.

Definition 4.3 A function $f : X \to Y$ between metric spaces is an *isometry* if

$$d(f(x_1), f(x_2)) = d(x_1, x_2)$$

for all $x_1, x_2 \in X$. If f is also bijective, then f is said to be an *isometric isomorphism* or a *global isometry*. If there exists an isometric isomorphism between X and Y, then we write $X \cong Y$ and say that X and Y are *isometric*.

A property of metric spaces is a *metric invariant* if whenever it holds for a metric space X it also holds for any other metric space isometric to X.

The following theorem is a rich source of metric spaces.

Theorem 4.1 *If V is a normed space, then defining $d : V \times V \to \mathbb{R}_+$ by*

$$d(x, y) = \|x - y\|$$

endows V with the structure of a metric space.

Proof Note first that the codomain of d is indeed $\mathbb{R}_+$ since the norm is always non-negative (in this case ∞ is never attained as the distance between points). For all vectors $x, y, z \in V$, symmetry follows by

$$d(x, y) = \|x - y\| = \|(-1) \cdot (y - x)\| = |-1| \cdot \|y - x\| = \|y - x\| = d(y, x),$$

while separation and the triangle inequality are just restatements of the conditions of positivity and the triangle inequality in the definition of normed space (refer to Definition 2.15, and with the aid of Proposition 2.13). $\qquad\square$

Example 4.1 Each of the normed spaces (and in particular the inner product spaces) introduced in Chap. 2 gives rise to a metric space, so we obtain the following examples of metric spaces.

The space $\mathbb{R}^n$, $n \geq 1$, with distance given by

$$d(x, y) = \left(\sum_{k=1}^{n} (x_k - y_k)^2 \right)^{1/2}$$

known as the *Euclidean metric* on $\mathbb{R}^n$. The space $\mathbb{C}^n$, $n \geq 1$, with distance given by

$$d(x, y) = \left(\sum_{k=1}^{n} |x_k - y_k|^2 \right)^{1/2}$$

where we may immediately note that $\mathbb{C}^n$ is isometric to $\mathbb{R}^{2n}$ with the Euclidean metric, a global isometry $f : \mathbb{C}^n \to \mathbb{R}^{2n}$ is given by

$$f(x_1 + i \cdot y_1, \ldots, x_n + i \cdot y_n) = (x_1, y_1, x_2, y_2, \ldots, x_n, y_n).$$

For all $1 \leq p < \infty$, every pre-L_p space, say of continuous real-valued functions on the closed interval $[a, b]$, is a metric space of functions with distance

$$d(x, y) = \left(\int_a^b dt \; |x(t) - y(t)|^p \right)^{1/p}.$$

In particular, $C([a, b], \mathbb{R})$ supports infinitely many different metric space structures, one for each $1 \le p < \infty$. The case $p = \infty$, i.e., with the L_∞ norm, gives rise to yet another distance function, namely

$$d(x, y) = \max_{a \le t \le b} |x(t) - y(t)|.$$

Similarly, the space $C(I, \mathbb{C})$ of continuous complex-valued functions on a closed interval $I = [a, b]$ supports an entire family of metric space structures. For each $1 \le p < \infty$ the space ℓ_p of absolutely p-power summable sequences of either real or complex numbers with distance given by

$$d(x, y) = \left(\sum_{k=1}^\infty |x_k - y_k|^p \right)^{1/p}$$

is yet another family of metric spaces. Of course the case $p = \infty$, i.e., the space ℓ_∞ of bounded sequences with distance given by

$$d(x, y) = \sup_{k \ge 1} |x_k - y_k|$$

is a metric space as well.

$\mathbb{R}^n$ and $\mathbb{C}^n$ support other metric structures, different than those presented above, as the following example shows.

Example 4.2 If we consider $\mathbb{R}^n$ as a subset of ℓ_p, with $1 \le p \le \infty$, by identifying $x = (x_1, \ldots, x_n)$ with $(x_1, \ldots, x_n, 0, 0, 0, \cdots)$, then $\mathbb{R}^n$ inherits the metric of ℓ_p and becomes a metric subspace of ℓ_p. In more detail, if $p = \infty$ then

$$d(x, y) = \max_{1 \le k \le n} |x_k - y_k|$$

and

$$d(x, y) = \left(\sum_{k=1}^n |x_k - y_k|^p \right)^{1/p}$$

for $1 \le p < \infty$. If $n > 1$ and $p \ne 2$, then each of these metrics is different than the Euclidean metric. A similar observation gives rise to metric structures on $\mathbb{C}^n$.

The following examples illustrate the subtleties that may arise with general metric spaces.

Example 4.3 Consider $\mathbb{R}^2$ with the Euclidean metric, i.e., the plane with its ordinary Euclidean metric structure. The unit circle $S = \{x \in \mathbb{R}^2 \mid \|x\| = 1\}$ has two natural metric structures on it. As a subspace of $\mathbb{R}^2$ the distance between two antipodal points in S is 2 (the length of the shortest straight line connecting the two points where the line is allowed to pass freely in the ambient space $\mathbb{R}^2$). On the other hand, it is possible to endow the unit circle S with the so called *intrinsic metric*, the one where the distance between two points is the length of the shorter of the two arcs connecting them. With that metric the distance between antipodal points is π (the length of a shortest geodesic connecting the two points, i.e. a "straight" line connecting the points where the line is not allowed to step outside of S and into the ambient space).

Other scenarios, perhaps more abstract, are also possible. For instance, given any set X, defining $d(x, x) = 0$ for all $x \in X$ and $d(x, y) = 1$ for all $x, y \in X$ with $x \neq y$ turns X into a metric space. In this metric space the distances between any two distinct points is 1. Thus, if X is infinite, then X is not isometric to any subspace of the Euclidean space $\mathbb{R}^n$, for any $n \geq 1$. Notice however that defining $d(x, y) = 0$ for all $x, y \in X$ fails to be a metric space (unless X has at most one element) since separation fails. It is however easily seen to be an example of a semimetric space.

Example 4.4 The set $\mathbb{R}_+$ can be given the metric structure

$$d(x, y) = \begin{cases} |x - y| & \text{if } x \neq \infty \neq y \\ 0 & \text{if } x = y = \infty \\ \infty & \text{otherwise.} \end{cases}$$

The metric axioms are easily verified. In fact, we will simply write $d(x, y) = |x - y|$ since the algebraic properties of this metric significantly resemble those of $|x - y|$ for real x, y.

Finally, we present one of many ways to endow the cartesian product $X \times Y$ of two metric spaces with a metric structure.

Definition 4.4 (*The Additive Metric Product*) For metric spaces X and Y the function $d : (X \times Y) \times (X \times Y) \to \mathbb{R}_+$ given by

$$d((x, y), (x', y')) = d_X(x, x') + d_Y(y, y')$$

is easily seen to be a distance function (as the reader may verify) and the metric space $(X \times Y, d)$ is called the *additive metric product* of X and Y.

Exercises

Exercise 4.1 Let $f : [0, \infty) \to [0, \infty)$ be a concave, strictly monotonically increasing function satisfying $f(0) = 0$. Suppose that (X, d) is a metric space and that d only attains finite values. Prove that $(X, f \circ d)$ is a metric space.

Exercise 4.2 Let (X, d_X) and (Y, d_Y) be two disjoint (i.e., $X \cap Y = \emptyset$) metric spaces and let $Z = X \cup Y$. Prove that $d : Z \times Z \to \mathbb{R}_+$ given by

$$d(z_1, z_2) = \begin{cases} d_X(z_1, z_2) & \text{if } z_1, z_2 \in X \\ d_Y(z_1, z_2) & \text{if } z_1, z_2 \in Y \\ \infty & \text{otherwise} \end{cases}$$

is a metric on Z. The metric space (Z, d) is then called the *coproduct* of X and Y.
Generalize this construction to give the coproduct of any number (finite or infinite)
of metric spaces.

Exercise 4.3 Let X be a metric space. Define the relation $\sim$ on X by $x \sim y$ precisely
when $d(x, y) < \infty$. Prove that $\sim$ is an equivalence relation. The equivalence class
$[x]$, for any point $x \in X$, is called the *galaxy* of x. Show that each galaxy, as a metric
subspace of X, has all distances finite and then show that X is the coproduct (see
previous exercise) of its galaxies.

Exercise 4.4 Suppose that (X, d) is a semimetric space and define the relation $\sim$ on
X by $x \sim y$ precisely when $d(x, y) = 0$. Prove that $\sim$ is an equivalence relation on
X and that $d([x], [y]) = d(x, y)$ on the quotient set $X/\sim$ is well defined and endows
it with the structure of a metric space.

Exercise 4.5 Let X be a set and $c: X \times X \to \mathbb{R}_+$ a function satisfying $c(x, x) = 0$
for all $x \in X$. Define the function $d: X \times X \to \mathbb{R}_+$ by

$$d(x, z) = \inf_{y_0, \ldots, y_n} \sum d(y_i, y_{i+1})$$

where the infimum is taken over all finite sequences of points $y_0, \ldots, y_n \in X$ with
$y_0 = x$ and $y_n = z$. Show that d is a semimetric on X.

Exercise 4.6 Let X be a set and write $\mathrm{Met}(X)$ for the set of all semimetric structures
d on X. Declare, for $d_1, d_2 \in \mathrm{Met}(X)$, that $d_1 \leq d_2$ when $d_1(x, y) \leq d_2(x, y)$ for all
$x, y \in X$.

1. Show that $\mathrm{Met}(X)$ with the given ordering is a poset.
2. For a family $\{d_i\}_{i \in I}$ of semimetrics on X prove that the point-wise infimum
   $d(x, y) = \inf_{i \in I}\{d_i(x, y)\}$ is the meet in $\mathrm{Met}(X)$ of $\{d_i\}_{i \in I}$.
3. Show that the construction from the previous exercise associates with the function
   c the meet in $\mathrm{Met}(X)$ of the family $\{e \in \mathrm{Met}(X) \mid c \leq e\}$ (with $c \leq e$ defined
   point-wise).
4. Show that $\mathrm{Met}(X)$ is a complete lattice. What happens if we took metrics instead
   of semimetrics?

Exercise 4.7 Consider $\mathbb{R}^2$ with the metric induced by the ℓ_p norm $\| - \|_p$. The unit
ball $\{x \in \mathbb{R}^2 \mid \|x\|_p \leq 1\}$ supports two metrics: the subspace metric and the intrinsic
one. For which values of p, $1 \leq p \leq \infty$, do these metrics coincide?

Exercise 4.8 Given any two metric spaces (X, d_X) and (Y, d_Y), define the function
$d_E : (X \times Y) \times (X \times Y) \to \mathbb{R}_+$ by

$$d_E((x_1, y_1), (x_2, y_2)) = \sqrt{d_X(x_1, x_2)^2 + d_Y(y_1, y_2)^2}.$$

Prove that $(X \times Y, d_E)$ is a metric space called the *Euclidean metric product* of X and Y. Generalize this construction to obtain the Euclidean metric product of any finite number of metric spaces.

Exercise 4.9 Prove that every isometry is injective.

Exercise 4.10 Let $f: X \to Y$ be an isometry between metric spaces. Show that if f is bijective, then its inverse is also an isometry.

Exercise 4.11 Prove the finite metric analogue of the Cantor-Schröder-Bernstein Theorem: if there exist isometries $f: X \to Y$ and $g: Y \to X$, then the metric spaces X and Y are isometric, provided X and Y are finite.

Exercise 4.12 Let X be a metric space with $d(x, y) = 1$ for all $x, y \in X$ with $x \neq y$. Prove that if $|X| = n$ and $n \geq 1$, then X is isometric to a subspace of $\mathbb{R}^{n-1}$ with the Euclidean metric. Prove no such isometric subspace exists in $\mathbb{R}^k$ for $k < n - 1$. Conclude that with the Euclidean metrics the spaces $\mathbb{R}^n$ and $\mathbb{R}^m$ are isometric if, and only if, $m = n$.

Exercise 4.13 For all metric spaces $X, Y,$ and Z prove that

1. $X \cong X$.
2. If $X \cong Y$, then $Y \cong X$.
3. If $X \cong Y$ and $Y \cong Z$, then $X \cong Z$.

Exercise 4.14 Prove that the induced metric in a normed space V satisfies the equality $d(x + z, y + z) = d(x, y)$ for all vectors $x, y, z \in V$. Namely, the induced metric in a normed space V is *translation invariant*.

Exercise 4.15 Consider the set R of all Riemann integrable functions $f : [a, b] \to \mathbb{R}$ and define $d : R \times R \to \mathbb{R}_+$ by $d(f, g) = \int_a^b dt \, |f(t) - g(t)|$. Is (R, d) a metric space?

Exercise 4.16 A *mid-point* for two distinct points x, z in a metric space X is a point $y \in X$ such that $d(x, y) = d(y, z) = d(x, z)/2$. Give examples of:

1. A metric space where every two distinct points admit a unique mid-point.
2. A metric space where every two distinct points admit a mid-point, and some admit two distinct mid-points.
3. A metric space where no two distinct points admit a mid-point.

4.2 Topology and Convergence in a Metric Space

This section is concerned with the topological structure of a metric space. Every metric space has a topology associated with it that is determined by the distance function.

A metric space is thus automatically a topological space and thus all of the concepts of topological spaces, including convergence, are meaningful in a metric space. After describing this so called induced topology we examine the concept of convergence in the metric spaces introduced above. It should be noted immediately that different metrics on the same set may induce the same topology. Topological spaces induced by a metric are called metrizable and we study some of their properties.

As motivation for the induced topology we follow a similar route to the one we took in Sect. 3.1 where the concept of topology was motivated as a means to capture continuity while avoiding arbitrary properties of the distance function.

Before we proceed we note that the familiar notion of continuity of a function applies (formally verbatim) to arbitrary metric spaces.

Definition 4.5 A function $f : X \to Y$ between metric spaces is *continuous* if for all $x \in X$ and $\varepsilon > 0$ there exists a $\delta > 0$ such that

$$d(x, x') < \delta \implies d(f(x), f(x')) < \varepsilon$$

for all $x' \in X$.

4.2.1 The Induced Topology

The topology induced by a metric is given in terms of open balls which are shown below to form a basis for a topology. It should be made clear at once that the resulting topology does not carry all of the metric information. In fact, it carries very little of it. There may be many radically different distance functions on a single set, all of which induce the same topology. Moreover, not every topology is necessarily induced by a metric. Thus replacing a metric by the induced topology looses a lot of information and misses many topological spaces. However, it is often convenient to be able to forget information, especially when it is not pertinent to the problem at hand.

Definition 4.6 Let X be a metric space, $x \in X$, and $\varepsilon > 0$. The *open ball* with *centre* x and *radius* ε is the set

$$B_\varepsilon(x) = \{y \in X \mid d(x, y) < \varepsilon\}.$$

Example 4.5 Consider the Euclidean metric on $\mathbb{R}^n$. The open ball $B_\varepsilon(x)$ in $\mathbb{R}$ is the open interval $(x - \varepsilon, x + \varepsilon)$, while in $\mathbb{R}^2$ it is the interior of the circle with centre x and radius ε, and in $\mathbb{R}^3$ it is the interior of a ball with centre x and radius ε.

In a general metric space an open ball need not look anything like a ball. We already had a look at some odd looking 'circles' in Fig. 2.1 located in Sect. 2.5 when $\mathbb{R}^2$ is equipped with the metric induced by the ℓ_p norm. Their interiors are the corresponding open balls in these spaces.

We show that the open balls in a metric space X form a basis for a topology.

Theorem 4.2 *The collection* $\{B_\varepsilon(x) \mid x \in X, \ \varepsilon > 0\}$ *of all open balls in a metric space X satisfies the conditions of Theorem 3.10.*

Proof Since $x \in B_\varepsilon(x)$ for all $x \in X$ and $\varepsilon > 0$, it follows at once that the open balls cover all of X. Next, suppose that $z \in B_{\varepsilon_1}(x) \cap B_{\varepsilon_2}(y)$ and we will demonstrate that $z \in B_\varepsilon(z) \subseteq B_{\varepsilon_1}(x) \cap B_{\varepsilon_2}(y)$ for $\varepsilon = \min\{\varepsilon_1 - d(x,z), \varepsilon_2 - d(y,z)\}$. Indeed, if $d(w,z) < \varepsilon$, then $d(w,x) \leq d(w,z) + d(z,x) < (\varepsilon_1 - d(x,z)) + d(z,x) = \varepsilon_1$, and so $w \in B_{\varepsilon_1}(x)$, and therefore $B_\varepsilon(z) \subseteq B_{\varepsilon_1}(x)$. Similarly, $B_\varepsilon(z) \subseteq B_{\varepsilon_2}(y)$. Noting that $\varepsilon > 0$, the proof is complete. $\qquad\square$

The following definition is now justified.

Definition 4.7 Given a metric space (X, d), the *induced topology* on the set X is the topology generated by the open balls in X. In particular, all open balls in X are open sets and a subset $U \subseteq X$ is open in the induced topology precisely when for all $x \in U$ there exists an $\varepsilon > 0$ such that $B_\varepsilon(x) \subseteq U$. Any topological space (X, τ) for which there exists a metric d on X inducing the given topology τ is called a *metrizable space*.

Remark 4.3 Every metric space is silently endowed with the induced topology, automatically making it a topological space.

Example 4.6 It is straightforward to see that the Euclidean distance $d(x, y) = |x - y|$ on $\mathbb{R}$ induces the Euclidean topology on $\mathbb{R}$. One may also readily confirm that the discrete topology on a set X is induced by the distance function $d : X \times X \to \mathbb{R}_+$ given by $d(x, y) = 1$ if $x \neq y$ and $d(x, x) = 0$, thus any discrete topological space is metrizable.

Examples of non-metrizable spaces are easily constructed once basic facts about metrizable spaces are observed. First though, we note the following.

Theorem 4.3 *A function $f : X \to Y$ between metric spaces is continuous in the sense of Definition 4.5 if, and only if, it is continuous with respect to the induced topologies.*

The proof is formally identical to the proof of Theorem 3.2 and is left for the reader as we turn to examine general properties of metrizable spaces.

Theorem 4.4 *Every metrizable space is Hausdorff.*

Proof Let X be a topological space whose topology is induced by a metric d. Given distinct points $x, y \in X$ let $\varepsilon = d(x, y)/2$, and notice that $\varepsilon > 0$. Then the open balls $B_\varepsilon(x)$ and $B_\varepsilon(y)$ separate x and y: if $z \in B_\varepsilon(x) \cap B_\varepsilon(y)$, then we would have $d(x, y) \leq d(x, z) + d(z, y) < 2\varepsilon = d(x, y)$, which is absurd. $\qquad\square$

By Theorem 3.12 convergent sequences in Hausdorff spaces have unique limits, leading to the following familiar result stated for general metric spaces.

Corollary 4.1 *In a metric space, a convergent sequence has a unique limit.*

Apart from being Hausdorff, every metrizable topology has other pleasant features influencing its topological behavior. It is worthwhile noticing that it is the interaction of a metric space with $\mathbb{R}_+$ (via the distance function) that is responsible for some of the familiar properties of the real numbers to reflect back to any metric space. This is shown in the next result.

Theorem 4.5 *Every metrizable space is first countable.*

Proof Let X be a topological space whose topology is induced by a metric d and let $x \in X$. We need to exhibit a countable local basis at x. The collection $\{B_{1/n}(x) \mid n \in \mathbb{N}\}$ is clearly countable. To show that it is a local basis, suppose that $x \in U$ for an open set $U \subseteq X$. Then there is an $\varepsilon > 0$ with $B_\varepsilon(x) \subseteq U$, and thus if we take any $n > 1/\varepsilon$, then $B_{1/n}(x) \subseteq B_\varepsilon(x) \subseteq U$, as required. $\qquad\square$

Theorem 3.9 yields the following corollaries.

Corollary 4.2 *A point x in a metric space X is a cluster point of a subset $S \subseteq X$ if, and only if, x is the limit of a sequence of points from S.*

Corollary 4.3 *A subset F of a metrizable space X is closed if, and only if, F contains all limit points of sequences in F.*

Corollary 4.4 (Heine's Continuity Criterion) *A function $f : X \to Y$ between two metrizable spaces is continuous if, and only if,*

$$x_m \to x_0 \implies f(x_m) \to f(x_0)$$

for all sequences $\{x_m\}_{m \geq 1}$ in X.

By Theorem 3.8 a second countable space is separable. For metric spaces the converse holds as well.

Theorem 4.6 *A separable metrizable space X is second countable.*

Proof Let d be a metric inducing the topology on X. We are given a countable and dense set $D = \{x_k\}_{k \in \mathbb{N}}$ in X and we need to exhibit a countable basis for X. In the proof of Theorem 4.5 we saw that the collection

$$\mathcal{B}_k = \{B_{1/n}(x_k) \mid n \in \mathbb{N}\}$$

is a local basis at x_k. We now show that

$$\mathcal{B} = \bigcup_{k \in \mathbb{N}} \mathcal{B}_k$$

is a basis for the topology of X, noticing that the result then follows since B, being a countable union of countable sets, is itself countable (see Sect. 1.2.13).

Let $U \subseteq X$ be an open subset and choose $x \in U$. There is then an $\varepsilon > 0$ such that $B_{2\varepsilon}(x) \subseteq U$. Choose $n \in \mathbb{N}$ such that $1/n < \varepsilon$. Since D is dense and $B_{1/n}(x)$ is open, it follows that $D \cap B_{1/n}(x)$ is not empty, and thus there is some m such that $x_m \in B_{1/n}(x)$, in other words, $d(x, x_m) < 1/n$. Now, the choice of n guarantees that $B_{1/n}(x_m) \subseteq B_{2\varepsilon}(x) \subseteq U$, while the choice of x_m guarantees that $x \in B_{1/n}(x_m)$. Noticing that $B_{1/n}(x_m) \in \mathcal{B}_m$, the above establishes that for every open set U and $x \in U$, there exists a $B \in \mathcal{B}$ with $x \in B \subseteq U$, showing that $\mathcal{B}$ is a basis for the topology on X. $\qquad\qquad\square$

4.2.2 Convergence in Metric Spaces

Every metric space is in particular a topological space and thus is equipped with a convergence notion. We investigate this convergence in the context of the main examples of metric spaces given above.

Proposition 4.1 *A sequence $\{x_m\}_{m\geq1}$ in a metric space X converges to $x_0 \in X$ if, and only if, $d(x_m, x_0) \to 0$ in $\mathbb{R}$ in the usual sense of convergence.*

Proof Suppose $x_m \to x_0$ in the topological sense. Given $\varepsilon > 0$ consider the open ball $B_\varepsilon(x_0)$, which is open in the induced topology. That means that there exists an $N \in \mathbb{N}$ such that $x_m \in B_\varepsilon(x_0)$ for all $m > N$. In other words, $d(x_m, x_0) < \varepsilon$ for all $m > N$, showing that $d(x_m, x_0) \to 0$. The converse is left for the reader. $\qquad\square$

Example 4.7 Consider $\mathbb{R}^n$ with the Euclidean metric

$$d(x, y) = \left(\sum_{k=1}^{n} (x_k - y_k)^2 \right)^{1/2}.$$

For a sequence $\{x^{(m)}\}_{m\geq1}$ of elements in $\mathbb{R}^n$ and a point $x \in \mathbb{R}^n$ the inequalities

$$|x_k^{(m)} - x_k| \leq \|x^{(m)} - x\|,$$

valid for all $1 \leq k \leq n$, show that $x^{(m)} \to x$ implies $x_k^{(m)} \to x_k$, for each component. The converse is also true, (since there are only finitely many components), and thus convergence in the Euclidean metric in $\mathbb{R}^n$ is equivalent to the usual convergence of the components in $\mathbb{R}$.

Example 4.8 Consider the space ℓ_p, $p \geq 1$ (see Sect. 2.5.4). Much as in the previous example, convergence in the ℓ_p norm implies the convergence of each of the components. However, as each vector now has infinitely many components, the converse does not necessarily hold. To see that, take the sequence $\{x^{(m)}\}_{m\in\mathbb{N}}$ where

$$x_k^{(m)} = \begin{cases} 1 & \text{if } m = k, \\ 0 & \text{if } m \neq k. \end{cases}$$

Each sequence $\{x_k^{(m)}\}_{m\in\mathbb{N}}$ is eventually constantly 0, and thus the only candidate for the limit of $\{x^{(m)}\}_{m\in\mathbb{N}}$ is the 0 vector. However,

$$d(x^{(m)}, 0) = \left(\sum_{i=1}^{\infty} |x_k^{(m)}|^p\right)^{1/p} = 1,$$

and thus $x^{(m)} \not\to 0$ in the metric sense, and therefore $\{x^{(m)}\}_{m\geq 1}$ does not converge at all in ℓ_p. The same sequence exhibits similar behaviour in ℓ_∞.

Example 4.9 Consider the space $C(I, \mathbb{R})$ of continuous real-valued functions on the closed interval $I = [a, b]$, with the metric induced by the L_∞ norm, i.e., for a sequence of functions $\{x_m\}_{m\geq 1}$ and a function x_0:

$$x_m \to x_0 \qquad \Longleftrightarrow \qquad \max_{t\in[a,b]} |x_m(t) - x_0(t)| \to 0.$$

Thus, for any fixed $\varepsilon > 0$ there exists an $N \in \mathbb{N}$ such that $|x_m(t) - x_0(t)| < \varepsilon$ for all $m > N$ and for all $t \in [a, b]$. In other words, x_m converges to x_0 uniformly on $[a, b]$. Since the converse is also true, as is easily seen by traversing the arguments backwards, we conclude that convergence in the L_∞ norm, is nothing but uniform convergence of functions on the interval $[a, b]$.

Example 4.10 For $1 \leq p < \infty$ consider the pre-L_p space $C(I, \mathbb{R})$, with $I = [a, b]$, i.e., with distance given by

$$d(x, y) = \left(\int_a^b dt \, |(x(t) - y(t)|^p\right)^{1/p}.$$

Convergence with respect to this metric is called L_p convergence. It is well-known, from elementary considerations of the Riemann integral, that if $x_m \to x_0$ uniformly, then x_m also converges to x_0 in the L_p norm. The converse is not generally true, as we exemplify for the L_2 norm. In $C([0, 1], \mathbb{R})$ consider the sequence

$$x_m(t) = \begin{cases} mt & \text{if } 0 \leq t \leq \dfrac{1}{m}, \\[2mm] 1 & \text{if } \dfrac{1}{m} < t \leq 1 \end{cases}$$

of functions. Consulting the diagram

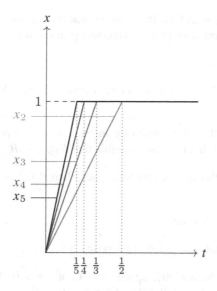

the computation

$$d(x_m, x)^2 = \int_0^1 dt \, (x_m(t) - 1)^2 = \int_0^{1/m} dt \, (mt - 1)^2 = \frac{1}{3m} \to 0$$

shows that x_m converges in L_2 to the constant function $x(t) = 1$. However, point wise $x_m(t)$ converges to the discontinuous function $\tilde{x}(t) = 1$ for $0 < t \le 1$ and $\tilde{x}(0) = 0$. As the uniform limit of continuous functions is continuous, we see that $\{x_m\}_{m \ge 1}$ does not converge with respect to the L_∞ norm.

Example 4.11   In fact, L_p convergence need not even imply point wise convergence at any t in the domain. Indeed, consider $C([0, 1], \mathbb{R})$ and, for simplicity, the L_1 norm. Let I_m be the following sequence of intervals. The first interval is $I_0 = [0, 1]$, the entire segment. The next two intervals are $I_1 = [0, 1/2]$ and $I_2 = [1/2, 1]$. The next three intervals are obtained by subdividing $[0, 1]$ into three equal segments, the next four intervals are obtained by subdividing $[0, 1]$ into four equal segments, and so on. Consider for each m the indicator function $x_m = \chi_{I_m}$. These functions are not continuous, but let us ignore this for now. With respect to the L_1 norm x_m converges to the constant function 0 since the integral $\int_0^1 dt \, \chi_m$ is the length of the segment that was considered in the m-th stage and the segments' sizes shrink to 0. However, for any $0 \le t \le 1$, the sequence $\{\chi_m(t)\}_{m \ge 1}$ contains a subsequence that is constantly 1 and a subsequence that is constantly 0, and thus does not converge. This example would thus serve our purpose, illustrating that L_1 convergence need not imply point wise convergence at any point, except that the functions are not continuous. This can easily be remedied, with minor changes in the details above, by sacrificing a little bit of the segment used to define x_m in order to add two linear segments to the graph of χ_m so as to obtain the graph of a continuous function x_m. This can be done in such a way as to only slightly affect the integrals, so that the L_1 convergence is unaltered, while retaining the point-wise non-convergence (see Exercise 4.32).

Exercises

Exercise 4.17 Let $\delta \leq \varepsilon$ and $x \in X$ a point in a metric space X. Show that $B_\delta(x) \subseteq B_\varepsilon(x)$.

Exercise 4.18 Show that in an arbitrary metric space it is possible that $B_\delta(x) = B_\varepsilon(x)$ while $\delta \neq \varepsilon$ and that it is also possible that $B_\delta(x) = B_\varepsilon(y)$ while $x \neq y$.

Exercise 4.19 Prove that if X is a normed space, then $B_\varepsilon(x) = B_\delta(y)$ implies $x = y$ and $\varepsilon = \delta$.

Exercise 4.20 Prove Theorem 4.3.

Exercise 4.21 Complete the proof of Proposition 4.1.

Exercise 4.22 Let X be a metric space, $x \in X$, and $\varepsilon > 0$. The *closed ball* with centre x and radius ε is the set $\overline{B}_\varepsilon(x) = \{y \in X \mid d(x, y) \leq \varepsilon\}$. Is the closure of the open ball $B_\varepsilon(x)$ with respect to the induced topology is the closed ball $\overline{B}_\varepsilon(x)$?

Exercise 4.23 Give an example of a non-second countable metric space.

Exercise 4.24 Prove Theorem 4.3.

Exercise 4.25 Let d_1 and d_2 be two distance functions on a set X such that there exist real numbers $\alpha, \beta > 0$ with

$$\alpha d_1(x, x') \leq d_2(x, x') \leq \beta d_1(x, x')$$

for all $x, x' \in X$. Prove that d_1 and d_2 induce the same topology on X.

Exercise 4.26 When is the cofinite topology on a set X metrizable?

Exercise 4.27 Prove that the coproduct of two metrizable spaces is metrizable. How far does this result generalize when considering more than two spaces?

Exercise 4.28 The metric analogue of the Cantor-Schröder-Bernstein Theorem fails: find non-isometric metric spaces X and Y together with isometries $f : X \to Y$ and $g : Y \to X$. Hint: a space consisting of countably many copies of a square is useful.

Exercise 4.29 Find a metrizable space with a non-metrizable quotient.

Exercise 4.30 Let $x^{(m)} \to x^{(0)}$ be a convergent sequence in ℓ_p for $p \geq 1$. Note that $x^{(m)}$ is thus a sequence and we denote by $\{x_k^{(m)}\}_{n \geq 1}$ the sequence of its k-th components. Show that $x_k^{(m)} \to x_k^{(0)}$. Make sure to treat the case $p = \infty$.

Exercise 4.31 Let $x_n \to x_0$ be a convergent sequence in the pre-L_p space $K_p, p \geq 1$, of continuous real valued functions on $[0, 1]$. Note that x_n is thus a function so for each $0 \leq t \leq 1$ there is the sequence $\{x_n(t)\}_{n\geq 1}$ obtained by evaluation at t. Show that if $p = \infty$, then $x_n(t) \to x_0(t)$ for all $0 \leq t \leq 1$. Demonstrate this implication need not hold when $1 \leq p < \infty$.

Exercise 4.32 Consider the function $f: [0, 1] \to \mathbb{R}$ given by $f(t) = 0$ if $t < 1/2$ and $f(t) = 1$ otherwise. Prove that for all $\varepsilon > 0$ there exists a continuous function $g: [0, 1] \to \mathbb{R}$ with $\int_0^1 dt \, |f(t) - g(t)| < \varepsilon$.

Informally, this shows that the given function f, though not continuous, can be approximated by a continuous function arbitrarily closely with respect to the L_1 norm. Show that the approximation can be obtained with respect to the L_p norm for all $1 \leq p < \infty$. Prove that a non-continuous function cannot be approximated arbitrarily closely with respect to the L_∞ norm.

Exercise 4.33 Show that the association of a topology to a metric space formally applies to semi-metric spaces too. Is such a topology always Hausdorff?

Exercise 4.34 Continuing the previous exercise recall the complete lattices $\text{Top}(X)$ from Exercise 3.41 and $\text{Met}(X)$ from Exercise 4.6. The fact that every semi-metric space has an associated topology states the existence of a function $O: \text{Met}(X) \to \text{Top}(X)$. Show that:

1. O is typically not injective.
2. O is typically not surjective.
3. O preserves the ordering.
4. $O(d_1 \wedge d_2) = O(d_1) \wedge O(d_2)$ for all semi-metrics d_1, d_2 on X.
5. O preserves finite meets in the sense that $O(\bigwedge S) = \bigwedge O(S)$, where $S \subseteq \text{Met}(X)$ is finite and $O(S) = \{O(d) \mid d \in S\}$.
6. O generally does not preserve arbitrary meets in the above sense for S infinite.

4.3 Non-expanding Functions and Uniform Continuity

Given two metric spaces X and Y the rich structure embodied in the metrics can be preserved to various degrees by functions $f: X \to Y$. In a sense the weakest preservation of structure one can impose is for f to be continuous while the strongest is for f to be an isometry. We look at two intermediate conditions for structure preservation, namely non-expansion and uniform continuity.

Definition 4.8 A function $f: X \to Y$ between metric spaces is *non-expanding* if

$$d(f(x), f(x')) \leq d(x, x')$$

for all $x, x' \in X$.

An important observation is that in any metric space (X, d) the distance function $d : X \times X \to \mathbb{R}_+$ is non-expanding. Recall the additive metric product structure (Definition 4.4) and the metric on $\mathbb{R}_+$ described in Example 4.4.

Lemma 4.1 *In any metric space X the distance function $d : X \times X \to \mathbb{R}_+$ is non-expanding.*

Proof Given (x, x'), $(y, y') \in X \times X$ we may assume without loss of generality that $d_X(x, x') \geq d_X(y, y')$. Let us further assume that both $d_X(x, x')$ and $d_X(y, y')$ are finite. It then follows that

$$d_{\mathbb{R}_+}(d(x, x'), d(y, y')) = |d_X(x, x') - d_X(y, y')| = d_X(x, x') - d_X(y, y').$$

Since

$$d_{X \times X}((x, x'), (y, y')) = d_X(x, y) + d_X(x', y')$$

we need to show that

$$d_X(x, x') \leq d_X(x, y) + d_X(y, y') + d_X(y', x')$$

which is indeed the case by the triangle inequality. Treating the cases where either $d_X(x, x') = \infty$ or $d_X(y, y') = \infty$ is similar.   □

Non-expansion is a rather strong property, quite close in strength to being an isometry. On the other side of the spectrum, closer to continuity, lies uniform continuity. We shall see that it is only slightly stronger than continuity.

Definition 4.9 A function $f : X \to Y$ between metric spaces is *uniformly continuous* if for every $\varepsilon > 0$ there is a $\delta > 0$ such that

$$d(x, x') < \delta \implies d(f(x), f(x')) < \varepsilon$$

for all $x, x' \in X$.

One can verify that if $f : X \to Y$ is non-expanding, then it is uniformly continuous, and that uniform continuity implies continuity. Neither converse implication is generally true. However, the Heine-Cantor Theorem states that continuity on a closed interval implies uniform continuity. The crucial property here is that a closed interval is compact. Indeed, we show that if X is compact, then continuity does imply uniform continuity. The proof we present relies on the topological concept known as the Lebesgue number of a covering and it requires a bit of preparation.

Definition 4.10 Let X be a metric space and $S \subseteq X$ a subset. The *diameter* of S is the extended real number

$$\text{diam}(S) = \sup_{x, y \in S} \{d(x, y)\}$$

(which may be ∞).

For instance, if $S = \{x, y\}$, then $\mathrm{diam}(S) = d(x, y)$. In $\mathbb{R}^n$ with the Euclidean metric $\mathrm{diam}(B_\varepsilon(x)) = 2\varepsilon$, for all $x \in \mathbb{R}^n$ and $\varepsilon > 0$. More generally, in an arbitrary metric space $\mathrm{diam}(B_\varepsilon(x)) \leq 2\varepsilon$ always holds and equality may fail.

Lemma 4.2 (Lebesgue Number Lemma) *Let X be a compact metric space. If $\{U_i\}_{i \in I}$ is an arbitrary open covering of X, then there exists a $\delta > 0$ such that for all subsets $S \subseteq X$ with $\mathrm{diam}(S) < \delta$ there is an $i \in I$ with $S \subseteq U_i$.*

Proof For each $x \in X$ there is an $i_x \in I$ with $x \in U_{i_x}$ and since U_{i_x} is open there is an $\varepsilon_x > 0$ such that $B_{2\varepsilon_x}(x) \subseteq U_{i_x}$. Clearly, the collection $\{B_{\varepsilon_x}(x)\}_{x \in X}$ covers X so by compactness there is a finite subcovering

$$X = \bigcup_{k=1}^{m} B_{\varepsilon_{x_k}}(x_k).$$

We show that

$$\delta = \min\{\varepsilon_{x_1}, \ldots, \varepsilon_{k_m}\},$$

which is clearly positive, is the required δ for the claim. Suppose then that $S \subseteq X$ is a subset with $\mathrm{diam}(S) < \delta$. We may assume $S \neq \emptyset$, so let us fix an element $y \in S$. Since $y \in X$ and X is covered by $\{B_{\varepsilon_x}(x)\}_{1 \leq k \leq n}$ we may find an $x = x_{k_0}$, with corresponding $\varepsilon = \varepsilon_{k_0}$, such that $y \in B_\varepsilon(x)$. It now follows, for any $y' \in S$, that $d(y, y') \leq \mathrm{diam}(S) < \delta$, and thus $y' \in B_\delta(y)$. However,

$$B_\delta(y) \subseteq B_\varepsilon(y) \subseteq B_{2\varepsilon}(x) \subseteq U_x$$

(as the reader is invited to verify) and thus $y' \in U_x$. As $y' \in S$ was arbitrary it follows that $S \subseteq U_x$, as required. $\qquad \square$

A δ as in the statement above is called a *Lebesgue number* for the covering.

Theorem 4.7 (Heine-Cantor) *A continuous function $f : X \to Y$ from a compact metric space X to an arbitrary metric space Y is uniformly continuous.*

Proof Given $\varepsilon > 0$ consider, for each $y \in Y$, the set

$$U_y = f^{-1}(B_{\frac{\varepsilon}{2}}(y))$$

which is open since f is continuous. Clearly, $x \in U_{f(x)}$ for all $x \in X$ and thus $\{U_y\}_{y \in X}$ is an open covering of X. We may now choose a Lebesgue number $\delta > 0$ for that covering. Suppose that

$$d(x, x') < \delta$$

for points $x, x' \in X$. Then, $\mathrm{diam}(\{x, x'\}) < \delta$ and thus $\{x, x'\} \subseteq f^{-1}(B_{\varepsilon/2}(y))$ for some $y \in Y$. That means that $d(f(x), y) < \varepsilon/2$ and that $d(f(x'), y) < \varepsilon/2$. Applying the triangle inequality yields

$$d(f(x), f(x')) \le d(f(x), y) + d(y, f(x')) < \varepsilon$$

as needed.   □

Since uniform continuity always entails continuity, the result above implies that for a function $f : X \to Y$ between metric spaces with a compact domain the concepts of continuity and uniform continuity coincide.

Exercises

Exercise 4.35 Prove that the identity function id : $X \to X$ on any metric space is uniformly continuous, and prove that the composition of two uniformly continuous functions is uniformly continuous. Repeat the exercise for non-expanding mappings instead of uniformly continuous ones.

Exercise 4.36 For $\mathbb{R}$ with the Euclidean metric show that $f : \mathbb{R} \to \mathbb{R}$ given by $f(x) = \alpha x$ is non-expanding if, and only if, $|\alpha| \le 1$.

Exercise 4.37 For $\mathbb{R}^n$ with the Euclidean metric and a diagonalizable $n \times n$ matrix M how that $f : \mathbb{R}^n \to \mathbb{R}^n$ given by $f(x) = Mx$ is non-expanding if, and only if, $|\lambda| \le 1$ for all eigenvalues λ of M.

Exercise 4.38 For a function $f : X \to Y$ between metric spaces show the implications

$$\text{isometry} \implies \text{non-expanding} \implies \text{uniformly continuous} \implies \text{continuous}$$

and that none of these implications is reversible.

Exercise 4.39 Let $f : \mathbb{R} \to \mathbb{R}$ be a differentiable function with $|f'(x)| \le 1$ for all $x \in \mathbb{R}$. Prove that f is non-expanding.

Exercise 4.40 Let X be a metric space and $\varepsilon > 0$. Show that if $d(x, y) < \varepsilon$, then $B_\varepsilon(y) \subseteq B_{2\varepsilon}(x)$.

Exercise 4.41 Prove that if $f : X \to Y$ is non-expanding, then $\text{diam}(f(S)) \le \text{diam}(S)$ for all $S \subseteq X$. Does the converse hold?

Exercise 4.42 Prove that if $f : X \to Y$ is uniformly continuous, then for every $\varepsilon > 0$ there exists a $\delta > 0$ such that $\text{diam}(f(S)) < \varepsilon$ for all $S \subseteq X$ with $\text{diam}(S) < \delta$. Does the converse hold?

Exercise 4.43 Prove that for all subsets S_1, S_2 of a metric space X, if $S_1 \subseteq S_2$, then $\text{diam}(S_1) \le \text{diam}(S_2)$. Does the converse hold?

Exercise 4.44 Prove that $\text{diam}(B_\varepsilon(x)) \le 2\varepsilon$ for all $\varepsilon > 0$ and for all points x in a metric space X. Demonstrate that $\text{diam}(B_\varepsilon(x))$ can be any number $0 \le t \le 2\varepsilon$.

Exercise 4.45 Show that in a normed space V with its induced metric, $\text{diam}(B_\varepsilon(x)) = 2\varepsilon$ for all $x \in V$ and $\varepsilon > 0$.

Exercise 4.46 Let V be a normed space. Prove that $\| - \| : V \to \mathbb{R}$ is non-expanding when $\mathbb{R}$ is given the Euclidean metric.

4.4 Complete Metric Spaces

This section is devoted to completeness in a metric space. A metric space X is said to be complete when every Cauchy sequence in it converges. Thus, in a sense, X is complete when every sequence in X that should converge actually does converge. The section starts with a discussion of the concept of Cauchy sequences in general metric spaces and then considers the completeness of $\mathbb{R}$. The reader may already be familiar with the completeness of the real numbers, a property that is commonly stated in the form of the least upper bound principle: every non-empty bounded above set of real numbers has a supremum (for more details see Sect. 1.2.19). We show below that the least upper bound property implies the metric completeness of $\mathbb{R}$, namely that any Cauchy sequence in $\mathbb{R}$ converges (when $\mathbb{R}$ is given the Euclidean metric). We then examine the metric spaces from the previous sections, together with an investigation of their completeness. Next we establish two fundamental theorems of complete metric spaces, namely the Banach Fixed Point Theorem and Baire's Theorem. The importance of these results is difficult to over emphasize, particularly in the context of this book. As will be shown in Chap. 6, the Banach Fixed Point Theorem can be used to iteratively solve certain differential equations. Some of the consequences of Baire's Theorem in the theory of Banach spaces will also be given. This chapter then closes with a description of a completion process, i.e., a way to turn any metric space into a complete one while altering it as little as possible. The construction we present is that of Cantor's using Cauchy sequences.

4.4.1 Complete Metric Spaces

Given a sequence $\{x_m\}_{m \geq 1}$ in a metric space X the statement

$$\exists x_0 \in X \ \forall \varepsilon > 0 \ \exists N \in \mathbb{N} : \ \forall m \geq N \ d(x_m, x_0) < \varepsilon$$

is nothing but the claim that $x_m \to x_0$. By interchanging the first two quantifiers we arrive at

$$\forall \varepsilon > 0 \ \exists x_\varepsilon \in X, \ N \in \mathbb{N} : \ \forall m \geq N \ d(x_m, x_\varepsilon) < \varepsilon$$

which we call the *global Cauchy condition*. It expresses something considerably weaker than the existence of a limit. Namely, to establish convergence one needs to manufacture an appropriate $N \in \mathbb{N}$ for any given $\varepsilon > 0$ *without* altering the limiting point. With the global Cauchy condition, on the other hand, one is first given an $\varepsilon > 0$, and may then manufacture the limit point x_ε and a corresponding $N \in \mathbb{N}$. The global Cauchy condition expresses the existence of a sort of varying limit of the sequence. In particular, if the sequence converges to x_0, then the global Cauchy condition is satisfied simply by taking $x_\varepsilon = x_0$.

A simple analysis of the situation reveals that the x_ε can be eliminated from the global Cauchy condition. Suppose that the global Cauchy condition is satisfied and let $\varepsilon > 0$ be given. There exist then an $x_\varepsilon \in X$ and an $N \in \mathbb{N}$ so that $d(x_m, x_\varepsilon) < \varepsilon/2$, for all $m \geq N$. It then follows that

$$d(x_k, x_m) \leq d(x_k, x_\varepsilon) + d(x_\varepsilon, x_m) \leq \frac{\varepsilon}{2} + \frac{\varepsilon}{2} = \varepsilon$$

for all $k, m \geq N$. Conversely, suppose that for all $\varepsilon > 0$ there exists an $N \in \mathbb{N}$ such that

$$d(x_k, x_m) < \varepsilon$$

for all $k, m \geq N$. Then by taking $x_\varepsilon = x_N$ we obtain

$$d(x_m, x_\varepsilon) = d(x_m, x_N) < \varepsilon$$

for all $m \geq N$, and so the global Cauchy condition is satisfied. This discussion is summarized as follows.

Definition 4.11 A sequence $\{x_m\}_{m \geq 1}$ in a metric space X is a *Cauchy sequence* if it satisfies the *Cauchy condition*:

$$\forall \varepsilon > 0 \, \exists N \in \mathbb{N} \, \forall k, m \geq N : d(x_k, x_m) < \varepsilon.$$

Equivalently, a sequence $\{x_m\}_{m \geq 1}$ is a Cauchy sequence if it satisfies the *global Cauchy condition*

$$\forall \varepsilon > 0 \, \exists x_\varepsilon \in X, \, N \in \mathbb{N} : \forall m \geq N \, d(x_m, x_\varepsilon) < \varepsilon.$$

The familiar fact that every convergent sequence in $\mathbb{R}$ (when endowed with the Euclidean metric) is a Cauchy sequence is now nothing but the remark above that the global Cauchy condition is trivially implied by the existence of a limit. The converse in $\mathbb{R}$ is well known to be true (and we prove that fact below) but for general metric spaces the converse may fail. To see that, start with any convergent sequence $x_m \to x_0$ in a metric space X satisfying $x_m \neq x_0$, for all $m \geq 1$, and consider the same sequence $\{x_m\}_{m \geq 1}$ in the subspace $X - \{x_0\}$. Clearly, $\{x_m\}_{m \geq 1}$ satisfies the Cauchy condition (indeed, it is a convergent sequence in X). However, it only converges in X; if it converged in $X - \{x_0\}$ too, say to the point $y \in X - \{x_0\}$, then it would also converge to y in X, contradicting uniqueness of limits in a metric space.

Definition 4.12 A metric space X in which every Cauchy sequence admits a limit *in X* is said to be a *complete metric space*.

As promised above we closely examine the completeness of the reals, which, as we will see, is a pivotal result.

Example 4.12 (*The completeness of* $\mathbb{R}$) The space $\mathbb{R}$ with the Euclidean metric $d(x, y) = |x - y|$ is a most fundamental example of a complete metric space. Recall

(Sect. 1.2.19) that the real numbers $\mathbb{R}$ are defined to be any Dedekind complete ordered field.

Given a Cauchy sequence $\{x_m\}_{m \geq 1}$ of real numbers we first note that $\{x_m\}_{m \geq 1}$ is bounded by some number M, i.e., $|x_m| < M$ for all $m \in \mathbb{N}$. Indeed, for $\varepsilon = 1$ there exists an $N \in \mathbb{N}$ such that $|x_k - x_m| < 1$ for all $k, m \geq N$, and thus we found a bounded tail of the sequence, and so the entire sequence is bounded. Let $S \subseteq \mathbb{R}$ be the subset consisting of all $x \in \mathbb{R}$ such that the inequality $x < x_m$ holds for almost all $m \in \mathbb{N}$ (meaning that it holds for all $m \in \mathbb{N}$ with at most finitely many exceptions). Clearly, for our chosen upper bound M we have that $-M \in S$ and $M \notin S$, and thus S is non-empty and bounded above. By the least upper bound principle in $\mathbb{R}$ the set S admits a supremum L. We will proceed to show that L is in fact the limit of the given sequence. Given $\varepsilon > 0$ let $N \in \mathbb{N}$ with $|x_k - x_m| < \varepsilon/2$ for all $k, m \geq N$. In particular, for $m = N$ we obtain

$$x_m - \frac{\varepsilon}{2} < x_k < x_m + \frac{\varepsilon}{2}$$

for all $k \geq N$. These inequalities imply, by definition of S, that

$$x_m - \frac{\varepsilon}{2} \in S, \quad \text{while} \quad x_m + \frac{\varepsilon}{2} \notin S.$$

Since L is the least upper bound of S it now follows that

$$x_m - \frac{\varepsilon}{2} \leq L \leq x_m + \frac{\varepsilon}{2}.$$

Combining all of the inequalities above, it follows that

$$|x_k - L| \leq |x_k - x_m| + |x_m - L| \leq \frac{\varepsilon}{2} + \frac{\varepsilon}{2} = \varepsilon$$

for all $k > N$, proving that $x_m \to L$. $\mathbb{R}$ is thus metrically complete.

Remark 4.4 The amount of work put into this example is quite adequately justified by noticing that the completeness of $\mathbb{R}$ is used in establishing the completeness of all of the forthcoming examples of complete metric spaces.

Remark 4.5 It should be noted that Dedekind completeness is a stronger property than metric completeness, in the following sense. While every Dedekind complete ordered field is metrically complete, there exist metrically complete ordered fields which are not Dedekind complete. None of these examples would be Archimedean though. These issues however are not particularly relevant to us here.

Example 4.13 In contrast to the situation with the real numbers, the metric space $\mathbb{Q}$ of rational numbers with $d(x, y) = |x - y|$ is not complete. Indeed, by the density of $\mathbb{Q}$ in $\mathbb{R}$, for every irrational α there is a sequence of rational numbers with $x_m \to \alpha$ in $\mathbb{R}$. Such a sequence is thus Cauchy in both $\mathbb{R}$ and $\mathbb{Q}$ but fails to converge in $\mathbb{Q}$.

Before examining the completeness property for the metric spaces introduced above we remark that completeness is a metric invariant. In fact, we establish a stronger result.

Proposition 4.2 (Uniformly Continuous Functions Preserve Cauchy Sequences) *If* $f : X \to Y$ *is a uniformly continuous function between metric spaces and* $\{x_m\}_{m \geq 1}$ *is a Cauchy sequence in X, then* $\{f(x_m)\}_{m \geq 1}$ *is a Cauchy sequence in Y.*

Proof Given $\varepsilon > 0$ let $\delta > 0$ be such that

$$d(x, x') < \delta \implies d(f(x), f(x')) < \varepsilon.$$

Since $\{x_m\}_{m \geq 1}$ is Cauchy, there exists an $N \in \mathbb{N}$ such that

$$k, m \geq N \implies d(x_k, x_m) < \delta.$$

Combining the two implications yields

$$k, m \geq N \implies d(f(x_k), f(x_m)) < \varepsilon,$$

as needed for showing that $\{f(x_m)\}_{m \geq 1}$ is Cauchy. $\square$

Corollary 4.5 *Let* $f : X \to Y$ *be an invertible continuous function between metric spaces such that* $f^{-1} : Y \to X$ *is uniformly continuous. If X is complete, then so is Y.*

Proof Let $\{y_m\}_{m \geq 1}$ be a Cauchy sequence in Y. Then $\{f^{-1}(y_m)\}_{m \geq 1}$ is a Cauchy sequence in X and since X is complete the sequence converges: $f^{-1}(y_m) \to x_0$. Since a continuous function perseveres limits of sequences, we may now apply f to conclude that $y_m \to f(x_0)$. $\square$

Corollary 4.6 *Completeness is a metric invariant.*

A very important property of complete metric spaces is that they allow to extend a uniformly continuous function from a subset to its closure.

Theorem 4.8 *Let* X, Y *be metric spaces with Y complete. If* $f : S \to Y$ *is a uniformly continuous function on a subset* $S \subseteq X$, *then f extends uniquely to a uniformly continuous function* $\hat{f} : \overline{S} \to Y$.

Proof Firstly, we show that if an extension exists, then it is unique. Let $x \in \overline{S}$ be an arbitrary point in the closure. There exists then a sequence $\{s_n\}_{n \geq 1}$ in S with

$$\lim_{n \to \infty} s_n = x$$

and as any extension of f to a uniformly continuous function $\hat{f}$ is in particular continuous we see that

$$\hat{f}(x) = \hat{f}(\lim_{n\to\infty} s_n) = \lim_{n\to\infty} \hat{f}(s_n) = \lim_{n\to\infty} f(s_n)$$

and is thus uniquely determined by f and the topologies of X and Y. We must now show that we can turn this observation into a definition, i.e., that

$$\hat{f}(x) = \lim_{n\to\infty} f(s_n)$$

is a well-defined uniformly continuous function for all $x \in \overline{S}$ and that it extends f. First, we note that the limit defining $\hat{f}$ is guaranteed to exist no matter which sequence in S is chosen to converge to x. This is so since such a sequence $\{s_n\}_{n\geq1}$ is Cauchy in X (since it converges) and thus the sequence $\{f(s_n)\}_{n\geq1}$ is Cauchy in Y (since f is uniformly continuous) and thus converges (since Y is complete). Suppose now that $t_n \to x$ for some other sequences in S. The triangle inequality and a limit argument then shows that

$$\lim_{n\to\infty} d(s_n, t_n) = 0$$

and uniform continuity of f gives us

$$\lim_{n\to\infty} d(f(s_n), f(t_n)) = 0.$$

Were it the case that $f(s_n) \to y$ while $f(t_n) \to z$ with $y \neq z$, then continuity of the distance function would yield

$$\lim_{n\to\infty} d(f(s_n), f(t_n)) = d(\lim_{n\to\infty} f(s_n), \lim_{n\to\infty} f(t_n)) = d(y, z) > 0$$

and thus a contradiction. We conclude that $y = z$ and so $\hat{f}(x)$ is independent of the choice of the sequence converging to x.

The uniform continuity of $\hat{f}$ follows again by the triangle inequality, and the fact that $\hat{f}$ extends f is trivial by considering, for $s \in S$, the constant sequence converging to it. ∎

We turn to some examples of completeness, or lack thereof, in our spaces of interest.

Example 4.14 (*Completeness of $\mathbb{R}^n$*) For the metric space $\mathbb{R}^n$ with the Euclidean metric

$$d(x, y) = \left(\sum_{k=1}^{n}(x_k - y_k)^2\right)^{1/2}$$

we already saw that $x^{(m)} \to x$ if, and only if, $x_k^{(m)} \to x_k$ for each $1 \leq k \leq n$. To show that $\mathbb{R}^n$ is complete, suppose that $\{x^{(m)}\}_{m\geq1}$ is a Cauchy sequence. Then, for each $1 \leq k \leq n$, the sequence $\{x_k^{(m)}\}_{m\geq1}$ is also Cauchy (since $|x_k^{(m)} - x_k^{(m')}| \leq d(x, y)$)

and since $\mathbb{R}$ is complete we conclude that $x_k^{(m)} \to x_k$ for all $1 \le k \le n$, and therefore that $x^{(m)} \to (x_1, \ldots, x_n)$. Thus, every Cauchy sequence in $\mathbb{R}^n$ converges, as required.

Example 4.15 (*Completeness of* $\mathbb{C}^n$) The metric space $\mathbb{C}^n$ with distance given by

$$d(x, y) = \left(\sum_{k=1}^{n} |x_k - y_k|^2 \right)^{1/2}$$

is complete. Since $\mathbb{C}^n$ is isometric to $\mathbb{R}^{2n}$ and completeness is a metric invariant, the completeness of $\mathbb{R}^{2n}$ implies that of $\mathbb{C}^n$.

Example 4.16 For a closed interval $I = [a, b]$ the space $C(I, \mathbb{R})$ of continuous functions on I with the metric structure induced by the L_∞ norm is complete. To see that recall that convergence with respect to this metric is uniform convergence of sequences of functions. If $\{x_m\}_{m \ge 1}$ is a Cauchy sequence of continuous real-valued functions on I, then one easily sees that for each $t \in [a, b]$ the sequence $\{x_m(t)\}_{m \ge 1}$ is a Cauchy sequence in $\mathbb{R}$ with respect to the Euclidean metric. As $\mathbb{R}$ is complete it follows that $x_m(t) \to x(t)$, showing that $\{x_m\}_{m \ge 1}$ converges point-wise to a function $x : [a, b] \to \mathbb{R}$, and in fact the convergence is easily seen to be uniform. To conclude we recall the well known fact that if a sequence of continuous functions converges uniformly to a function x, then x is continuous. Thus, $x \in C(I, \mathbb{R})$ and is the uniform limit of the given sequence.

Example 4.17 In contrast to the L_∞ case, for all $1 \le p < \infty$ the pre-L_p space $C([a, b], \mathbb{R})$ is not complete. For simplicity let us consider the case $[a, b] = [0, 1]$. Let

$$x_m(t) = \begin{cases} 1 & \text{if } 0 \le x \le \frac{1}{2}, \\ mx - \frac{1}{2m} - 1 & \text{if } \frac{1}{2} \le x \le \frac{1}{2} + \frac{1}{m}, \\ 0 & \text{if } \frac{1}{2} + \frac{1}{m} \le x \le 1. \end{cases}$$

Then, for $k \ge m$

$$\|x_m - x_k\|_p^p = \int_{\frac{1}{2}}^{\frac{1}{2} + \frac{1}{n}} dt \, |x_m(t) - f_k(t)|^p \le \frac{1}{n}$$

and thus $\{x_m\}_{m \ge 1}$ is Cauchy. However, if this sequence converged to a continuous function $x : [0, 1] \to \mathbb{R}$, then

$$\int_0^{\frac{1}{2}} dt \, |x(t) - x_m(t)|^p \le \|x - x_m\|_p^p \to 0$$

and thus $x(t) = 1$ for all $0 \le t < \frac{1}{2}$. Similarly $x(t) = 0$ for all $\frac{1}{2} < t \le 1$, but then x is not continuous.

Example 4.18 Going back to the complete space $C(I, \mathbb{R})$ with the metric obtained from the L_∞ norm (for which convergence is equivalent to uniform convergence) consider the subspace $P \subseteq C(I, \mathbb{R})$ consisting only of the polynomial functions. As a subspace of the complete space $C(I, \mathbb{R})$ the space P is not complete. To see that, recall that the polynomials

$$p_m(t) = \sum_{k=0}^{m} \frac{t^k}{k!}$$

converge uniformly to the exponential function e^t, which is not a polynomial. Thus the sequence $\{p_m\}_{m \geq 1}$ is Cauchy (since it converges in $C(I, \mathbb{R})$) but it does not converge to a polynomial and thus does not converge in P at all (since if it would, then it would have two different limits in $C(I, \mathbb{R})$)

Example 4.19 The space ℓ_p (of all absolutely p-power summable sequences of, say, real numbers) is complete. To see that, let $\{x^{(m)}\}_{m \geq 1}$ be a Cauchy sequence in ℓ_p. Since $|x_k^{(m)} - x_k^{(n)}| \leq \|x^{(m)} - x^{(n)}\|_p$, for all $k \geq 1$, it follows readily that $\{x_k^{(m)}\}_{m \geq 1}$ is a Cauchy sequence in $\mathbb{R}$, and thus it converges to a point $x_k \in \mathbb{R}$. Thus $\{x^{(m)}\}_{m \geq 1}$ converges component wise to $x = (x_1, \ldots, x_k, \ldots)$. The completeness of ℓ_p will follow by showing that the convergence is also in the ℓ_p norm. Firstly, to verify that $x \in \ell_p$, note that since $\{x^{(m)}\}_{m \geq 1}$ is Cauchy, it is bounded, say by M. Now, for all $K \geq 1$ and $m \geq 1$

$$\left(\sum_{k=1}^{K} (x_k^{(m)})^p \right)^{1/p} \leq \|x^{(m)}\|_p \leq M.$$

Allowing m to tend to ∞, and then for K to tend to ∞ shows that $\|x\|_p \leq M$, and thus $x \in \ell_p$. Finally, to show that $x_m \to x$ in the ℓ_p norm, let $\varepsilon > 0$ be given. Let $N \in \mathbb{N}$ with

$$\|x^{(m)} - x^{(n)}\|_p < \varepsilon$$

for all $m, n \geq N$. We now have, for all $K \geq 1$ and $m, n \geq N$, that

$$\left(\sum_{k=1}^{K} (x_k^{(m)} - x_k^{(n)})^p \right)^{1/p} \leq \|x^{(m)} - x^{(n)}\|_p < \varepsilon$$

and thus, letting n tend to infinity, and then K tend to infinity, it follows that

$$\|x^{(m)} - x\|_p < \varepsilon$$

for all $m > N$, as required.

4.4.2 Banach's Fixed Point Theorem

A fixed point for a function $f : X \to X$ is a point $x_0 \in X$ with $f(x_0) = x_0$. Often, a problem can be rephrased in terms of finding a fixed point for a suitable function. Fixed point theory is thus a useful tool when solving practical problems. We present a geometric condition on a function on a metric space that guarantees the existence of a unique fixed point. Unlike many existence proofs, this one is constructive and thus can be used in practice to obtain approximations for solutions.

The geometric notion we require is that of a contraction.

Definition 4.13 A function $f : X \to Y$ between metric spaces is called a *contraction* if there exists a real number $0 \le \alpha < 1$ such that

$$d(f(x_1), f(x_2)) \le \alpha d(x_1, x_2)$$

for all $x_1, x_2 \in X$.

We note at once that a function need not be a contraction in order to posses fixed points. An extreme example is the identity function $\mathrm{id} : X \to X$ on any metric space. It is not a contraction yet every point is a fixed point of it.

Theorem 4.9 (Banach Fixed Point Theorem) *If X is a complete non-empty metric space and $f : X \to X$ is a contraction, then f has a unique fixed point $x_0 \in X$.*

Proof To prove the existence of a fixed point let x_1 be an arbitrary element in X and consider the sequence $x_2 = f(x_1), x_3 = f(x_2)$, and in general $x_{m+1} = f(x_m)$. If this sequence converges to some limit $x_0 \in X$, then, since continuous functions preserve limits of sequences, we will have that

$$f(x_0) = f(\lim_{m \to \infty} x_m) = \lim_{m \to \infty} f(x_m) = \lim_{m \to \infty} x_{m+1} = \lim_{m \to \infty} x_m = x_0$$

and so x_0 is a fixed point of f. To show that the sequence does indeed converge we invoke the completeness of X and proceed to show that the sequence is Cauchy. Since the function f is a contraction we know that there exists an $0 \le \alpha < 1$ such that

$$d(f(x), f(y)) \le \alpha d(x, y)$$

for all $x, y \in X$. We need to estimate the quantities $d(x_k, x_m)$ for the sequence defined above, and we notice first that for all $x, y \in X$ the triangle inequality yields

$$d(x, y) \le d(x, f(x)) + d(f(x), f(y)) + d(f(y), y)$$
$$\le d(x, f(x)) + \alpha d(x, y) + d(f(y), y)$$

and thus

$$d(x, y) \leq \frac{d(x, f(x)) + d(y, f(y))}{1 - \alpha}.$$

Incidentally, this already shows that if x and y are both fixed points of f, then $d(x, y) = 0$, and thus $x = y$, so the uniqueness clause is proven.

The proof is completed by a straightforward application of the latter inequality that leads to

$$d(x_k, x_m) \leq \frac{\alpha^k + \alpha^m}{1 - \alpha} d(x_0, x_1)$$

from which the fact that $\{x_m\}_{m \geq 1}$ is a Cauchy sequence is clear.   □

4.4.3 Baire's Theorem

The next result has numerous important implications in analysis in general and in functional analysis and Banach space theory in particular. Baire's theorem can be interpreted as stating that a non-empty complete metric space can not be *small*. But, size is a multifaceted concept. For instance, the set $\mathbb{Q}$ of rational numbers is countable while $\mathbb{R}$ is uncountable and thus, in terms of cardinality of sets, $\mathbb{Q}$ is negligible compared to $\mathbb{R}$. However, from a topological point of view, since the closure of $\mathbb{Q}$ is all of $\mathbb{R}$, the set $\mathbb{Q}$ is quite large. The opposite situation is that of a *nowhere dense* set, namely a set S whose closure has empty interior, and is thus a (topologically speaking) very small set. Sets which are countable unions of nowhere dense sets are called *meager* sets. The claim of Baire's Theorem is that a non-empty complete metric space is never meager.

Definition 4.14 A subset $S \subseteq X$ in a topological space X is said to be *nowhere dense* if the interior of its closure is empty. Equivalently, S is nowhere dense when $\overline{S}$ contains no non-empty open subsets.

Nowhere dense sets are, in a sense, very small. If x is not an isolated point, i.e., $\{x\}$ is not open, then $\{x\}$ is nowhere dense. A subset of a nowhere dense subset is nowhere dense, and a finite union of nowhere dense subsets is itself dense. However, a countable union of nowhere dense subsets need not be nowhere dense. For instance, in $\mathbb{R}$ with the Euclidean topology no point is isolated and thus every singleton $\{x\}$ is nowhere dense Consequently, every finite subset of $\mathbb{R}$ is nowhere dense but $\mathbb{Q}$, a countable union of singletons, is not nowhere dense. In fact, it is everywhere dense, i.e., its closure is the entire space $\mathbb{R}$.

Definition 4.15 A subset $S \subseteq X$ in a topological space X is said to be *meager* if it is a countable union of nowhere dense subsets.

Clearly, every nowhere dense subset is meager, but not vice-versa ($\mathbb{Q}$ is meager in $\mathbb{R}$ but not nowhere dense). It is left to the reader to establish that a subset of a meager set is itself meager, and that a countable union of meager subsets is meager.

Since X is a subset of itself we may speak of X itself being meager. Armed with this new form of measuring sizes, here is the main theorem.

Theorem 4.10 (Baire) *A complete non-empty metric space is not meager.*

Proof Suppose that X is a complete, non-empty, and meager metric space. By the definition of meagerness we may write

$$X = \bigcup_{m=1}^{\infty} S_m$$

where each $S_m \subseteq X$ is nowhere dense. By noting that the closure of a nowhere dense subset is still nowhere dense, we may further assume (by taking closures) that each S_m is closed. Since X is non-empty X is not nowhere dense, and thus $S_m \neq X$ for all $m \geq 1$.

From the non-empty open subset $X \setminus S_1$ we extract a point $x_1 \in X$ and $\varepsilon_1 > 0$ with

$$B_{\varepsilon_1}(x_1) \subseteq X \setminus S_1.$$

Since S_2 does not contain any open subset it certainly does not contain $B_{\varepsilon_1}(x)$. Thus the set $B_{\varepsilon_1}(x_1) \cap (X \setminus S_2)$, which is open, is not empty. Therefore, there exists a point $x_2 \in X$ and $\varepsilon_2 > 0$ with

$$B_{\varepsilon_2}(x_2) \subseteq B_{\varepsilon_1}(x_1) \cap (X \setminus S_2).$$

Continuing in this manner we obtain, for each $m \geq 1$, a point $x_m \in X$ and $\varepsilon_m > 0$ such that

$$B_{\varepsilon_{m+1}}(x_{m+1}) \subseteq B_{\varepsilon_m}(x_m) \cap (X \setminus S_{m+1}).$$

To continue we need to optimize the radii. Generally, $\overline{B_{\delta/2}}(x) \subseteq B_\delta(x)$ holds and so we can make the stronger assumption that

$$\overline{B_{\varepsilon_{m+1}}(x_{m+1})} \subseteq B_{\varepsilon_m}(x_m) \cap (X \setminus S_{m+1}).$$

as well as that that $\varepsilon_m < 1/m$ holds for all $m \geq 1$. We now have a sequence $\{B_{\varepsilon_m}(x_m)\}_{m \geq 1}$ of nested balls with rapidly shrinking radii and that implies (by Exercise 4.56) that there exists a (unique) point

$$x_0 \in \bigcap_{m \geq 1} B_{\varepsilon_m}(x_m).$$

Now, since $\{S_m\}_{m \geq 1}$ covers X we must have $x_0 \in S_{m_0}$ for some $m_0 \geq 1$. But then $x_0 \notin B_{\varepsilon_{m_0}}(x_{m_0})$, a contradiction. $\qquad \square$

Corollary 4.7 *A complete non-empty metric space X with no isolated points must be uncountable.*

Proof If X were countable, then since

$$X = \bigcup_{x \in X} \{x\}$$

and each $\{x\}$ is nowhere dense it would follow that X is meager.   □

Corollary 4.8 *The set $\mathbb{R}$ of real numbers is uncountable.*

Proof When endowed with the Euclidean metric the space $\mathbb{R}$ is complete and has no isolated points.   □

4.4.4   Completion of a Metric Space

If X is a metric space that is not complete, then it is because there exist Cauchy sequences in it that fail to converge in X. Quite often though the space X is seen to be embedded inside a larger metric space that is complete. This is the situation for instance for $\mathbb{Q}$ as a metric subspace of $\mathbb{R}$ with the Euclidean metric. There are however other situation where it is not a-priori clear that the metric space in question embeds in a complete one. For instance, the space $C(I, \mathbb{R})$ with the metric obtained from the L_∞ norm is complete but is incomplete with respect to the L_p norm for $1 \le p < \infty$. In that case it is not evident what an ambient complete metric space looks like.

It is natural to ask if any metric space can be completed. That is, whether every metric space embeds in a potentially (much) larger complete metric space. Of course, one would have to set some expectations from such a completion, primarily that it is not too wasteful. In more detail, we are willing to accept that the completion may need to be larger than the original space, but we do not want it to be any larger than it must be. The following definition formalizes this expectation.

Definition 4.16 Let X be a metric space. A complete metric space Y together with an isometry $\iota : X \to Y$ is said to be a *metric completion* (or simply a *completion*) of X if $\iota(X)$ is dense in Y. It is common to speak of Y as a completion of X, leaving $\iota : X \to Y$ implicit, and it is also common to write $\hat{X}$ for a completion.

Example 4.20 Clearly, the inclusion $\iota : \mathbb{Q} \to \mathbb{R}$ expresses $\mathbb{R}$ as a completion of $\mathbb{Q}$, when both sets are endowed with the metric $d(x, y) = |x - y|$.

We record the following convenient and important property of the completion.

Theorem 4.11 *Let $f : X \to Y$ be a uniformly continuous function between metric spaces, with Y complete. If $\iota : X \to \hat{X}$ is a completion, then there exists a unique uniformly continuous function $F : \hat{X} \to Y$ that extends f, i.e., such that $f = F \circ \iota$.*

Proof Apply Theorem 4.8 to the image $\iota(X)$ which is by definition dense in X, so its closure is X.

Remark 4.6 The ability to extend a uniformly continuous function to a completion of the domain does not hold for arbitrary continuous functions. For instance, take the set $S = \{1/n \mid n \in \mathbb{N}\}$ as a metric subspace of $\mathbb{R}$ under the Euclidean metric. Topologically, it is a discrete space and thus any function $f : S \to Y$, to any metric space Y, is continuous. Such a function merely corresponds to a sequence in Y. Now, let $\hat{S} = S \cup \{0\}$ and notice that with the inclusion function $\iota : S \to \hat{S}$ we obtain a completion of S. The space $\hat{S}$ is not discrete and a continuous function $F : \hat{S} \to Y$ corresponds to a sequence in Y together with a limit point of it. In other words, while continuous functions $f : S \to Y$ correspond to arbitrary sequences in Y, continuous functions $F : \hat{S} \to Y$ correspond to convergent sequences in Y. Clearly, there are plenty of examples of complete metric spaces Y where not every sequence converges, and thus not every continuous function $f : S \to Y$ extends to a continuous function $F : \hat{S} \to Y$.

We pose and answer two natural questions about completions of a metric space X: does a completion always exist and if so, is it unique. Regarding uniqueness, suppose that $\iota : X \to Y$ is a completion. Given any bijection $g : Y \to Y'$, with Y' an arbitrary set, the metric structure of Y can be transferred to a metric structure on Y' by defining $d : Y' \times Y' \to \mathbb{R}_+$ by $d(y_1', y_2') = d(g^{-1}(y_1'), g^{-1}(y_2'))$, and then g is a global isometry. Informally, there is no essential difference between Y and Y', and thus it is expected that Y' can also serve as a completion for X. This is indeed the case, as the reader can check that the composition $g \circ \iota : X \to Y'$ is a completion. It thus follows that a completion is never unique (except when it is empty), but, more importantly, we are guided to ask a more refined question: given two completions $\iota : X \to Y$ and $\iota' : X \to Y'$ for the same metric space X, are Y and Y' isometric? The affirmative answer is embodied in the following result.

Theorem 4.12 *Let X be a metric space. If $\iota : X \to Y$ and $\iota' : X \to Y'$ are completions of X, then there exists a global isometry $\rho : Y \to Y'$.*

Proof Since $\iota' : X \to Y'$ is an isometry, it is in particular uniformly continuous and thus ι' extends uniquely to a uniformly continuous function $\rho : Y \to Y'$ satisfying $\rho \circ \iota = \iota'$. Reversing the roles of ι and ι', the same argument yields a uniformly continuous function $\rho' : Y' \to Y$ with $\rho' \circ \iota' = \iota$. Let us now consider $\rho' \circ \rho : Y \to Y$, for which we have that $\rho' \circ \rho \circ \iota = \rho' \circ \iota' = \iota$. Thus, $\rho' \circ \rho : Y \to Y$ is a uniformly continuous function that extends ι. However, quite trivially, the identity function $\text{id}_Y : Y \to Y$ satisfies $\text{id}_Y \circ \iota = \iota$, namely it too is a uniformly continuous function extending ι. But, such an extension was proven above to be unique, so we conclude that $\rho' \circ \rho = \text{id}_Y$. A similar argument reveals that $\rho \circ \rho' = \text{id}_{Y'}$, and thus that ρ is bijective and ρ' is its inverse. Finally, due to the continuity of the distance function, for all $y_1, y_2 \in Y$, choosing sequences $\{a_m\}_{m \geq 1}$ and $\{b_m\}_{m \geq 1}$ with $\iota(a_m) \to y_1$ and $\iota(b_m) \to y_2$,

$$d_Y(y_1, y_2) = \lim_{m \to \infty} d_Y(\iota(a_m), \iota(b_m)) = \lim_{m \to \infty} d_X(a_m, b_m)$$

while

$$d_{Y'}(\rho(y_1), \rho(y_2)) = \lim_{m \to \infty} d_{Y'}((\rho \circ \iota)(a_m), (\rho \circ \iota)(b_m))$$

$$= \lim_{m \to \infty} d_{Y'}(\iota'(a_m), \iota'(b_m)) = \lim_{m \to \infty} d_X(a_m, b_m).$$

In short then, $d_Y(y_1, y_2) = d_{Y'}(\rho(y_1), \rho(y_2))$, namely ρ is an isometry, and, as seen above, a global one. $\square$

The result above shows that all completions of a metric space X are isometric and thus that the completion is *unique up to an isomorphism*. We cannot hope for anything more than that.

Having settled the uniqueness question we turn to the existence of a completion of an arbitrary metric space X. Since a completion is only unique up to an isometry, and thus if one (non-empty) completion of X exists, then infinitely many different completions of X exist, it should not be surprising that there are in the literature different constructions of embeddings $\iota : X \to \hat{X}$, yielding different (yet isometric) completions. The one we present is modelled after Cantor's construction of the real numbers from the rationals.

We fix a metric space X and construct a completion for it in a series of steps, leaving it to the reader to fill in the proofs. If X is not complete, then there exists a Cauchy sequence $\{x_m\}_{m \geq 1}$ that fails to have a limit in X. Intuitively, completing X entails adjoining to X formal limits for all such Cauchy sequences. If we denote the formal limit of $\{x_m\}_{m \geq 1}$ by $[\{x_m\}]$, then we are led to define $\hat{X}$ as the union

$$X \cup \{[\{x_m\}] \mid \{x_m\}_{m \geq 1} \text{ is a Cauchy sequence in } X \text{ that fails to converge}\}.$$

However, two different Cauchy sequences $\{x_m\}_{m \geq 1}$ and $\{y_m\}_{m \geq 1}$ may need to converge to the same point (for instance, there are many different sequences of rationals that converge to, e.g., $\sqrt{2}$), and thus an equivalence relation needs to be introduced. First, though, a simplifying observation: we may identify an element $x \in X$ with the constant sequence $\{x_m = x\}_{m \geq 1}$ (which is clearly Cauchy). Thus, instead of the union of X with the formal limits of limit-less Cauchy sequences, we expect a completion of X to be formed as the set

$$\hat{X} = \{[\{x_m\}] \mid \{x_m\}_{m \geq 1} \text{ is a Cauchy sequence in } X\}$$

of equivalence classes for a suitable equivalence relation on the set of all Cauchy sequences in X. The following steps realize this plan.

1. Let $\tilde{X}$ be the set of all Cauchy sequences in X. Define $\tilde{d} : \tilde{X} \times \tilde{X} \to \mathbb{R}_+$ by

$$\tilde{d}(\{x_m\}, \{y_m\}) = \lim_{m \to \infty} d(x_m, y_m).$$

2. Prove that the limit defining $\tilde{d}((x_m), (y_m))$ always exists. (Hint: The metric function $d : X \times X \to \mathbb{R}_+$ is uniformly continuous and $\mathbb{R}$ is complete.)
3. Prove that $(\tilde{X}, \tilde{d})$ is a semimetric space.

4. Recall Exercise 4.4 and let $\hat{X}$ be the metric space thus obtained from $\tilde{X}$.
5. Prove that $(\hat{X}, d)$ is complete. (Hint: avoid the temptation of taking the diagonal of a Cauchy sequence of Cauchy sequences. Instead, use the Cauchy condition whenever you can.)
6. Note that for all $x \in X$ the constant sequence $\{x_m = x\}_{m \geq 1}$ is Cauchy, thereby defining a function $\iota : X \to \hat{X}$. Prove that it is an isometry.
7. Prove that $\iota(X)$ is dense in $\hat{X}$.

Remark 4.7 Any metric space X has a completion, and thus, if non-empty, X has infinitely many different (yet isometric) completions. It is important to realize that there is no distinguished completion that deserves to be called *the* completion of X. Any particular construction of a completion for X is just that: one model among many. Theoretically, since isometric metric spaces are essentially the same, it does not matter which completion model is used; they are all isometric. However, some models may offer syntactic advantages over others.

Exercises

Exercise 4.47 Prove the very last estimate in the proof of Theorem 4.9 and use it to show the relevant sequence is Cauchy.

Exercise 4.48 Demonstrate that the Banach Fixed Point Theorem may fail for a contraction $f : X \to X$ if X is not complete.

Exercise 4.49 Show that a contraction $f : X \to X$ on a metric space is necessarily non-expanding. Demonstrate that the converse may fail.

Exercise 4.50 Show that the Banach Fixed Point Theorem may fail for a non-expanding function $f : X \to X$; a fixed point need not exist nor, if one exists, need it be unique.

Exercise 4.51 Suppose $f : X \to Y$ is a bijective uniformly continuous function between metric spaces. Show that f^{-1} need not be uniformly continuous.

Exercise 4.52 Prove that a subsequence of a Cauchy sequence is Cauchy.

Exercise 4.53 Prove that a Cauchy sequence in a metric space X converges if, and only if, it has a convergent subsequence.

Exercise 4.54 Suppose that X is a non-empty metric space with the property that any contraction $f : C \to C$ on a non-empty closed subset $C \subseteq X$ has a fixed point. We shall establish a sort of converse to the Banach Fixed Point Theorem, namely that in that case X is complete. Let thus $\{s_m\}_{m \geq 1}$ be a Cauchy sequence in X and the goal is to show that it converges. By the previous exercise it suffices to find a convergent subsequence of $\{s_m\}_{m \geq 1}$, a fact that will be used repeatedly. Let $C = \{s_m \mid m \geq 1\}$ be the set of values of the sequence.

1. Establish the result in case C is finite. Proceed under the assumption that C is infinite.

2. Argue that we may assume $s_n \neq s_m$ for all $m \neq n$.
3. Consider the function $f : C \to C$ given by $f(s_m) = s_{m+1}$. It need not be a contraction but show that it can be arranged to be a contraction by passing ahead of time to a suitable subsequence.
4. From the assumption on X and the details above conclude that X is not closed.
5. The closure of C contains an extra point not in C. Show that it is the limit of the sequence.

Exercise 4.55 In a metric space we say that a sequence $\{B_{\varepsilon_m}(x_m)\}_{m \geq 1}$ of balls has *shrinking radii* if $B_{\varepsilon_{m+1}}(x_{m+1}) \subseteq B_{\varepsilon_m}(x_m)$ and that it has *rapidly shrinking radii* if $\overline{B_{\varepsilon_{m+1}}(x_{m+1})} \subseteq B_{\varepsilon_m}(x_m)$.

Consider $\mathbb{R}$ with the Euclidean metric and the sequence $\{B_{1/n}(1/n)\}_{n \geq 1}$ of balls. Verify that this is a sequence of balls with vanishing radii yet the intersection of these balls is empty.

Consider now the space $X = \mathbb{R} \setminus \{0\}$ as a metric subspace of $\mathbb{R}$ and the sequence of balls $\{B_{1/n^2}(1/n)\}$. Verify that this is a sequence of balls with rapidly shrinking radii yet the intersection of these balls is empty.

Exercise 4.56 See previous exercise for terminology. Let X be a metric space and $\{B_{\varepsilon_m}(x_m)\}_{m \geq 1}$ a sequence of balls in it with $B = \bigcap_{m \geq 1} B_{\varepsilon_m}(x_m)$ the intersection of the balls.

1. Prove that if the sequence of balls has shrinking radii, then B contains at most one point.
2. Let $\{s_m\}_{m \geq 1}$ be an arbitrary sequence with $s_m \in B_{\varepsilon_m}(x_m)$. Show that if the sequence of balls has shrinking radii, then the sequence $\{s_m\}_{m \geq 1}$ is Cauchy.
3. Prove that if X is complete and the radii shrink rapidly, then $B \neq \emptyset$ and thus consists of a single point.

Exercise 4.57 For $k \geq 1$ and $I = [a, b]$ a non-degenerate closed interval, show that the metric space $C^k(I, \mathbb{R})$ of functions with a continuous k-th derivative with the metric induced by the L_∞ norm is not complete.

Exercise 4.58 Prove that a subset of a nowhere dense set is nowhere dense and that a finite union of nowhere dense sets is nowhere dense.

Exercise 4.59 Prove that a subset of a meager set is meager and that a countable union of meager sets is meager. Is an arbitrary union of meager sets necessarily meager?

Exercise 4.60 Show that the closure of a nowhere dense subset remains nowhere dense. Is the closure of a meager set necessarily meager?

Exercise 4.61 Show that $\mathbb{R} \setminus \mathbb{Q}$ is not a meager subset of $\mathbb{R}$ with respect to the Euclidean topology.

Exercise 4.62 Show that it is possible for an uncountable set to be meager. Can a finite set fail to be meager?

Exercise 4.63 Consider the topological space $\mathbb{Q}$ as a subspace of $\mathbb{R}$ endowed with the Euclidean topology. Prove that $\mathbb{Q}$ is metrizable but not by any complete metric.

Exercise 4.64 Let $\{f_m : \mathbb{R} \to \mathbb{R}\}_{m \geq 1}$ be a countable family of continuous functions (with respect to the Euclidean topology on $\mathbb{R}$). Suppose that for every point $x \in \mathbb{R}$ there exist distinct functions f_k, f_m such that $f_k(x) = f_m(x)$. Prove that there exists an interval (a, b), with $a < b$, and distinct functions f_k, f_m such that $f_k(x) = f_m(x)$ for all $x \in (a, b)$.

Exercise 4.65 Let X and Y be metric spaces and consider respective completions $\iota_X : X \to \hat{X}$ and $\iota_Y : Y \to \hat{Y}$. Recall that the product of metric spaces can be metrized by either the Euclidean product metric or the additive product metric. Check, in each case, if the function $\iota : X \times Y \to \hat{X} \times \hat{Y}$, given by $\iota(x, y) = (\iota_X(x), \iota_Y(y))$, is a completion of $X \times Y$.

Exercise 4.66 Continuing Remark 4.6 give an explicit continuous function $f : S \to \mathbb{R}$ that fails to extend continuously to $\hat{S}$. By Theorem 4.11 any uniformly continuous function $f : S \to \mathbb{R}$ does extend continuously to $\hat{S}$. In this particular case there is a simple reason embodied in recognizing what it means for a function $f : S \to \mathbb{R}$ to be uniformly continuous. Clarify what that reason is.

Exercise 4.67 In Theorem 4.11 an extension $\hat{f}$ of a uniformly continuous function f is constructed. Prove that if f is non-expanding, then so is its extension $\hat{f}$.

4.5 Compactness and Boundedness

Compactness is a topological property and since a metric space is naturally endowed with a topology, compactness is meaningful in any metric space. The notion of completeness on the other hand is a metric property with no topological counterpart. Nonetheless, completeness and compactness are not independent properties. The space $\mathbb{R}$ with the Euclidean metric is complete but not compact, its subset $[0, 1]$ is both complete and compact, and $(0, 1)$ is neither complete nor compact. The final possibility though, as we show below, is implied; if a metric space is compact, then it must be complete. In this section we show the remarkable fact that compactness in a metric space is equivalent to completeness together with a property known as total boundedness. The latter is a purely metric property and thus the result shows that the topological property of compactness is equivalent (for metric spaces, of course) to the conjunction of the purely metric conditions of completeness and total boundedness.

The reader is familiar with the notion of bounded sets in the Euclidean space $\mathbb{R}^n$. In a general metric space there are two notions of boundedness. These two notions coincide in the Euclidean spaces. Recall that for a subset $S \subseteq X$ of a metric space X, its diameter is given by $\mathrm{diam}(S) = \sup_{x,y \in S}\{d(x, y)\}$, which may be ∞.

Definition 4.17 A subset S in a metric space X is *bounded* if its diameter is finite. A subset S is *totally bounded* if no two points in it are infinite distance apart and for every $\varepsilon > 0$ one can cover S by finitely many sets of diameter smaller than ε.

The reader is invited to show that every totally bounded set is bounded, but that the converse need not hold. It is also evident that if $S' \subseteq S \subseteq X$ and S is bounded (respectively totally bounded), then S' is bounded (respectively totally bounded). Since X is a subset of itself, we may speak of the metric space X as being bounded or totally bounded.

The following result, whose proof is left for the reader, has an important corollary that is pivotal for the proof of the main theorem below.

Proposition 4.3 *The equality* $\mathrm{diam}(S) = \mathrm{diam}(\overline{S})$ *holds for all subsets S of a metric space X.*

Corollary 4.9 *Let X be a totally bounded metric space, $\mathcal{U} = \{U_i\}_{i \in I}$ an open covering of X, and $\varepsilon > 0$. If X can not be covered by finitely many elements from $\mathcal{U}$, then there exists a non-empty closed set $F \subseteq X$ with $\mathrm{diam}(F) < \varepsilon$ that cannot be covered by finitely many elements from $\mathcal{U}$.*

Proof Since X is totally bounded it can be covered by finitely many subsets of diameter smaller than ε. By the preceding proposition, we may take the closures of these sets without altering their diameters and thus we may assume they are all closed. If each of these sets was covered by finitely many elements from $\mathcal{U}$, then (by taking the union) X itself would be covered by finitely many elements from $\mathcal{U}$, contrary to the assumption on X. Thus at least one of these closed sets is the desired one. $\qquad \square$

Theorem 4.13 *For a metric space X the following conditions are equivalent.*

1. *X is compact.*
2. *X is sequentially compact, i.e., every sequence $\{x_m\}_{m \geq 1}$ admits a convergent subsequence $\{x_{m_k}\}_{k \geq 1}$.*
3. *X is complete and totally bounded.*

Proof We prove first that if X is compact, then X is sequentially compact, and so let $\{x_m\}_{m \geq 1}$ be a sequence in X. If the set $S = \{x_m \mid m \in \mathbb{N}\}$ is finite, then some of its elements must repeat infinitely often in the given sequence. In other words, the sequence contains a constant subsequence, which certainly converges. Otherwise S is an infinite subset in a compact topological space and thus, by Theorem 3.17, it admits a cluster point $x_0 \in X$ and by Corollary 4.2 this cluster point is the limit of a sequence of elements from S.

Next we show that if X is sequentially compact, then it is complete and totally bounded. Suppose then that $\{x_m\}_{m \geq 1}$ is a Cauchy sequence in X. By sequential compactness we may extract a subsequence $\{x_{m_k}\}_{k \geq 1}$ converging to a point $x_0 \in X$. But convergence of a subsequence of a Cauchy sequence forces the sequence itself to converge (to the same limit). So every Cauchy sequence converges, namely X is complete.

To see that X is also totally bounded, assume for the contrary that it is not. Then there exists an $\varepsilon_0 > 0$ such that

$$X \neq \bigcup_{k=1}^{m} B_{\varepsilon_0}(x_k)$$

for any choice of points $x_1, \ldots, x_m \in X$, $m \geq 1$. We now construct the following sequence. Since $X \neq \emptyset$ (otherwise it is totally bounded), choose an arbitrary element $x_1 \in X$. Suppose that we have chosen points $x_1, \ldots, x_m \in X$ with the property that $d(x_i, x_j) \geq \varepsilon_0$ for all $1 \leq i \neq j \leq m$. We can now choose an arbitrary element

$$x_{m+1} \in X \setminus \bigcup_{k=1}^{m} B_{\varepsilon_0}(k_k)$$

and the condition $d(x_i, x_j) \geq \varepsilon_0$ now holds for all $1 \leq i \neq j \leq m + 1$. In this way we obtain the infinite sequence $\{x_m\}_{m \geq 1}$ which, by sequential compactness, has a convergent subsequence. In particular, $\{x_m\}_{m \geq 1}$ has a Cauchy subsequence. But by the construction of $\{x_m\}_{m \geq 1}$, no such subsequence exists, and thus we obtain a contradiction.

Finally, we prove that if X is complete and totally bounded, then X is compact. Suppose that $\mathcal{U}$ is an open covering of X that does not admit a finite sub-covering. Our aim will be to construct a non-convergent Cauchy sequence and thus obtain a contradiction.

Let us call a non-empty subset $F \subseteq X$ a *problematic* set if it is closed and cannot be covered by finitely many elements from $\mathcal{U}$ (so X itself is problematic). By Corollary 4.9 (applied to X with $\varepsilon = 1$) there exists a problematic set F_1 with $\text{diam}(F_1) < 1$. Applying the same corollary again (to F_1 and with $\varepsilon = 1/2$) there exists a problematic set $F_2 \subseteq F_1$ with $\text{diam}(F_2) < 1/2$. Continuing inductively in this manner we obtain a sequence $F_1 \supseteq F_2 \supseteq F_3 \supseteq \cdots \supseteq F_m \supseteq \cdots$ of non-empty closed sets with $\text{diam}(F_m) < 1/m$, all of which are problematic. Let $x_m \in F_m$ be an arbitrary element in F_m. The sequence $\{x_m\}_{m \geq 1}$ is clearly Cauchy since $x_k \in F_m$ for all $k \geq m$ and thus $d(x_k, x_{k'}) < 1/m$ for all $k, k' > m$. To see that the sequence cannot converge assume that $x_k \to x_0$ for some $x_0 \in X$. Recalling that every closed set contains its limit points we see that $x_0 \in F_m$ for all $m \geq 1$ (since $x_k \in F_m$ for all $k \geq m$). Further, since $\mathcal{U}$ covers X there exists an element $U \in \mathcal{U}$ with $x_0 \in U$. Since U is open there exists a $\delta > 0$ with $B_\delta(x) \subseteq U$. Let m now be large enough so that $1/m < \delta$. We then have that $x_0 \in F_m$ and $\text{diam}(F_m) < \delta$ and thus $F_m \subseteq B_\delta(x_0) \subseteq U$. We thus found an element in $\mathcal{U}$ which covers F_m, contrary to F_m being a problematic set. It is thus shown that $\{x_m\}_{m \geq 1}$ is a non-convergent Cauchy sequence, contradicting the completeness of X. The proof is complete. $\qquad\square$

Exercises

Exercise 4.68 Exercise 3.83 establishes Weierstrass' Theorem for functions on $[a, b]$. Prove the stronger result: a continuous function $f: X \to \mathbb{R}$ on a compact metric space X attains a minimum and a maximum.

Exercise 4.69 Let (X, d) be a metric space and let $d_t : X \times X \to \mathbb{R}_+$ be given by $d_t(x, y) = d(x, y)$ if $d(x, y) < 1$ and $d_t(x, y) = 1$ otherwise. Prove that (X, d_t) is a metric space inducing the same topology X as d.

Exercise 4.70 Continuing the preceding exercise, prove that in (X, d_t) every subset is bounded while the totally bounded subsets in (X, d) and (X, d_t) coincide.

Exercise 4.71 Prove that a totally bounded subset S in a metric space X is bounded but that the converse may fail.

Exercise 4.72 Prove that if $\{x_m\}_{m \geq 1}$ is a Cauchy sequence in a metric space X and $\{x_{m_k}\}_{k \geq 1}$ is a convergent subsequence converging to x_0, then $x_m \to x_0$.

Exercise 4.73 Prove that a metric subspace S of $\mathbb{R}^n$ with the Euclidean metric is complete if, and only if, S is closed in $\mathbb{R}^n$.

Exercise 4.74 In $\mathbb{R}^n$ with the Euclidean metric, prove that a subset is bounded if, and only if, it is totally bounded. Conclude that the compact subsets are precisely the closed and bounded ones.

Exercise 4.75 Prove Proposition 4.3.

Exercise 4.76 Prove that any subset of a totally bounded metric space X is totally bounded and that a closed subset of a complete metric space is complete. Deduce that a closed subset of a compact metric space is compact.

Exercise 4.77 Prove that the unit sphere $\{x \in \mathbb{R}^n \mid \|x\| = 1\}$ is compact in $\mathbb{R}^n$ (here the norm and the topology are the Euclidean ones).

Exercise 4.78 Prove that the unit sphere $\{x \in c_0 \mid \|x\|_\infty = 1\}$ is not compact in c_0 with the induced topology (c_0 is the space of sequences, say of real numbers, converging to 0 and the norm is the ℓ_∞ norm).

Exercise 4.79 Let $f : X \to X$ be an isometry of a metric space X. Demonstrate that if X is not compact, then f need not be surjective. However, for compact X prove f is surjective. Hint: Assume, to the contrary, that $y \in X \setminus f(X)$ exists and consider iteratively applying f to y.

Exercise 4.80 Prove the analogue of the Cantor-Schröder-Bernstein Theorem for compact metric spaces: if there exist isometries $f : X \to Y$ and $g : Y \to X$, then the metric spaces X and Y are isometric, provided X and Y are compact. Hint: Use the previous exercise.

Further Reading

The adventurous reader is urged to delve into Fréchet's original thesis [4] for a fascinating glimpse into the mind of the father of the subject matter of this chapter. Of the numerous introductory level textbooks on the subject the reader may wish to consult [6] for an introduction to metric spaces very much in line with the presentation as given in this chapter, or the more advanced [7] for a more comprehensive introduction to analysis in general. The reader intrigued by Baire's Theorem will find a historical account and numerous examples of its uses in [5]

For a glimpse of the overwhelming ubiquity of metric spaces in mathematics see the Encyclopedia of Distances [2]. The encyclopedia presents an almost uncountable list of examples of metric spaces as well as various generalizations of metric spaces. Some of the generalizations (e.g., [9]) are inspired by problems in physics. Other generalizations (e.g., [3]) give rise to a unifying perspective on the relationship between metric spaces and topology (see [1]). For an investigation in the opposite direction, i.e., metric spaces with special properties, see [8].

References

1. Cook, D.S., Weiss, I.: The topology of a qauntale valued metric space. Fuzzy Sets and Systems (2020)
2. Deza, M.M., Deza, E.: Encyclopedia of Distances, 2nd edn., p. xviii+650. Springer, Heidelberg (2013)
3. Flagg, R.C.: Quantales and continuity spaces. Algebra Universalis **37**(3), 257–276 (1997)
4. Fréchet, M.R.: Sur Quelque Points du Calcul Fonctionnel. Rendic. Circ. Mat., Palermo **22**, 1–74 (1906)
5. Hawtrey Jones, S.: Applications of the Baire category theorem. Real Anal. Exch. **23**(2), 363–394 (1999)
6. O'Searcoid, M.: Elements of Abstract Analysis. Springer Undergraduate Mathematics Series, p. xii+298. Springer-Verlag London, Ltd., London (2002)
7. O'Searcoid, M.: Metric Spaces. Springer Undergraduate Mathematics Series, p. xx+304. Springer, London (2007)
8. Rammal, R., Toulouse, G., Virasoro, M.A.: Ultrametricity for physicists. Rev. Mod. Phys. **58**(3), 765–788 (1986)
9. Skakala, J., Visser, M.: Bi-metric pseudo-Finslerian spacetimes. J. Geom. Phys. **61**(8), 1396–1400 (2011)

Chapter 5
A Non-classical View of the Classical Spaces

The classical spaces ℓ_p of sequences and L_p of functions are central to analysis. Defining the sequence spaces and studying their basic properties is elementary but even just properly defining the function spaces is much more delicate and traditionally requires a dose of measure theory. Measure theory is itself a tool of central importance but it is technically involved and this text will not do it justice if attempting to develop it. Instead, the approach we take in this chapter builds upon the metric machinery developed thus far. By so doing we are able to quickly obtain precise definitions and results at the cost of not having an explicit model. For that reason we take the time to present a conceptual framework in some detail. In particular, the intention is not to survey the main properties of the classical spaces but rather to show their precise relationship to familiar finite-dimensional spaces. The pros and cons of this approach are highlighted throughout the chapter.

Section 5.1 introduces metric and topological notions and emphasizes the normed spaces notation. Section 5.2 concentrates on finite-dimensional spaces with the two-fold aim of proving that topologically all norms are equivalent in that case and to highlight the building blocks of the classical spaces. Section 5.3 directly builds upon the finite-dimensional spaces to describe two countable-dimensional spaces and to formalize their construction as colimits of finite-dimensional spaces. Section 5.4 introduces the classical spaces ℓ_p and L_p, utilizing the material of the previous section in order to emphasize similarity and differences between them. Finally, Sect. 5.5 addresses the separability of the classical spaces and the structure of linear functionals on them.

5.1 Metric and Topological Considerations

We introduce some metric and topological aspects of general interest, contextualized to normed spaces.

© Springer Nature Switzerland AG 2021
C. Alabiso and I. Weiss, *A Primer on Hilbert Space Theory*, UNITEXT for Physics,
https://doi.org/10.1007/978-3-030-67417-5_5

5.1.1 The Induced Topology and Equivalent Norms

Let V be a normed linear space. Recall that the induced topology consists of all sets $U \subseteq V$ such that for all $x \in U$ there exists $\varepsilon > 0$ with

$$B_\varepsilon(x) \subseteq U$$

where

$$B_\varepsilon(x) = \{y \in V \mid \|x - y\| < \varepsilon\}$$

is the open ball of radius ε with center x. The homogeneity of $\| - \|$ implies the element-wise scaling of open balls: for $\alpha > 0$

$$\alpha \cdot B_\varepsilon(x) = B_{\alpha\varepsilon}(x)$$

where the meaning of the set on the left-hand side is

$$\alpha \cdot B_\varepsilon(x) = \{\alpha \cdot y \mid y \in B_\varepsilon(x)\}.$$

In words: scaling an ε-ball by α results in a ball about the same center whose radius got scaled by α. The translation invariance of the metric $d(x, y) = \|x - y\|$ induced by the norm manifests itself in a translation property of open balls, i.e.,

$$B_\varepsilon(x) + w = B_\varepsilon(x + w)$$

for any vector $w \in V$ (with the evident meaning of the set on the left-hand side). In words: translating an open ball by w results in another open ball of the same radius whose center got translated by w.

Since the open balls determine the topology, i.e., a set is open precisely when it is a union of open balls, translation invariance implies that a set is open precisely when it is a union of translates of open balls with center 0. By the scaling property of open balls the collection of open balls centered at 0 is precisely the collection of all positive scalings of the unit open ball $B_1(0)$. It is worth stressing out thus that the entire topology induced by the norm $\| - \|$ is determined by just a single open ball. Let us phrase this property more precisely.

Theorem 5.1 *Let V be a normed space. Then a subset $U \subseteq V$ is open if, and only if, there exists a family of vectors $\{w_x\}_{x \in U}$ and positive scalars $\{\varepsilon_x\}_{x \in U}$ such that*

$$U = \bigcup_{x \in U} \varepsilon_x \cdot B + w_x$$

where $B = \{y \in V \mid \|y\| < 1\}$ is the open unit ball in V.

Proof Suppose that U be can be written as such a union. If each of the sets $\varepsilon_x \cdot B + w_x$ can be shown to be open, then U is open as the union of open sets. But $\varepsilon_x \cdot B + w_x =$

$B_{\varepsilon_x}(w_x)$ which is an open ball and thus certainly an open set. In the other direction, suppose U is open. Thus, by definition, for each $x \in U$ there exists $\varepsilon_x > 0$ such that

$$x \in B_{\varepsilon_x}(x) \subseteq U$$

and so

$$U = \bigcup_{x \in U} B_{\varepsilon_x}(x).$$

Noting that $B_{\varepsilon_x}(x) = B_{\varepsilon_x}(0) + x = \varepsilon_x \cdot B + x$ we see that we can take $w_x = x$ to complete the proof. $\square$

A sequence $\{x_n\}_{n \geq 1}$ of vectors in a normed space V converges to $x \in V$ precisely when

$$\lim_{n \to \infty} \|x_n - x\| = 0$$

and the limit x, if it exists, is unique. The sequence is Cauchy if for all $\varepsilon > 0$ there exists $n \geq 1$ such that

$$\|x_m - x_k\| < \varepsilon$$

for all $m, k > n$. Every convergent sequence is Cauchy and if V is complete, then the converse holds as well.

A given linear space may support more than one norm and so it becomes interesting to compare the induced topologies. In particular, we seek conditions on the norms that will tell us that the topologies they induce coincide.

Proposition 5.1 *Let $\| - \|_1$ and $\| - \|_2$ be two norms on V.*

1. *If there exists a constant $c > 0$ with $\|x\|_1 = c\|x\|_2$ for all $x \in V$, then the two norms induce the same topology on V.*
2. *If $\|x\|_1 \leq \|x\|_2$ for all $x \in V$, then the topology induced by $\| - \|_1$ is coarser than the one induced by $\| - \|_2$. In other words, if a set $U \subseteq V$ is open with respect to $\| - \|_1$, then it is open with respect to $\| - \|_2$.*

Proof

1. By the stated condition

$$\{x \in V \mid \|x\|_1 < 1\} = \{x \in V \mid \|x\|_2 < 1/c\} = \frac{1}{c}\{x \in V \mid \|x\|_2 < 1\}$$

namely the respective unit open balls are each other's scalings. Therefore, the collection of scalings of one ball agrees with the scalings of the other ball. By Theorem 5.1 the two norms generate the same topology.
2. If U is open with respect to $\| - \|_1$ and $x \in U$, then $B_\varepsilon(x) \subseteq U$ for some $\varepsilon > 0$ where the open ball is with respect to $\| - \|_1$. But the condition implies that the open ball $B_\varepsilon(x)$ with respect to $\| - \|_2$ is contained in the same ball $B_\varepsilon(x)$ with respect to $\| - \|_1$, and thus U is also open with respect to $\| - \|_2$.

$\square$

Definition 5.1 Two norms $\| - \|_1$ and $\| - \|_2$ on the same linear space V are *equivalent norms if there exist constants $\alpha, \beta > 0$ such that*

$$\alpha \|x\|_1 \leq \|x\|_2 \leq \beta \|x\|_1$$

for all $x \in V$.

Proposition 5.2 *Two norms on a linear space V are equivalent if, and only if, they determine the same topology.*

Proof Repeatedly using Proposition 5.1 we see that $\| - \|_1$ and $\alpha \| - \|_1$ induce the same topology while the first inequality implies that $\alpha \| - \|_1$ induces a coarser topology than $\| - \|_2$, so the topology induced by $\| - \|_2$ is coarser. Further, the second inequality implies that $\| - \|_2$ induces a coarser topology than $\beta \| - \|_1$ while $\beta \| - \|_1$ and $\| - \|_1$ induce the same topology, so the topology induced by $\| - \|_1$ is coarser. The two topologies must then coincide. The converse is left as an exercise. $\square$

Proposition 5.3 *Equivalence of norms is an equivalence relation on the set of all norms on a given linear space V.*

Proof The proof can be given directly in terms of Definition 5.1 by finding suitable constants to establish reflexivity, symmetry, and transitivity. A more conceptual alternative uses Proposition 5.2, e.g., to show transitivity, if $\| - \|_1$ is equivalent to $\| - \|_2$, then they determine the same topology τ. If $\| - \|_2$ is also equivalent to $\| - \|_3$, then they determine the same topology ρ. The two topologies are determined by the common norm $\| - \|_2$ and thus $\tau = \rho$. As $\| - \|_1$ and $\| - \|_3$ determine the same topology, they are equivalent norms. $\square$

For the comparison of different norms on a given linear space it is convenient to know that the norm function $\| - \| \colon V \to \mathbb{R}$ is continuous with respect to itself (and the Euclidean topology on $\mathbb{R}$). This fact is part of the following general observation stating that pretty much all the structure maps in a normed space are continuous.

Theorem 5.2 *Let V be a normed space endowed with its induced metric structure.*

1. *The norm function $\| - \| \colon V \to \mathbb{R}$ is non-expanding with respect to the Euclidean metric on $\mathbb{R}$.*
2. *For each $w \in V$ the translation function $x \mapsto x + w$ is an isometry.*
3. *For each scalar $\alpha \in K$ the function $A_\alpha \colon V \to V$ given by $A_\alpha(x) = \alpha x$ satisfies $\|A_\alpha(x)\| \leq |\alpha| \cdot \|x\|$ for all $x \in V$.*

In particular, all of the functions are uniformly continuous.

Proof The first claim is the well-known inequality

$$|\|x\| - \|y\|| \leq \|x - y\|$$

called the reverse triangle inequality. Its proof is immediate from the triangle inequality. The second claim is a tautology and the third claim is simply the homogeneity property of the norm:

$$\|A_\alpha(x)\| = \|\alpha x\| = |\alpha| \|x\|.$$

$\square$

5.1.2 Nets and Dense Sets

The size of a space can be measured in many different ways, one of which is the size of a dense subset, an essentially topological notion. A metric notion of size is given by nets, which we introduce and compare to density.

Definition 5.2 A *net* in a normed space V is a set $X \subseteq V$ of vectors for which there exists $\varepsilon > 0$ with

$$\|x_1 - x_2\| \geq \varepsilon$$

for all distinct $x_1, x_2 \in X$.

Recall that a set $Y \subseteq V$ is dense if $\overline{Y} = V$ where the closure is taken with respect to the induced topology. In a metric space density of Y is equivalent to every vector $x \in V$ in the ambient space being a limit of elements from Y. Thus, in a normed space, Y is dense precisely when for all $x \in V$ and $\varepsilon > 0$

$$\|x - y\| < \varepsilon$$

for at least one vector $y \in Y$.

Intuitively, a net is spacious and thus a large one implies a large ambient space. A dense subset fills up most of the ambient space and thus a small one implies a small ambient space. This intuition is made formal as follows.

Proposition 5.4 *In a normed space V the cardinality of a net is not greater than the cardinality of a dense set.*

Proof Let $X \subseteq V$ be a net, so there exists $\varepsilon > 0$ with

$$\|x_1 - x_2\| \geq \varepsilon$$

for all distinct $x_1, x_2 \in X$. Let $Y \subseteq V$ be a dense subset and pick any $x \in X$. There is then at least one $y \in Y$ with

$$\|x - y\| < \frac{\varepsilon}{2}$$

and choosing any such vector defines a function $f \colon X \to Y$. The claim about the cardinalities of the sets will follow by showing that f is injective so suppose $f(x_1) = f(x_2) = y$ for some $x_1, x_2 \in X$. Then

$$\|x_1 - f(x_1)\| \le \frac{\varepsilon}{2} \quad \text{and} \quad \|x_2 - f(x_2)\| < \frac{\varepsilon}{2}$$

by the construction of f. By the triangle inequality

$$\|x_1 - x_2\| = \|x_1 - y + y - x_2\| \le \|x_1 - y\| + \|y - x_2\| < \frac{\varepsilon}{2} + \frac{\varepsilon}{2} = \varepsilon$$

while if $x_1 \ne x_2$ then

$$\|x_1 - x_2\| \ge \varepsilon$$

so we must have that $x_1 = x_2$, as needed.   □

5.1.3   Point-to-Set Distance and Closed Sets

The closure of a set $S \subseteq V$ in a normed space is its closure in the induced topology, i.e., the smallest closed set containing S. We take this opportunity to introduce another perspective on closed sets in a metric space.

Given a point $x \in X$ and a subset $S \subseteq X$ in a metric space it is natural to define the point-to-set distance to be the infimum of all distances from x to each element in the subset S. By slight abuse of notation we extend the distance notation to include point-to-set scenarios.

Definition 5.3 The *point-to-set distance* from a point $x \in X$ to a subset $S \subseteq X$ of a metric space X is

$$d(x, S) = \inf_{s \in S} d(x, s).$$

In low-dimensional spaces it is fairly simple to visualize point-to-set distances. For instance, in $\mathbb{R}^2$ with the Euclidean metric one can easily draw the relevant open ball $B_\varepsilon(0)$ and witness that $d(t, B_\varepsilon(0)) = d(0, t) - \varepsilon$ for any t outside of the ball $B_\varepsilon(0)$. Note in particular that for any t on the boundary of the ball, i.e., $d(0, t) = \varepsilon$, the point-to-set distance satisfies $d(t, B_\varepsilon(0)) = 0$ even though $t \notin B_\varepsilon(0)$. Intuitively, such a point lying at 0 distance from the set S is as close as possible to S. Topologically, this characterizes the closure of S under the topology induced by the metric.

Lemma 5.1 *Let X be a metric space, $x \in X$ a point, and $S \subseteq X$ a subset. The condition $d(x, S) = 0$ is equivalent to $x \in \overline{S}$. In words: x has zero distance to S if, and only if, x belongs to the closure of S.*

Proof Suppose first that $x \in \overline{S}$ and we show that $d(x, S) = 0$. Consider an arbitrary $\varepsilon > 0$ and the open ball $B_\varepsilon(x)$. Since it is an open set it intersects S so there is a point $s \in S$ with $d(x, s) < \varepsilon$. But then

$$d(x, S) \le d(x, s) < \varepsilon$$

and as $\varepsilon > 0$ was arbitrary it follows that $d(x, S) = 0$. In the other direction, suppose that $d(x, S) = 0$ and consider an arbitrary open set with $x \in U$. In order to show that $x \in \overline{S}$ we must show that $U \cap S \neq \emptyset$. Since U is open there is $\varepsilon > 0$ with $B_\varepsilon(x) \subseteq U$. But then, since

$$\varepsilon > 0 = d(x, S) = \inf_{s \in S} d(x, s)$$

there is $s \in S$ with $\varepsilon \geq d(x, s)$. In other words, s belongs to the open ball $B_\varepsilon(x)$ and so we found a point common to it and S, as was the aim.   □

Corollary 5.1 *A subset $C \subseteq X$ in a metric space is closed if, and only if,*

$$x \in C \iff d(x, C) = 0$$

for all $x \in X$.

Directly in terms of the norm, the closure of S in V is the set of all vectors $x \in V$ that are zero distance from S, i.e.,

$$\overline{S} = \{x \in V \mid d(x, S) = 0\}$$

where

$$d(x, S) = \inf_{s \in S} \|x - s\|.$$

We illustrate this characterization of closedness in terms of distance as follows. Note that a sequence x_n converges in V to x implies that $d(x, S) = 0$ where $S = \{x_n \mid n \in \mathbb{N}\}$ is the set of values of the sequence. Conversely, if $d(x, S) = 0$, then there exists a sequence in S converging to x.

5.1.4 The Operator Norm

The norm $\|x\|$ in a normed space represents the length of the vector x. Given an operator A between normed spaces it is natural to study the way the operator affects the norm, that is how $\|x\|$ and $\|Ax\|$ are related. If the operator does not alter the norm of any vector by more than some constant factor, then it is said to be bounded. This class of operators is studied more extensively in Chap. 6. Linear operators will typically be denoted by A or B and their values on vectors x by $A(x)$ or simply Ax.

Definition 5.4 An operator $A : V \to W$ between normed spaces is called a *bounded operator* if there exists a real number M such that

$$\|Ax\| \leq M\|x\|$$

for all $x \in V$.

Equivalently, an operator A is bounded if it maps bounded sets in the domain to bounded sets in the codomain (where a set S of vectors in a normed space is bounded if there exists a positive M such that $\|s\| \leq M$ for all $s \in S$).

Example 5.1 A simple example was already encountered in Theorem 5.2; the function $A_\alpha \colon V \to V$, clearly a linear operator, is shown there to be bounded. For an unbounded example consider the space c_{00} with the ℓ_p norm $\| - \|_p$ for any $1 \leq p \leq \infty$. Let $A \colon c_{00} \to c_{00}$ be given by

$$A(\xi_1, \ldots, \xi_n, \ldots) = (\xi_1, 2\xi_2, 3\xi_3, \ldots, n\xi_n, \ldots)$$

and note that it is a linear operator. It is seen to be unbounded by considering the standard basis vectors since $\|A(e_n)\|_p = n$ while $\|e_n\| = 1$.

The situation in the finite dimensional case is very convenient.

Proposition 5.5 *Any linear operator $A \colon V \to W$ between normed spaces where V is finite-dimensional is bounded.*

The proof is deferred to the exercises.

The interplay between the linear structure and the topology in a normed space implies a strong unification of concepts.

Theorem 5.3 *The following conditions for a linear operator $A \colon V \to W$ between normed spaces are equivalent.*

1. *A is uniformly continuous.*
2. *A is continuous.*
3. *A is continuous at $x = 0$.*
4. *A is bounded.*

Proof Obviously, uniform continuity implies continuity, and continuity implies continuity at $x = 0$. Suppose now that A is continuous at $x = 0$. Then there exists a $\delta > 0$ such that $\|Ax\| \leq 1$ for all $x \in V$ with $\|x\| < 2\delta$. Now, given an arbitrary $y \in V$, if $y = 0$, then $Ay = 0$ while if $y \neq 0$, then, noting that $\|\delta y/\|y\|\| = \delta < 2\delta$, one obtains that

$$\|Ay\| = \|A\frac{\delta\|y\|y}{\delta\|y\|}\| = \frac{\|y\|}{\delta}\|A\frac{\delta y}{\|y\|}\| \leq \frac{1}{\delta}\|y\|,$$

showing that A is bounded by $1/\delta$. Finally, if A is bounded by some $M > 0$, then $\|Ax - Ax'\| = \|A(x - x')\| \leq M\|x - x'\|$, for all $x, x' \in V$, and thus uniform continuity follows easily. $\qquad\square$

For a bounded operator A the inequality $\|Ax\| \leq M\|x\|$ sets an upper bound on how much A can distort the length of vectors in the domain. It is natural to consider the infimum over all such upper bounds M and define that to be the norm of the operator A.

Definition 5.5 Let $A : V \to W$ be a linear operator. The *operator norm* or simply the *norm* of A is defined to be

$$\|A\| = \inf\{M \geq 0 \mid \forall x \in V \quad \|Ax\| \leq M\|x\|\}$$

adopting the convention that $\inf \emptyset = \infty$.

Thus, a linear operator is bounded if, and only if, its norm is finite.

Example 5.2 Continuing Example 5.1, directly evaluating the relevant infimum shows that $\|A_\alpha\| = |\alpha|$.

The homogeneity of the norm together with continuity considerations give rise to the following expressions for the operator norm.

Proposition 5.6 *For a bounded linear operator $A : V \to W$ between normed spaces the following expressions are all equal to $\|A\|$.*

1. $\sup\{\|Ax\| \mid x \in V, \quad \|x\| \leq 1\}$.
2. $\sup\{\|Ax\| \mid x \in V, \quad \|x\| < 1\}$.
3. $\sup\{\|Ax\| \mid x \in V, \quad \|x\| = 1\}$.
4. $\sup\{\|Ax\|/\|x\| \mid x \in V, \quad x \neq 0\}$.

The proof is left for the reader.

In general, any bounded operator, simply since it is continuous, is determined by its values on a dense subset. However, not any bounded operator on a dense subset can be extended to the full domain. The next result shows that when the codomain is complete this difficulty disappears.

Theorem 5.4 *Let V be a normed space, $\mathcal{D} \subseteq V$ a dense linear subspace of V, and $\mathcal{B}$ a complete normed space. If $A : \mathcal{D} \to \mathcal{B}$ is a bounded linear operator, then there exists a unique linear operator $B : V \to \mathcal{B}$ such that $Bx = Ax$ for all $x \in \mathcal{D}$ and $\|B\| = \|A\|$. Such a B is called an extension of A.*

Proof Given $x_0 \in V$ let $\{x_m\}_{m \geq 1}$ be a sequence in $\mathcal{D}$ with $x_m \to x_0$. Since a bounded linear operator is uniformly continuous we may apply the extension principle of Theorem 4.8 to obtain a function $B : V \to \mathcal{B}$ that extends A. In particular, for every $x \in V$ and a sequence in $\mathcal{D}$ with $x_m \to x$ we have $Ax_m \to Bx$. This extension is unique so it remains to verify the required properties of B.

That B is linear is seen as follows. If $x, y \in V$ and $x_m \to x$ and $y_m \to y$ are as above, then it follows that $x_m + y_m \to x + y$ and thus

$$B(x + y) = \lim_{m \to \infty} A(x_m + y_m) = \lim_{m \to \infty} Ax_m + \lim_{m \to \infty} Ay_m = Bx + By.$$

A similar argument shows that $B(\alpha x) = \alpha(Bx)$. As for showing that $\|A\| = \|B\|$, note first that $\|A\| \leq \|B\|$ just by the fact that B is an extension of A. Next, for $x \in V$ and $x_m \to x_0$ as above, one has

$$\|Ax_m\| \le \|A\|\|x_m\|$$

and passing to the limit, and remembering that the norm is continuous, we obtain

$$\|Bx\| \le \|A\|\|x\|$$

showing that $\|B\| \le \|A\|$, and thus the desired equality holds.   □

Another consequence of the extension principle of uniformly continuous functions to complete spaces implies that the completion of a normed space V inherits the structure and becomes itself a normed space.

Theorem 5.5 *Let V be a normed space with its induced metric structure, and let $\iota : (V, d) \to (\hat{V}, \hat{d})$ be a completion of it. Then the linear structure of V extends to $\hat{V}$ and so does the norm, making $\hat{V}$ a normed space. Moreover, $\hat{V}$ is complete in its induced norm.*

Proof The proof is very similar to the proof of Theorem 5.4. The first step is to note that by Theorem 5.2 the structure mappings of V are all uniformly continuous so each can be extended as long as the codomain is complete. In more detail, the norm function $\| - \| : V \to \mathbb{R}$ is uniformly continuous and so Theorem 4.8 provides us with an extension $\widehat{\| - \|} : \hat{V} \to \mathbb{R}$ since $\mathbb{R}$ is complete. The linear structure of V extends to a linear structure on $\hat{V}$ via the inclusion $V \to \hat{V}$, as follows. For any scalar α, the function $A_\alpha : V \to \hat{V}$ from Example 5.1 is uniformly continuous thus extends to $\hat{V}$ and provides it with a scalar product. Vector addition is extended similarly and all of the linear space axioms and norm axioms then follow by a limit argument as was done in the proof of Theorem 5.4.   □

Finally, we give here the way of comparing two different normed spaces.

Definition 5.6 A function $A : V \to W$ between two normed spaces is a *linear isometry* if A is a linear isomorphism that respects the norms in the sense that $\|Ax\| = \|x\|$ for all $x \in V$.

One can easily verify that a function $A : V \to W$ is a linear isometry if, and only if, it is a linear isomorphism that is also an isometry with respect to the induced metrics on V and W.

Exercises

Exercise 5.1 Let $\| - \|$ be a norm on either $K = \mathbb{R}$ or $K = \mathbb{C}$. Prove that there exists $\alpha > 0$ such that $\|x\| = \alpha |x|$ for all x where $| - |$ is the absolute value norm.

Exercise 5.2 Show that if $\| - \|_1$ and $\| - \|_2$ are norms on a given linear space, then $\alpha \| - \|_1 + \beta \| - \|_2$ is a norm for all $\alpha, \beta > 0$.

Exercise 5.3 Complete the proof of Proposition 5.2 by showing that if two norms on V induce the same topology, then the norms are equivalent.

Exercise 5.4 Two norms $\| - \|_1$ and $\| - \|_2$ on a linear space V are *strongly equivalent* if there exists $\alpha > 0$ such that $\|x\|_1 = \alpha \cdot \|x\|_2$ for all $x \in V$. Prove that $\dim(V) = 1$ if, and only if, all norms on V are strongly equivalent.

Exercise 5.5 Prove the following for an arbitrary normed space V.

1. A finite set $S \subseteq V$ is always a net and never dense.
2. A subset of a net is a net and a superset of a dense set is dense.
3. The union $N \cup F$ of a net and a finite set is a net.
4. The set difference $D \setminus F$ of a dense set and a finite set is dense.
5. The union of two nets need not be a net.
6. The intersection of two dense sets need not be dense.

Which properties remain true if V is replaced by an arbitrary metric space?

Exercise 5.6 Show that the point-to-set distance satisfies $d(x, y) = d(x, \{y\})$ and $d(x, S) = 0$ if $x \in S$.

Exercise 5.7 Show that for the point-to-set distance $d(x, T) \leq d(x, S)$ whenever $S \subseteq T$.

Exercise 5.8 We showed in Lemma 5.1 that $\overline{S} = \{x \in X \mid d(x, S) = 0\}$. Show further that the boundary of S is $\partial S = \{x \in X \mid d(x, S) = 0 \text{ and } d(x, X \setminus S) = 0\}$ and that the interior of S is $\mathrm{int}(S) = \{x \in X \mid d(x, X \setminus S) > 0\}$.

Exercise 5.9 Find a finite-dimensional normed space V, a point $x \in V$, and a set $S \subseteq V$ such that the distance $d(x, S)$ is attained at infinitely many different points of S.

Exercise 5.10 Define the set-to-set distance

$$\inf_{s \in S, t \in T} \|s - t\|$$

for subsets S, T of a normed space V. Show that the triangle inequality may fail so this does not yield a metric function on $\mathcal{P}(V)$.

Exercise 5.11 Show that in a normed space V if $\{x_n\}_{n \geq 1}$ converges to x, then $d(x, \{x_n \mid n \in \mathbb{N}\}) = 0$ but the converse may fail.

Exercise 5.12 Suppose that $d(x, S) = 0$ for some point x and a subset S of a normed space V. Show that there exists a sequence $\{x_n\}_{n \geq 1}$ converging to x.

Exercise 5.13 Consider the space ℓ_∞ of bounded sequences in $\mathbb{C}$ (or $\mathbb{R}$) with the supremum norm $\| - \|_\infty$. For the point-to-set distance show, for all $x \in \ell_\infty$, that $d(x, c_{00}) = 0$ if, and only if, $x \in c_0$. In other words: $\overline{c_{00}} = c_0$.

Exercise 5.14 For $1 \leq p \leq \infty$ compute the operator norm of the evaluation mapping $\ell_p \to \mathbb{C}$ given by $x \mapsto x_k$, the k-th component of x.

Exercise 5.15 Let $A \colon V \to W$ be a linear operator between normed spaces. Show that A is bounded if, and only if, the image $A(S)$ under A of the unit sphere $S = \{x \in V \mid \|x\| = 1\}$ is bounded in B. Conclude that if V is finite dimensional, then A is automatically bounded.

Exercise 5.16 Prove Proposition 5.6.

Exercise 5.17 A bijective continuous function $f \colon X \to Y$ between topological spaces need not have a continuous inverse. This can be easily seen by considering X and Y to have the same underlying set but different topologies. It is more interesting that a bijective continuous function $f \colon X \to X$ from a topological space to itself need not have a continuous inverse. Use the space c_{00} to demonstrate that by constructing a bounded linear operator on it whose inverse is unbounded.

Exercise 5.18 For two vectors $x, y \in c_{00}$, say,

$$x = (x_1, \ldots, x_n, 0, 0, \ldots) \quad y = (y_1, \ldots, y_m, 0, 0, \ldots)$$

define their *concatenation* to be

$$x \star y = (x_1, \ldots, x_n, y_1, \ldots, y_m, 0, 0, \ldots).$$

For fixed $x \in c_{00}$ consider the function $A_x \colon c_{00} \to c_{00}$ given by $A_x(y) = x \star y$ is a linear operator. Similarly, for $y \in c_{00}$ show that $B_y \colon c_{00} \to c_{00}$ given by $B_y(x) = x \star y$ is a linear operator. Compute their operator norms. You should observe a significant difference between $\|A_x\|$ and $\|B_y\|$.

Exercise 5.19 Prove the following properties in $\mathbb{R}^2$ with respect to the ℓ_2 norm, i.e., its Euclidean structure:

1. The distance from a point to a nonempty closed set is attained.
2. The distance from a point to the end-points of a segment is larger than the distance to the mid-point of the segment.
3. The distance to a non-empty convex set, if attained, is attained uniquely.

Which of these properties remain true in an arbitrary normed space?

Exercise 5.20 Refer to the proof of Theorem 5.5. Prove the triangle inequality for the extension of the norm function $\| - \|$ to $\hat{V}$. Explain in detail how vector addition extends from V to $\hat{V}$.

Exercise 5.21 In the spirit of Theorem 5.5 prove that if V is a linear space equipped with an inner product, and thus with an induced norm, then the inner product extends to an inner product on the completion $\hat{V}$. Show first that the inner product is uniformly continuous in a suitable way.

5.2 Finite-Dimensional Spaces

In elementary linear algebra the finite-dimensional spaces $\mathbb{R}^n$ and $\mathbb{C}^n$ are familiar both as inner product spaces and with various other norms introduced such as $\| - \|_1$ and $\| - \|_\infty$. These are thus in particular examples of metric spaces and therefore carry an induced topology. Contrary to the infinite-dimensional situation the compatibility of the linear structure and the topology is as good as one can desire: all such spaces are complete, all linear operators between them are continuous, every subspace is closed, and so on. We will not prove these facts here since the details of the proofs are not of particular interest. It suffices to remember that topological considerations are rarely an issue for finite-dimensional linear spaces.

What this section does demonstrate is an even stronger fact: all norms on a given finite-dimensional linear space are equivalent, i.e., they induce the same topology. In the context of applications to physics, the infinite-dimensional spaces are vital for quantum mechanics while their less exotic finite-dimensional brothers are the foundations of quantum computation and quantum information theory. However, finite-dimensional spaces serve as the building blocks of many of the infinite-dimensional ones used in quantum mechanics.

5.2.1 All Norms are Equivalent

In the finite-dimensional case topological issues rarely play any role. Part of the reason is that the induced topology in a finite-dimensional normed space is independent of the norm.

Theorem 5.6 *For $n \geq 1$ all norms on $\mathbb{R}^n$ are equivalent and all norms on $\mathbb{C}^n$ are equivalent.*

Proof The details in each case are similar so we only give the details for $\mathbb{R}^n$. First, by Proposition 5.3 norm equivalence is an equivalence relation and so it suffices to show that an arbitrary norm $\| - \|$ on $\mathbb{R}^n$, that we now fix, is equivalent to the particular norm

$$\|x\|_1 = \sum_{i=1}^{n} |\xi_i|$$

where we write $x = (\xi_1, \ldots, \xi_n)$ for a typical vector in $\mathbb{R}^n$. Using the standard basis $e_1, \ldots, e_n \in \mathbb{R}^n$ we compute

$$\|x\| = \| \sum_{i=1}^{n} \xi_i e_i \| \leq \sum_{i=1}^{n} |\xi_i| \|e_i\|$$

so if we denote

$$\beta = \max_{1 \le i \le n} \|e_i\|$$

then

$$\|x\| \le \beta \sum_{i=1}^{n} |\xi_i| = \beta \|x\|_1$$

for all $x \in \mathbb{R}^n$. By Proposition 5.1 the topology induced by $\| - \|$ is coarser than the one induced by $\| - \|_1$. This implies that $\| - \| : \mathbb{R}^n \to \mathbb{R}$ is continuous with respect to the norm $\| - \|_1$. To see that, recall that what we need to show is that if $U \subseteq \mathbb{R}$ is open, then $\| - \|^{-1}(U)$ is open in $\mathbb{R}^n$ with respect to $\| - \|_1$. But, any norm is continuous with respect to itself and thus $\| - \|^{-1}(U)$ is open in $\mathbb{R}^n$ with respect to $\| - \|$, and any open set with respect to $\| - \|$ is also open with respect to $\| - \|_1$.

By Weierstrass' Theorem (Exercise 4.68) any real-valued continuous function on a non-empty compact set attains its minimum. The unit sphere $S = \{x \in \mathbb{R} \mid \|x\|_1 = 1\}$ is a closed and bounded set and thus, by Exercise 4.74, S is compact (this is where the finite-dimensionality assumption is crucial, cf. Exercise 5.24). Therefore, $\| - \| : S \to \mathbb{R}$ attains its minimum, say

$$\|x_m\| = \alpha$$

where

$$\alpha = \min\{\|x\| \mid x \in S\}$$

and $x_m \in S$. Let $y \in \mathbb{R}^n$ be arbitrary and non-zero and write $x = y/\|y\|_1$. Then $x \in S$ and so

$$\alpha = \|x_m\| \le \|x\| = \frac{\|y\|}{\|y\|_1}$$

and thus

$$\alpha \|y\|_1 \le \|y\|.$$

Finally we conclude that

$$\alpha \|y\|_1 \le \|y\| \le \beta \|y\|_1$$

holds for all $y \in \mathbb{R}^n$ (since for $y = 0$ it certainly holds). To conclude that $\| - \|$ and $\| - \|_1$ are equivalent it remains to show that $\alpha > 0$. If that was not the case then, as a minimum of non-negative values, $\alpha = 0$. But then $\|x_m\| = 0$ holds which axiomatically forces $x_m = 0$ and thus $\|x_m\|_1 = 0$, contrary to $x_m \in S$. The two norms are thus equivalent. □

Proposition 5.7 *A linear space V is finite dimensional if, and only if, all norms on it are equivalent.*

Proof Theorem 5.6 establishes that all norms on $\mathbb{R}^n$ and $\mathbb{C}^n$ are equivalent. The proof can easily be adapted to show that all norms on a finite-dimensional linear space are equivalent. For the converse we show that an infinite-dimensional linear space V carries at least two inequivalent norms. By the Axiom of Choice there exists a Hamel basis T for V. Thus each vector $x \in V$ can be written uniquely as

$$x = \sum_{t \in T} \alpha_t(x) \cdot t$$

where $\{\alpha_t(x)\}_{t \in T}$ are scalars that depend on x and, for each x, only finitely many of them are non-zero. Therefore the definitions

$$\|x\|_1 = \sum_{t \in T} |\alpha_t(x)| \quad \text{and} \quad \|x\|_\infty = \max_{t \in T}\{|\alpha_t(x)|\}$$

are justified as the sum is finite and the maximum exists. The verification that $\| - \|_1$ and $\| - \|_\infty$ are norms on V is straightforward and so it remains to see why they are not equivalent. Since V is infinite dimensional the set T is infinite and we may list infinitely many distinct vectors in it, say, $\{t_i\}_{i \in \mathbb{N}}$. Clearly, $\|x\|_\infty \leq \|x\|_1$ so suppose $\|x\|_1 \leq \beta \|x\|_\infty$ is true for some constant $\beta > 0$ and all $x \in V$. In particular then, for

$$x_n = \sum_{i=1}^{n} t_i$$

we have

$$\|x_n\|_1 = n \quad \text{and} \quad \|x_n\|_\infty = 1$$

and so

$$n \leq \beta$$

must hold for all $n \in \mathbb{N}$. But this is absurd. $\qquad\square$

5.2.2 Finite-Dimensional Sequence and Function Spaces

A restricted case of the sequence spaces ℓ_p defined in Definition 2.16 yields, for every $1 \leq p < \infty$, a norm on $\mathbb{C}^n$ given by

$$\|x\|_p^p = \sum_{i=1}^{n} |\xi_i|^p$$

for all $(\xi_1, \ldots, \xi_n) \in \mathbb{C}^n$ and

$$\|x\|_\infty = \max\{|\xi_1|, \ldots, |\xi_n|\}$$

when $p = \infty$. We know that all of these norms on any given $\mathbb{C}^n$ are equivalent and thus induce the same topology on $\mathbb{C}^n$ (necessarily the Euclidean topology). In particular, for all $1 \le p, q \le \infty$ there exist constants $\alpha, \beta > 0$ with

$$\alpha \|x\|_q \le \|x\|_p \le \beta \|x\|_q$$

for all $x \in \mathbb{C}^n$. We can in fact obtain precise values for these constants.

Theorem 5.7 *For all $1 \le p < q \le \infty$ and $n \ge 1$ the inequalities*

$$\|x\|_q \le \|x\|_p \le n^{1/p - 1/q} \|x\|_q$$

hold for all $x \in \mathbb{C}^n$.

Proof The first inequality is left as an exercise. We address the second inequality and consider the case $q < \infty$. In the following computation the inequality is justified by Hölder's Inequality (Theorem 2.15)

$$\sum_{i=1}^{n} |\xi_i|^p = \sum_{i=1}^{n} |\xi_i| \cdot 1 \le \left(\sum_{i=1}^{n} (|\xi_i|^p)^{q/p} \right)^{\frac{p}{q}} \cdot \left(\sum_{i=1}^{n} 1^{q/(q-p)} \right)^{1 - \frac{p}{q}} = \left(\sum_{i=1}^{n} |\xi_i|^q \right)^{\frac{p}{q}} \cdot n^{1 - \frac{p}{q}}$$

and thus for $x = (\xi_1, \ldots, \xi_n)$

$$\|x\|_p \le n^{1/p - 1/q} \cdot \left(\sum_{i=1}^{n} |\xi_i|^q \right)^{\frac{1}{q}} = n^{1/p - 1/q} \|x\|_q$$

as needed. The case $q = \infty$ is left for the reader.   □

Spaces of polynomials were considered in Example 2.23. We write here $\mathbb{P}_n$ for the linear space of all polynomials of degree strictly smaller than n with complex coefficients. It is a linear space of dimension n and thus isomorphic to $\mathbb{C}^n$. Fix an interval $I = [a, b]$, and with reference to Sect. 2.5.5, we obtain for all $1 \le p < \infty$ the L_p norm on $\mathbb{P}_n$ by

$$\|x\|_p^p = \int_a^b dt \, |x(t)|^p$$

and

$$\|x\|_\infty = \max_{a \le t \le b} |x(t)|$$

when $p = \infty$. Again, we know that all of these norms induce the same topology on $\mathbb{P}_n$. Abstractly, since $\mathbb{P}_n \cong \mathbb{C}^n$, the two spaces are essentially the same and topologically the norms are indistinguishable. But there is a significant difference that we now clarify.

For all $n \ge 1$ and $1 \le k \le n$ let $\mathrm{ev}_k \colon \mathbb{C}^n \to \mathbb{C}$ be the evaluation of the k-th component function

$$\mathrm{ev}_{n,k}(\xi_1, \ldots, \xi_n) = \xi_k.$$

For $t_0 \in [a, b]$ let $\mathrm{ev}_{n,t_0} : \mathbb{P}_n \to \mathbb{C}$ be the evaluation at t_0 function

$$\mathrm{ev}_{n,t_0}(x) = x(t_0).$$

It is immediate that these are linear operators. The norms of these operators reveal important subtleties.

Proposition 5.8 *Let $1 \le p \le \infty$ and consider $\mathbb{C}^n$ with the ℓ_p norm $\| - \|_p$. Then, given $k \ge 1$, there exists an M such that $\|\mathrm{ev}_{n,k}\| \le M$ for all $n \ge k$.*

Proof Clearly each operator is bounded since it is defined on a finite-dimensional space. The uniform bound is easily seen to be $M = 1$ by direct computation, left as an exercise. □

Stated differently, the operators $\{\mathrm{ev}_{n,k}\}_{n \ge k}$ are bounded uniformly. The situation is very different for the spaces of polynomials.

Proposition 5.9 *Let $1 \le p \le \infty$ and consider $\mathbb{P}_n$ with the L_p norm. Then, given $t_0 \in [a, b]$, the operators $\{\mathrm{ev}_{n,t_0}\}_{n \ge 1}$ are bounded but only uniformly bounded for $p = \infty$.*

Proof The fact that each operator is bounded is, again, immediate from finite dimensionality. In the case $p = \infty$ it is immediate that $\|\mathrm{ev}_{n,t_0}\| = 1$. For $1 \le p < \infty$ it suffices to find, for each $n \ge 1$, a vector $x_n \in \mathbb{P}_n$ that evaluates to 1 at t_0 while $\|x\|_p \le 1/n$. Assuming for simplicity, that $[a, b] = [0, 1]$, $t_0 = 1$, and $p = 1$ we can take $x_n(t) = t^{n-1}$ by a direct evaluation of the norm. It is left for the reader to handle the general case. □

We conclude this section by noting that while on any finite-dimensional linear space all norms are equivalent in the topological sense they certainly are not equivalent metrically: a finite-dimensional linear space typically supports infinitely many non-isometric norms. A glimpse is offered in the exercises.

Exercises

Exercise 5.22 Prove the norms in the proof of Proposition 5.7 are indeed norms.

Exercise 5.23 Complete the proof of Theorem 5.7, verify the case $q = \infty$, and show that the given constants are optimal. To prove that $\|x\|_p \le \|x\|_q$ show first that for all $0 < t \le 1$ one has $(x + y)^t \le x^t + y^t$ for all $x, y > 0$ by elementary techniques. Then consider $t = p/q$.

Exercise 5.24 Show, for all $1 \le p \le \infty$, that the unit sphere $\{x \in \ell_p \mid \|x\| = 1\}$ is closed and bounded but not compact. Hint: The sphere is not even sequentially compact.

Exercise 5.25 Show that $\| - \|_p$, for all $1 \leq p \leq \infty$, is permutation invariant on $\mathbb{C}^n$, i.e., $\|(\xi_1, \ldots, \xi_n)\| = \|(\xi_{\tau(1)}, \ldots, \xi_{\tau(n)})\|$ for all permutations τ of $\{1, 2, \ldots, n\}$. Find other norms with the same property.

Exercise 5.26 Referring to Proposition 5.8 show that $\|\mathrm{ev}_{n,k}\| = 1$ for all $1 \leq k \leq n$.

Exercise 5.27 Fill in the missing details in the proof of Proposition 5.9.

Exercise 5.28 Prove that a subset $S \subseteq V$ in a finite-dimensional normed space is compact if, and only if, it is closed and bounded.

Exercise 5.29 Show that the closed unit ball $\{x \in V \mid \|x\| \leq 1\}$ is compact in any finite-dimensional normed space V. (The converse, namely that compactness of the the unit ball in V implies V has finite dimension, is true but harder to prove.)

Exercise 5.30 Show that $\mathbb{R}^2$ with the norm $\| - \|_1$ is isometrically isomorphic to $\mathbb{R}^2$ with the norm $\| - \|_\infty$.

Exercise 5.31 Prove that if $n \geq 3$, then $\mathbb{R}^n$ with the norm $\| - \|_1$ is not isometrically isomorphic to $\mathbb{R}^n$ with the norm $\| - \|_\infty$. Hint: Count corners.

Exercise 5.32 The spaces $\mathbb{C}^n$ and $\mathbb{P}_n$ are isomorphic since they have the same dimension. An 'obvious' isomorphism is given by choosing the basis $e_1, \ldots, e_n$ in $\mathbb{C}^n$ and mapping e_k to the polynomial $t \mapsto t^{k-1}$, and then extending linearly. What is the norm of the resulting isomorphism?

5.3 Countable-Dimensional Sequence and Function Spaces

We move beyond the finite-dimensional spaces to the next cadinality. This is done through the formalism of colimits thus emphasizing the link to the finite-dimensional spaces and setting up a convenient bridge to transfer information from the familiar finite-dimensional case to infinite dimensions.

Recall the finite-dimensional normed spaces $\mathbb{C}^n$ and $\mathbb{P}_n$ from Sect. 5.2.2.

5.3.1 The Space c_{00} as a Colimit

The space c_{00} of all sequences that are eventually 0 is, in some sense, the simplest infinite-dimensional space. We can precisely pinpoint this intuition in the concept of a colimit—a fundamental concept in category theory. We shall not develop any of the general theory here but rather present the concept just for our cases of interest. Recall that for concreteness we are assuming all linear spaces are over $\mathbb{C}$. Consider the diagram

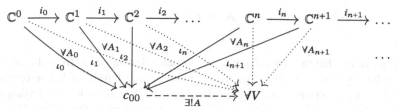

where the top row is infinite and contains all the spaces $\mathbb{C}^n$, $n \geq 0$, (by convention $\mathbb{C}^0 = \{0\}$, the trivial linear space) and each of the arrows is a linear operator with $i_n : \mathbb{C}^n \to \mathbb{C}^{n+1}$ given by

$$i_n(\xi_1, \ldots, \xi_n) = (\xi_1, \ldots, \xi_n, 0).$$

Such a collection of linear spaces and linear operators is called a *diagram of linear spaces*. Of the two spaces appearing at the bottom part of the diagram let us focus first on c_{00}. Each of the arrows leading to it is a linear operator with $\iota_n : \mathbb{C}^n \to c_{00}$ given by

$$\iota_n(\xi_1, \ldots, \xi_n) = (\xi_1, \ldots, \xi_n, 0, 0, \ldots).$$

We note that each triangle in the diagram consisting of solid arrows *commutes* in the sense that for all $n \geq 0$

$$\iota_n = \iota_{n+1} \circ i_n$$

so that tracing the triangle along one side or along the other two, following the directions of the arrows, results in the same overall process. The verification of the commutativity of the triangle is done by computing on a generic element:

$$\iota_n(\xi_1, \ldots, \xi_n) = (\xi_1, \ldots, \xi_n, 0, 0, \ldots)$$
$$\iota_{n+1}(i_n(\xi_1, \ldots, \xi_n)) = \iota_{n+1}(\xi_1, \ldots, \xi_n, 0) = (\xi_1, \ldots, \xi_n, 0, 0, \ldots)$$

leading to the same result (in words, ι appends infinitely many 0's while i appends only one, but appending first one 0 and then infinitely many more is the same as just appending infinitely many to start with). The linear space c_{00} together with the solid arrows from the diagram to it satisfying the commutativity of each triangle is called a *cone from the diagram to c_{00}*. We now turn to the portion of the diagram that contains the ∀ symbols. It states the condition that is required for the dashed arrow to exist: for any other cone from the diagram to an arbitrary linear space V the dashed arrow exists. In more detail, it reads as: for all linear spaces V and linear operators $A_n : \mathbb{C}^n \to V$, one for each $n \geq 0$, such that each triangle commutes, namely

$$A_n = A_{n+1} \circ i_n$$

for all $n \geq 0$, a unique dashed arrow exists. The precise meaning of the dashed arrow is that it is a linear operator that factorizes the cone to V via the cone to c_{00}. In more detail, it is a linear operator $A : c_{00} \to V$ such that

$$A_n = A \circ \iota_n$$

for all $n \geq 0$.

Let us verify that the space c_{00} has the property that any cone from the diagram to an arbitrary linear space V, given by a compatible family of linear operators $A_n : \mathbb{C}^n \to V$, factorizes uniquely via a dashed arrow $A : c_{00} \to V$ as in the diagram. Firstly, note that while no single $\iota_n : \mathbb{C}^n \to c_{00}$ is surjective, the entire collection $\{\iota_n\}_{n \geq 1}$ is surjective in the sense that for all $x \in c_{00}$ there exists $n \geq 1$ and $x_n \in \mathbb{C}^n$ with $x = \iota_n(x_n)$. Therefore, the condition $A_n = A \circ \iota_n$ forces the definition of A to be

$$A(x) = A(\iota_n(x_n)) = A_n(x_n)$$

so if an extension exists it is unique and given by this formula. Before verifying that this definition indeed gives rise to a linear operator we must verify it is well defined, but this is a direct consequence of the compatibility among the A_n (see Exercise 5.33).

It remains to show that A is a linear operator. We repeatedly use the fact that if $x = \iota_n(x_n)$, then $A(x) = A_n(x_n)$. If $x = \iota_n(x_n)$, then $\alpha x = \iota_n(\alpha x_n)$ by homogeneity of ι_n, and so

$$A(\alpha x) = A_n(\alpha x_n) = \alpha A_n(x_n) = \alpha A(x)$$

so A is homogeneous. For additivity, given $x = \iota_n(x_n)$ and $y = \iota_m(y_m)$ the compatibility in the cone condition implies that we may assume $m = n$, and then $x + y = \iota_n(x_n + y_n)$ so

$$A(x + y) = A_n(x_n + y_n) = A_n(x_n) + A(y_n) = A(x) + A(y)$$

as required.

This universal property of c_{00} with respect to the diagram of inclusions $\mathbb{C}^n \to \mathbb{C}^{n+1}$ as above is said to make c_{00} the *colimit of the diagram*.

5.3.2 The Space $\mathbb{P}$ as a Colimit

The space $\mathbb{P}$ of all polynomials, considered as functions over the fixed interval $[a, b]$, is another infinite-dimensional linear space enjoying a straightforward definition. Just as c_{00} is built as a colimit of a diagram of the finite-dimensional building blocks $\mathbb{C}^n$, so is $\mathbb{P}$, the space of all polynomials, a colimit of a diagram consisting of the finite-dimensional spaces $\mathbb{P}_n$. Since the details are very similar we proceed more rapidly. The point to pay attention to is the emphasis we give to the arrows rather than to the linear spaces themselves.

Consider the diagram

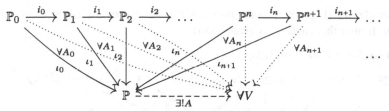

where the top row is infinite and contains all the polynomial spaces $\mathbb{P}^n$, $n \geq 0$, and each of the arrows is a linear operator. By convention, $\mathbb{P}_0 = \{0\}$, consisting of the 0 function. Since $\mathbb{P}_n \subseteq \mathbb{P}_{n+1}$ we may simply take each $i_n : \mathbb{P}_n \to \mathbb{P}_{n+1}$ to be the inclusion function. Moreover, $\mathbb{P}_n \subseteq \mathbb{P}$ and so we may also take each $\iota_n : \mathbb{P}_n \to \mathbb{P}$ to be given by inclusion. In such a way we obtain a cone from the diagram to $\mathbb{P}$ since obviously

$$\iota_n = \iota_{n+1} \circ i_n$$

as either side of the equation is simply the inclusion function.

The rest of the diagram deciphers in exactly the same way as was the case for c_{00} and the verification of the relevant claims is just as easy (if not easier). In detail, given any other cone from the diagram to an arbitrary linear space V, namely

$$A_n = A_{n+1} \circ \iota_n$$

for all $n \geq 0$, a unique linear operator $A : \mathbb{P} \to V$ exists that factorizes the cone to V via the cone to $\mathbb{P}$ in the sense that

$$A_n = A \circ \iota_n$$

for all $n \geq 0$.

The formal similarity between c_{00} and $\mathbb{P}$ is now in place: c_{00} is the colimit of the diagram of inclusions $\mathbb{C}^n \to \mathbb{C}^{n+1}$ and $\mathbb{P}$ is the colimit of the diagram of inclusions $\mathbb{P}_n \to \mathbb{P}_{n+1}$. Note that the most natural choice of an isomorphism $\mathbb{P}_n \to \mathbb{C}^n$ is itself compatible with the diagram structures and so one may consider the diagrams as a whole to be isomorphic entities (see Exercise 5.35). Consequently, the respective colimits are isomorphic, i.e., $\mathbb{P} \cong c_{00}$, as can easily be seen directly. Nonetheless, we shall see below that different choices of norms on the building blocks in each case lead to radically different structures on the colimits.

5.3.3 Norm Extension

The colimit expressions for c_{00} and $\mathbb{P}$ is a bridge from the finite-dimensional building blocks to the infinite-dimensional linear space. One would expect the transfer of structure along this bridge. This is indeed possible.

Consider the diagram from the previous section expressing c_{00} as a colimit and assume further that each $\mathbb{C}^n$ is given a norm.

Theorem 5.8 *Consider the diagram*

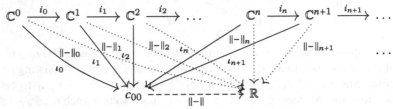

where $\{\| - \|_n \colon \mathbb{C}^n \to \mathbb{R}\}_{n \geq 0}$ *is a family of norms. If the given norms are compatible in the sense that* $\| - \|_n = \|i_n(-)\|_{n+1}$, *then there exists a unique norm* $\| - \| \colon c_{00} \to \mathbb{R}$ *such that* $\|\iota_n(-)\| = \| - \|_n$. *Moreover, in such a situation the operators* ι_n *are all of norm 1.*

Proof The arguments are very similar to those given when showing the universal property of c_{00}. We start again with the observation that the collection $\{\iota_n\}_{n \geq 1}$ is surjective in the sense that for all $x \in c_{00}$ there exists $n \geq 0$ and $x_n \in \mathbb{C}^n$ with $x = \iota_n(x_n)$. Therefore, the condition $\|\iota_n(-)\| = \| - \|_n$ forces the definition of $\| - \|$ to be

$$\|x\| = \|\iota_n(x_n)\| = \|x_n\|_n$$

so if an extension exists it is unique and given by this formula. The requirement that the norms are compatible, i.e., $\| - \|_n = \|i_n(-)\|_{n+1}$, implies independence from the precise value of n when writing $x = \iota_n(x_n)$ so $\|x\|$ is well defined. Armed with the knowledge that $\|x\| = \|x_n\|_n$ whenever $x = \iota_n(x_n)$ we can proceed to verify the norm axioms. We shall only show

$$\|x + y\| \leq \|x\| + \|y\|$$

leaving the rest as an exercise. We choose $m, n \geq 1$, $x_n \in \mathbb{C}^n$, and $y_m \in \mathbb{C}^m$ with $x = \iota_n(x_n)$ and $y = \iota_m(y_m)$. The same argument as above allows us to assume that $m = n$. We can therefore compute as follows

$$\|x + y\| = \|\iota_n(x_n) + \iota_n(y_n)\| = \|\iota_n(x_n + y_n)\| = \|x_n + y_n\|_n \leq$$
$$\leq \|x_n\|_n + \|y_n\|_n = \|\iota_n(x)\| + \|\iota_n(y)\| = \|x\| + \|y\|$$

as required. With the norm in place verifying that each ι_n has norm 1 is immediate. $\square$

5.3.4 Operator Extension and Boundedness

The expression of c_{00} as a colimit means a linear operator out of c_{00} is equivalently described as a compatible family of linear operators on the finite-dimensional linear spaces $\mathbb{C}^n$. We have seen that if each $\mathbb{C}^n$ is a normed space and the norms are suitably compatible, then the norms extend to c_{00}. In the presence of such norms it is then natural to consider the boundedness of the operators involved. In the statement of the next result refer to the diagram

$$\cdots \xrightarrow{\imath_{n-1}} (\mathbb{C}^n, \| - \|_n) \xrightarrow{\imath_n} (\mathbb{C}^{n+1}, \| - \|_{n+1}) \xrightarrow{\imath_{n+1}} \cdots$$

$$\begin{array}{c} \imath_{n+1} \\ \imath_n \downarrow \qquad \searrow \qquad \forall A_n \qquad \downarrow \forall A_{n+1} \\ (c_{00}, \| - \|) \dashrightarrow^{\exists! A} (V, \| - \|) \end{array}$$

which is the description of c_{00} as a colimit fortified with norms.

Theorem 5.9 *Consider c_{00} as a colimit where each $\mathbb{C}^n$ is endowed with a norm, the norms are compatible, and have been extended to a norm on c_{00} as in Theorem 5.8. Suppose further that compatible linear operators $A_n \colon \mathbb{C}^n \to V$ to some normed space V are given and let $A \colon c_{00} \to V$ be the linear operator they extend to. Then*

$$\|A\| = \sup_{n \geq 0} \|A_n\|$$

namely A is bounded if, and only if, the supremum is finite and in that case the supremum is the operator norm of A.

Proof The proof relies on the fact that $\|x\| = \|x_n\|_n$ and $A(x) = A_n(x_n)$ whenever $x = \imath_n(x_n)$ independently of n. We write $M = \sup_{n \geq 0} \|A_n\|$. First we have

$$\|A(x)\| = \|A_n(x_n)\| \leq \|A_n\| \|x_n\|_n = \|A_n\| \|x\| \leq M \|x\|$$

and so $\|A\| \leq M$. Now fix $\varepsilon > 0$ and choose $n \geq 0$ with $\|A_n\| > M - \varepsilon$. Then choose $x_n \in \mathbb{C}^n$ with $\|A_n(x_n)\| > (M - \varepsilon)\|x_n\|_n$. It follows for $x = \imath_n(x_n)$ that

$$\|A(x)\| = \|A_n(x_n)\| > (M - \varepsilon)\|x_n\|_n = (M - \varepsilon)\|x\|$$

and so $\|A\| \geq M - \varepsilon$. As ε was arbitrary we conclude that $\|A\| \geq M$ and as $\|A\| \leq M$ was already established the desired equality follows. $\qquad \square$

The formal similarly between c_{00} and $\mathbb{P}$, each being a colimit of a diagram of inclusions, implies a norm extension and operator extension results for spaces of polynomials. We leave it to the reader to draw the relevant diagrams accompanying the claim.

Theorem 5.10 *If each $\mathbb{P}_n$, $n \geq 0$, is endowed with a norm and the norms are compatible with respect to the inclusions $\mathbb{P}_n \to \mathbb{P}_{n+1}$, then there is a unique norm on $\mathbb{P}$*

*extending the given norms, and then the canonical operators ι_n are of norm 1. More-
over, if compatible operators $A_n: \mathbb{P}_n \to V$ are given, then the induced operator
$A: \mathbb{P} \to V$ satisfies*

$$\|A\| = \sup_{n \geq 0} \|A_n\|.$$

Proof A careful consideration of the proofs of the analogous results about c_{00} reveals
that they were formulated entirely in terms of properties of the arrows appearing in
the diagram expressing c_{00} as a colimit. The arrows in the diagram for $\mathbb{P}$ as a colimit
satisfy the same properties and thus the same proof is still valid.   □

This diagrammatic reasoning is at once a labor saving formalism and a conceptual
tool. We shall use the results of this section in the remaining chapter and note here
that the categorical approach can be utilized much further than we are doing (see
[3]).

Exercises

Exercise 5.33 With reference to any of the colimit diagrams, show that if $\{A_n\}_{n \geq 0}$
is a compatible collection of linear operators and $x = \iota_n(x_1) = \iota_m(x_2)$ for some
$m, n \geq 0$, then $A_n(x_1) = A_m(x_2)$. Conclude that in the relevant diagrammatic arguments one can take $m = n$.

Exercise 5.34 Fill in the detail in the discussion expressing $\mathbb{P}$ as a colimit in
Sect. 5.3.2.

Exercise 5.35 Consider the linear isomorphism $\Psi_n: \mathbb{C}^n \to \mathbb{P}_n$ determined by send-
ing the vector e_k to the polynomial p_k with $p_k(t) = t^{k-1}$. Show that for this family
of isomorphisms

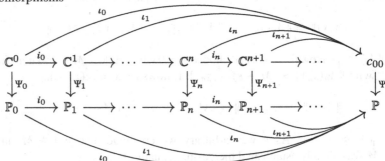

each square commutes, i.e, $\Psi_{n+1} \circ i_n = i_n \circ \Psi_n$. Further, show that there exists a
unique isomorphism $\Psi: c_{00} \to \mathbb{P}$ between the colimits satisfying $\Psi \circ \iota_n = \iota_n \circ \Psi_n$,
for all $n \geq 0$.

Exercise 5.36 Referring to the diagram in Theorem 5.8 show that the compatibility
of the family of norms implies that $\|x_n\|_n = \|x_m\|_m$ whenever $\iota_n(x_n) = \iota_m(x_m)$.

Exercise 5.37 Complete the details in the proof of Theorem 5.8, i.e., show that the
extended function $\| - \|: c_{00} \to \mathbb{R}$ is a norm and that each ι_n has norm 1.

Exercise 5.38 Using the techniques of this section determine, for each family of functions below, those that extend to a bounded linear operator on the colimit.

1. $A_n \colon \mathbb{C}^n \to \mathbb{C}$ given by $A_n(x) = (1/n) \sum_{k=1}^n \xi_k$.
2. $A_n \colon \mathbb{P}^n \to \mathbb{P}$ given by $A_n(x)(t) = x(-t)$.
3. Fix $y \in c_{00}$. $A_n \colon \mathbb{C}^n \to c_{00}$ given by $A_n(x) = y \star x$ (see Exercise 5.18).
4. Fix $y \in c_{00}$. $A_n \colon \mathbb{C}^n \to c_{00}$ given by $A_n(x) = x \star y$.

5.4 The Classical Sequence and Function Spaces

The previous section, where c_{00} was described as the colimit of the diagram of inclusions $\mathbb{C}^n \to \mathbb{C}^{n+1}$ and $\mathbb{P}$ was presented as the colimit of the similar diagram of inclusions $\mathbb{P}_n \to \mathbb{P}_{n+1}$, was designed to enable the following presentation of the classical sequence spaces ℓ_p and function spaces L_p. Remember that the polynomials in the polynomial spaces are considered as functions on some fixed interval $[a, b]$.

5.4.1 From Finite Dimensions to the Classical Spaces

Let $1 \le p \le \infty$ and endow each of $\mathbb{C}^n$ and $\mathbb{P}_n$ with its $\| - \|_p$ norm. It is easily seen that the norms on the diagram of the spaces $\mathbb{C}^n$ are compatible and thus, by Theorem 5.8, they extend to a norm $\| - \|_p$ on c_{00}. Similarly, the norms on the diagram of the spaces $\mathbb{P}_n$ are compatible and thus extend to a norm $\| - \|_p$ on $\mathbb{P}$ by Theorem 5.10. With these norms the spaces c_{00} and $\mathbb{P}$ are not complete.

5.4.1.1 The Case $p \ne \infty$

We define the classical spaces ℓ_p and L_p for $1 \le p < \infty$ by means of the machinery developed in this chapter.

Definition 5.7 Let $1 \le p < \infty$. The classical sequence space ℓ_p is the completion of c_{00} with respect to the norm $\| - \|_p$ obtained by extension of the norms $\| - \|_p$ on the spaces $\mathbb{C}^n$. The classical function space $L_p = L_p([a, b], \mathbb{C})$ is the completion of $\mathbb{P}$ with respect to the norm $\| - \|_p$ obtained by extension of the norms $\| - \|_p$ on the spaces $\mathbb{P}_n$.

This sort of definition describes a mathematical object not by construction but by its relation to other spaces. In particular, such an object is only defined up to isomorphism (by Theorem 4.12 any two completions of the same metric space are isometrically isomorphic). The conjunction of a particular model of the mathematical object together with an understanding of its relation to other objects is most fruitful, offering different perspectives. We illustrate that for the ℓ_p and L_p spaces as just defined. Since ℓ_p is now defined as the completion of c_{00} with respect to $\| - \|_p$ two facts are obvious.

Theorem 5.11 *For $1 \le p < \infty$ the space ℓ_p is a complete normed space and c_{00} is dense in it.*

Proof By Theorem 5.5 the linear structure of c_{00} and its norm extend uniquely to the completion, making it a complete normed space. By the very definition of the completion of a metric space (Definition 4.16) c_{00} is dense in its completion ℓ_p. The fact that any metric space admits a completion was outlined toward the end of Sect. 4.4.4. □

One may follow the instructions for the construction of the completion of c_{00} and obtain a more explicit description of ℓ_p. However, the result would be very cumbersome to work with. It is therefore of great interest that a much simpler model exists, allowing us, equivalently, to define ℓ_p by means of p-power summable sequences as in Example 2.25. The tradeoff is that with that particular model Theorem 5.11 ceases to be obvious and requires a more elaborate proof.

Theorem 5.12 *Consider the sequence model of ℓ_p with the norm $\| - \|$ defined explicitly as an infinite sum, i.e., the vectors in ℓ_p are sequences x for which*

$$\sum_{n=1}^{\infty} |\xi_n|^p$$

converges and the norm is given by

$$\|x\|_p = \left(\sum_{n=1}^{\infty} |\xi_n|^p \right)^{1/p}$$

for all vectors $x \in \ell_p$. Then ℓ_p is complete and c_{00} is dense in it.

Proof The completeness of ℓ_p was shown in Example 4.19. Density of c_{00} in ℓ_p amounts to the property that tails of convergent series can be made arbitrarily small: given a vector $(\xi_1, \ldots, \xi_k, \ldots) \in \ell_p$ the vectors $x_n = (\xi_1, \ldots, \xi_n, 0, 0, \ldots) \in c_{00}$ converge to x precisely because $\|x - x_n\|_p^p$ is a tail of the convergent series

$$\sum_{n=1}^{\infty} |\xi_n|^p$$

and so vanish to 0 as n tends to ∞. □

Let us now turn to the L_p spaces. Since L_p was defined as the completion of $\mathbb{P}$ two facts are again self-evident.

Theorem 5.13 *For $1 \le p < \infty$ the space $L_p = L_p([a, b], \mathbb{C})$ is a complete normed space and $\mathbb{P}$ is dense in it.*

Proof Formally identical to the proof of Theorem 5.11.   □

For ℓ_p it is the case that a model of the completion of c_{00} exists in terms of infinite sequences. One would expect, or at least hope, that for L_p a model of the completion of $\mathbb{P}$ exists in terms of functions. This is the case but the situation is complicated: for fixed k the evaluation functionals on $\mathbb{C}^n$ are uniformly bounded (Proposition 5.8) and thus extend (by Theorem 5.9) to an evaluation functional on c_{00} which then extends to an evaluation functional on ℓ_p (by Theorem 5.4) but for fixed t_0 the evaluation functionals on $\mathbb{P}_n$ are not uniformly bounded (Proposition 5.9) so that L_p lacks evaluation functionals. This situation suggests that the vectors in L_p are not quite functions. Therefore the following analogue of Theorem 5.12 is considerably deeper and unavoidably subtler.

Theorem 5.14 *There exists a model of $L_p = L_p([a, b], \mathbb{C})$ whose elements are equivalence classes of functions on $[a, b]$. The representatives are called Lebesgue measurable functions, the equivalence relation is agreement modulo sets of measure zero, and the norm is given in terms of the Lebesgue integral. In particular, L_p is complete and $\mathbb{P}$ is dense in it.*

Each of the claims made requires a healthy dose of measure theory and falls well outside the scope of this book.

5.4.1.2 The Case $p = \infty$

In the discussion of the classical spaces thus far the case $p = \infty$ was excluded. To be clear, the completion of c_{00} with respect to $\| - \|_\infty$ certainly exists, as does the completion of $\mathbb{P}$, but these are not the spaces ℓ_∞ and L_∞, respectively. The reason, in a sense, is that these completions are too small. It is still interesting to identify explicit models for these completions and observe a difference when compared to the case $p \neq \infty$.

Theorem 5.15 *The space c_0 of sequences of complex numbers that converge to 0 is the completion of c_{00} with respect to the norm $\| - \|_\infty$.*

Proof What needs to be shown is that c_0 is complete and that c_{00} is dense in c_0, all with respect to the norm $\| - \|_\infty$. Both claims are elementary observations regarding the definition of the limit of a sequence. In some more detail, if $\{x_n\}_{n \geq}$ is a Cauchy sequence in c_0, then by evaluation at k we obtain a Cauchy sequence of complex numbers $\{\mathrm{ev}_k(x_n)\}_{n \geq 1}$, which thus converges (since $\mathbb{C}$ is complete). We therefore have a candidate vector x for the limit of the sequence $\{x_n\}_{n \geq 1}$, i.e., the sequence x satisfying $\mathrm{ev}_k(x_n) \to \mathrm{ev}_k(x)$. Showing that $x \in c_0$ and that $x_n \to x$ follows easily from the rather strong consequences of the norm $\| - \|_\infty$. Finally, for the density of c_{00}, an arbitrary $x \in c_0$ is the limit of its truncations, i.e., assuming $x = (\xi_1, \xi_2, \ldots)$ setting

$$x_n = (\xi_1, \ldots, \xi_n, 0, 0, \ldots)$$

the fact that $\xi_n \to 0$ implies at once that

$$\|x - x_n\|_\infty \to 0$$

as needed.   □

As expected, the function space offers a greater challenge. The completion of $\mathbb{P}$ with respect to $\| - \|_\infty$ is familiar but proving so is a significant theorem.

Theorem 5.16 *The completion of $\mathbb{P}$ with respect to $\| - \|_\infty$ is the space $C([a, b], \mathbb{C})$ of continuous functions on $[a, b]$.*

Proof The fact that $C([a, b], \mathbb{C})$ is complete is shown just as the case for real-valued functions in Example 4.16. The density of $\mathbb{P}$ in $C([a, b], \mathbb{C})$ is, precisely, the Weierstrass approximation theorem—a standard result in analysis but not an elementary one.   □

Remark 5.1 To the reader familiar with holomorphic functions we note that if instead of complex polynomials on $[a, b]$ one considers complex polynomials on, say, the unit disk in $\mathbb{C}$, then the completion of $\mathbb{P}$ is the space of all continuous complex-valued function on the unit disk that are holomorphic in its interior. The much larger space of continuous complex-valued functions on the disk is the completion of a variant of $\mathbb{P}$, namely the space of complex polynomials in both z and $\bar{z}$. Both of these are important spaces which, as we see, also arise in a similar fashion to the spaces we are considering.

At this point we observe a significant difference between ℓ_p and c_0 for all $1 \le p < \infty$, providing a first indication that c_0 is too small.

Definition 5.8 A sequence $\{x_n\}_{n \ge 1}$ in ℓ_p is *advancing* if there exists a sequence $\{\xi_k\}_{k \ge 1}$ in $\mathbb{C}$ such that

$$x_n = (\xi_1, \dots, \xi_n, 0, 0, \dots)$$

for all $n \ge 1$.

Proposition 5.10 *For $1 \le p < \infty$ every bounded advancing sequence $\{x_n\}_{n \ge 1}$ in ℓ_p converges, but the same is not true in c_0.*

Proof Let $\xi = \{\xi_k\}_{k \ge 1}$ be the sequence realizing the advancement of the given sequence. The assumption on the sequence being bounded implies the existence of M with

$$\|x_n\|_p^p = \sum_{k=1}^{n} |\xi_k|^p < M$$

for all $n \ge 1$. In other words, the partial sums of

$$\sum_{k=1}^{\infty} |\xi_k|^p$$

are bounded and so the series converges. Therefore, the vector

$$\xi = (\xi_1, \xi_2, \ldots, \xi_n, \ldots)$$

belongs to ℓ_p and clearly

$$\lim_{n \to \infty} \|x_n - \xi\|_p = 0$$

so $x_n \to \xi$ in ℓ_p. A counterexample in c_0 is, for instance, given by

$$x_n = (1, 1, 1, \ldots, 1, 0, 0, \ldots)$$

with n 1s followed by 0s: clearly $\|x_n\|_\infty = 1$ so the sequence is bounded but $\|x_n - x_m\|_\infty = 1$ for all $n \neq m$ so the sequence is not Cauchy and thus fails to converge. □

At the heart of the argument in the proof is the property that in $\mathbb{R}$ a bounded monotone sequence converges. This fact is equivalent to the completeness of $\mathbb{R}$. In a sense, ℓ_p is large enough to include any sequence whose initial segments all belong to ℓ_p. The space c_0 is not large enough to possess this property.

Remark 5.2 A similar phenomenon distinguishes qualitatively between L_p and L_∞. The details require measure theory.

Let us summarize the results above.

Theorem 5.17 *For all* $1 \leq p < \infty$ *the spaces* ℓ_p *and* L_p, *each with its* $\| - \|_p$ *norm, is complete. Each sequence space* ℓ_p *is the completion of the colimit of the diagram of inclusions* $\mathbb{C}^n \to \mathbb{C}^{n+1}$ *as normed spaces. There exists an explicit model of* ℓ_p *in terms of sequences. Each function space* L_p *is the completion of the colimit of the diagram of inclusions* $\mathbb{P}_n \to \mathbb{P}_{n+1}$ *as normed spaces. There exists an explicit model of* L_p *in terms of equivalence classes of functions. The only spaces in which the norm is induced by an inner product are* ℓ_2 *and* L_2.

Proposition 5.10 is a first indication that c_0 is too small and does not quite fit in the tower of ℓ_p spaces, and similarly that $C([a, b], \mathbb{C})$ does not belong to the family of L_p spaces. We complete the picture now with a discussion of ℓ_∞ and L_∞. By now we should expect the function space case to be more demanding. The space ℓ_∞ is the space of all bounded sequences of complex numbers with the supremum norm (see Sect. 2.5.4). The space L_∞ is, just like the L_p spaces, not quite a space of functions due to a lack of evaluation functionals. The vectors in L_∞ are equivalence classes of essentially bounded functions and the norm is given by the essential supremum. These notions, again, require some measure theory and are thus not included here.

5.4.2 Summation and Integration

Consider on $\mathbb{C}^n$ the summation function $\sigma_n \colon \mathbb{C}^n \to \mathbb{C}$ given by

$$\sigma_n(\xi_1, \ldots, \xi_n) = \sum_{k=1}^{n} \xi_k$$

which is clearly a linear operator and thus bounded with respect to any one of the norms $\| - \|_p$, $1 \leq p \leq \infty$ (or any other norm).

Proposition 5.11 *The collection $\{\sigma_n\}_{n \geq 0}$ of all summation operators is uniformly bounded if, and only if, $p = 1$.*

Proof Suppose $p = 1$. If $x = (\xi_1, \ldots, \xi_n)$ is an arbitrary vector in $\mathbb{C}^n$, then

$$|\sigma_n(x)| \leq \sum_{i=1}^{n} |\xi_i| = \|x\|_1$$

and so $\|\sigma_n\| \leq 1$, for all $n \geq 0$, so the summation operators are uniformly bounded. When $1 < p < \infty$ denote by $\mathbf{1}_n$ the vector $(1, 1, \ldots, 1) \in \mathbb{C}^n$. Then $\|\mathbf{1}_n\|_p = n^{1/p}$ while $\sigma_n(\mathbf{1}_n) = n$ so

$$|\sigma_n(\mathbf{1}_n)| = n^{1-1/p} \|\mathbf{1}_n\|_p$$

showing that $\|\sigma_n\| \geq n^{1-1/p}$. Therefore, the summation operators are not uniformly bounded. The case $p = \infty$ is similar. $\qquad\square$

Clearly the summation operators, for any given $1 \leq p \leq \infty$, are compatible and thus form a cone from the diagram of inclusions $\mathbb{C}^n \to \mathbb{C}^{n+1}$ to $\mathbb{C}$ (see Sect. 5.3.1). By the universal property of c_{00} as a colimit the summation operators extend to a summation operator $c_{00} \to \mathbb{C}$. However, by Theorem 5.9 and Proposition 5.11, only for $p = 1$ is the extension bounded, and so one obtains a unique extension of the finite summation operators to a summation operator $\sigma_\infty : \ell_1 \to \mathbb{C}$ by Theorem 5.4. In terms of the explicit model of ℓ_1 this is a familiar linear functional.

Theorem 5.18 *The linear functional $\sigma_\infty : \ell_1 \to \mathbb{C}$ obtained as the unique extension of the finite summation operators $\sigma_n : \mathbb{C}^n \to \mathbb{C}$ is the usual sum of an infinite series when ℓ_1 is modelled as the set of absolutely summable sequences.*

Proof The proof is essentially the elementary fact that absolute convergence of a series implies convergence. $\qquad\square$

Moving the discussion to the function spaces, observe that definite integration yields a linear functional

$$\int_a^b dt - : \mathbb{P}_n \to \mathbb{C}$$

and these are clearly compatible with the inclusions $\mathbb{P}_n \to \mathbb{P}_{n+1}$. Again, each of these is bounded since the linear spaces are finite-dimensional and the only question is for which value of p are they uniformly bounded.

Proposition 5.12 *The collection of the integration operators*

$$\left\{ \int_a^b dt - : \mathbb{P}_n \to \mathbb{C} \right\}_{n \geq 0}$$

is uniformly bounded if, and only if, $p = 1$.

Proof The analysis is similar in spirit to the proof of Proposition 5.11 and is left for the reader.   □

The discussion leading to the existence of the summation functional σ_∞ on ℓ_1 can now be repeated verbatim (by virtue of the formally identical way in which ℓ_1 and L_1 have been constructed) to conclude the existence of an integration functional on L_1. As always, its explicit description is considerably more challenging than was the case with σ_∞.

Theorem 5.19 *In the measure-theoretic model of L_1 the unique extension of the Riemann integral from $\mathbb{P}$ to L_1 is precisely the Lebesgue integral.*

Proof The proof requires a substantial amount of measure theory.   □

We conclude this section by referring to [1] where a relatively recent characterization of the ℓ_p and the L_p norms is given as the the only ones that are permutation invariant and multiplicative with respect to a suitably defined product.

Exercises

Exercise 5.39 Let $c_c \subseteq \mathbb{C}^\infty$ be the space of eventually constant sequences. Show that its closure in ℓ_∞ with respect to $\| - \|_\infty$ is c, the space of convergent sequences. Define $l : c_c \to \mathbb{C}$, for a sequence $x = (x_1, x_2, \ldots, x_n, t, t, t, \ldots)$ that is eventually equal to t, by $l(x) = t$. Prove that l is non-expanding with respect to the d_∞ metric on c_c and the Euclidean metric on $\mathbb{C}$. Conclude that l admits a unique non-expanding extension to a function $c \to \mathbb{C}$. Show that this function is the operation of taking the limit of a convergent sequence.

Exercise 5.40 In the discussion following Theorem 5.13 the evaluation functionals $\text{ev}_{n,k} : \mathbb{C}^n \to \mathbb{C}$ are referenced. However, if $n < k$, then no such evaluation functional exists. There are two ways to resolve this problem for any fixed $k \geq 1$: (1) keep the colimit diagram of c_{00} as it is and define $\text{ev}_{n,k} : \mathbb{C}^n \to \mathbb{C}$ for $n < k$ to be identically zero; (2) remove an initial segment from the diagram defining c_{00} so it starts with $\mathbb{C}^k$ instead of $\mathbb{C}^0$. Show that these two approaches lead to isomorphic solutions.

Exercise 5.41 Consider the function $A : \ell_1 \to \ell_1$ given by

$$A(x)_n = \frac{n^\alpha}{e^n} x_n$$

where $\alpha \geq 1$ is a parameter. Show that A is a bounded linear operator on ℓ_1 and compute its norm.

Exercise 5.42 Prove Theorem 5.18.

Exercise 5.43 Continuing Exercise 5.38, in terms of the explicit model of the completion, give a description for the extension to the completion when it exists.

Exercise 5.44 For a subset $S \subseteq \mathbb{N}$ of the natural numbers let $X_S \subseteq \ell_\infty$ consist of all sequences with $x_s = 0$ for all $s \in S$. Show that X_S is a closed linear subspace of ℓ_∞. Such a closed subspace is called *complemented* if there exists a closed linear subspace Y such that $\ell_\infty = X_S \oplus Y$. Show that X_S is complemented and that unless $\dim(X_S)$ is finite $X_S \cong \ell_\infty$. (In 1967 Lindenstrauss proved that all closed infinite-dimensional complementary subspaces of ℓ_∞ are isomorphic to ℓ_∞.)

Exercise 5.45 For $1 \le p \le \infty$ let $\ell_e = \{x \in \ell_p \mid x_{2n} = 0, n \ge 1\}$ and $\ell_o = \{x \in \ell_p \mid x_{2n+1} = 0, n \ge 1\}$. Show that $\ell_p = \ell_e \oplus \ell_o$. Show that each of the subspaces is closed by finding a suitable bounded linear operator $A \colon \ell_p \to \ell_p$ for which $\ell_e = \text{Ker}(A)$ and similarly for ℓ_o.

Exercise 5.46 In the previous exercise we obtained a decomposition of ℓ_p in terms of sequences concentrated in even and odd places. In this exercise we look for a similar decomposition for the L_p spaces for $1 \le p < \infty$. The difficulty is that we do not have an explicit model for L_p so we proceed with the tools we have. We refer to the construction of L_p as the completion of the colimit of the polynomial space inclusions $\mathbb{P}_n \to \mathbb{P}_{n+1}$.

1. Let $A_n \colon \mathbb{P}_n \to \mathbb{P}$ be given by $A(x)(t) = x(-t) - x(t)$. Show that A is a bounded linear operator.
2. Show that $\{A_n\}_{n \ge 0}$ forms a cone to $\mathbb{P}$ and conclude a bounded extension $A \colon \mathbb{P} \to \mathbb{P}$ exists.
3. Conclude that an extension of A to L_p exists. What is its norm?
4. Define $L_e = \text{Ker}(A)$ and show it is a closed linear subspace of L_p.
5. Similarly define L_o and show that $L_p = L_e \oplus L_o$.

5.5   Separability and Duality

Separability tells us that topologically a space is not too big. Every point in a separable space is accessible as the limit of a sequence constructed out of a repository of countable size. Duality refers to the structure of bounded linear functionals on the space. Using the expressions of the classical spaces as we have presented them we address the separability of the classical spaces and analyse some of the structure of their bounded linear functionals.

5.5.1   Separability

We establish the separability of ℓ_p and L_p for $1 \le p < \infty$.

Proposition 5.13 *For all $n \geq 0$ the space $\mathbb{C}^n$ is separable with respect to any norm.*

Proof By Theorem 5.6 all norms on $\mathbb{C}^n$ are equivalent and since separability is a topological invariant we may assume any particular norm on $\mathbb{C}^n$, so we choose the L_∞ norm $\| - \|_\infty$. We now exploit the elementary fact that $\mathbb{Q}$ is dense in $\mathbb{R}$. It implies at once that $\mathbb{Q} + i\mathbb{Q} = \{a + ib \mid a, b \in \mathbb{Q}\}$ is dense in $\mathbb{C}$. It can now easily be checked directly that $(\mathbb{Q} + i\mathbb{Q})^n$ is dense in $\mathbb{C}^n$ with respect to $\| - \|_\infty$. Finally, $(\mathbb{Q} + i\mathbb{Q})^n$ is countable as a finite cartesian product of countable sets. □

Corollary 5.2 *All finite-dimensional normed spaces are separable.*

Proof We address the complex case. Any n-dimensional linear space V is isomorphic to $\mathbb{C}^n$ and since in the finite-dimensional case all linear operators are continuous it follows that $\mathbb{C}^n$ and V are homeomorphic no matter which norms are used. In particular, the separability of $\mathbb{C}^n$ implies that of V. □

Having established that separability is automatic in finite dimensions we turn to c_{00} and $\mathbb{P}$. Each is built out of finite-dimensional spaces and so the building blocks are separable. The diagram expressing c_{00} and $\mathbb{P}$ as colimits is itself countable and that allows to transfer the separability from the building blocks to the colimit.

Proposition 5.14 *The spaces c_{00} and $\mathbb{P}$ are separable with respect to the norm $\| - \|_p$ for all $1 \leq p \leq \infty$.*

Proof Let us fix a value for p and proceed to exhibit a countable subset in c_{00} dense with respect to $\| - \|_p$. By the separability of $\mathbb{C}^n$ there is a countable set $S_n \subseteq \mathbb{C}^n$ dense with respect to $\| - \|_p$. Using the canonical operators $\iota_n \colon \mathbb{C}^n \to c_{00}$ let

$$S = \bigcup_{n \geq 0} \iota_n(S_n)$$

be the union of the images of the given dense sets. The fact that the norm on c_{00} was built by extension of the norms on $\mathbb{C}^n$ implies at once that S is dense in c_{00}. As a countable union of countable sets S is countable and thus c_{00} is separable. The argument for $\mathbb{P}$ is formally identical since the ingredients used above were the separability of the finite-dimensional building blocks, the compatibility of the norms, and the countability of the diagram. □

The final step to the classical spaces is the completion and so we note the following.

Proposition 5.15 *The completion of a separable metric space is separable.*

Proof Let X be a metric space, $\hat{X}$ its completion, and $S \subseteq X$ a dense subset. We shall show that S is also dense in $\hat{X}$, implying the result. Let $y \in \hat{X}$ be arbitrary and fix $\varepsilon > 0$. Since X is dense in $\hat{X}$ there exists $x \in X$ with $d(x, y) < \varepsilon/2$. Since S is dense in X there exists $s \in S$ with $d(s, x) < \varepsilon/2$. The triangle inequality implies that $d(s, y) < \varepsilon$. So, for all $y \in \hat{X}$ and $\varepsilon > 0$ there exists $s \in S$ with $d(s, y) < \varepsilon$, as needed. □

Putting these results together we obtain our objective.

Theorem 5.20 *The spaces ℓ_p and L_p for $1 \le p < \infty$ as well as c_0 and $C([a, b], \mathbb{C})$ are separable.*

Proof Each space is the completion of a space obtained as the colimit of a countable diagram of separable spaces.   □

5.5.2 Duality

Bounded linear functionals $V \to \mathbb{C}$ are of importance for the study of V. Using the machinery developed thus far we look into the structure of bounded linear functionals on classical spaces, starting with the finite-dimensional ones.

Let us consider $\mathbb{C}^n$ with the $\| - \|_p$ norm for some $1 \le p \le \infty$. Any linear functional $F \colon \mathbb{C}^n \to \mathbb{C}$ is automatically bounded. Such an operator F is necessarily of the form

$$F(x) = \langle \Phi, x \rangle$$

for a unique $\Phi \in \mathbb{C}^n$ where the inner product is the standard one. To see why this is so note that using the standard basis $e_1, \ldots, e_n \in \mathbb{C}^n$, for all $x \in \mathbb{C}^n$

$$F(x) = F\left(\sum_{i=1}^n \xi_i e_i\right) = \sum_{i=1}^n \xi_i F(e_i)$$

so that $\Phi = (\overline{F(e_1)}, \ldots, \overline{F(e_n)})$.

There is thus a bijective correspondence $F \mapsto \Phi$ between bounded functionals on $\mathbb{C}^n$ and $\mathbb{C}^n$ itself. Up to this point the choice of norm was irrelevant since all norms are in any case equivalent. With respect to the norm $\| - \|_p$ we now determine the operator norm $\|F\|$ in terms of its representing vector Φ. Let q satisfy $1/p + 1/q = 1$ and note that Hölder's Inequality (Theorem 2.15) then implies that

$$|F(x)| = |\langle \Phi, x \rangle| \le \|\Phi\|_q \|x\|_p$$

which thus shows that $\|F\| \le \|\Phi\|_q$. In fact, one can show that equality holds.

A similar observation holds for the spaces $\mathbb{P}_n$. To see how note that the vector representing F above was obtained by evaluating F on an orthonormal basis. If one starts with a functional F on $\mathbb{P}_n$, then one can similarly represent it in the same way by evaluating on an orthonormal basis. Obtaining an orthonormal basis in this case is a simple matter of applying the Gram-Schmidt process with respect to the standard inner product on $\mathbb{P}_n$ given by integration. When $[a, b] = [-1, 1]$, starting with the basis $1, t, t^2, \ldots, t^{n-1}$ results in the famous Legendre polynomials. However, no particular choice needs be made to know that any functional F on $\mathbb{P}_n$ is of the form

$$F(x) = \int dt\ \overline{(\Phi(t))}x(t)$$

for a unique $\Phi \in \mathbb{P}_n$. The fact that then $\|F\| = \|\Phi\|_q$ follows in precisely the same way as above from Hölder's Inequality. Let us summarize.

Proposition 5.16 *The linear operators F on $\mathbb{C}^n$ (respectively $\mathbb{P}_n$) are in bijective correspondence with $\mathbb{C}^n$ (respectively $\mathbb{P}_n$) and the operator norm $\|F\|$ with respect to $\| - \|_p$ corresponds to the norm $\| - \|_q$.*

Having handled the finite-dimensional linear spaces we turn now to ℓ_p for $1 \leq p < \infty$, so let $F: \ell_p \to \mathbb{C}$ be a bounded linear functional. We immediately make use of the boundedness of F, and hence its uniform continuity, to obtain a significant simplification. Since c_{00} is dense in ℓ_p it follows that F is uniquely determined by its restriction to c_{00}. In other words, to determine the structure of bounded functionals on ℓ_p it suffices to study the bounded functionals on c_{00} with respect to the norm $\| - \|_p$.

We proceed by examining $F: c_{00} \to \mathbb{C}$. Using the fact that c_{00} is the colimit of the diagram of inclusions $\mathbb{C}^n \to \mathbb{C}^{n+1}$ we know that such an operator F is uniquely determined by the family of operators $F \circ \iota_n : \mathbb{C}^n \to \mathbb{C}$, which we call F_n. But now we are back to the finite-dimensional case and so each F_n is represented by a vector $\Phi_n \in \mathbb{C}^n$ and moreover $\|F_n\|_p = \|\Phi_n\|_q$. The family $\{F_n : \mathbb{C}^n \to \mathbb{C}\}_{n \geq 0}$ can be arbitrary as long as it is compatible (so as to form a cone) and uniformly bounded (so as to extend to a bounded operator on the colimit). Up to this point we were only referring to a single diagram expressing c_{00} as a colimit of inclusions $\mathbb{C}^n \to \mathbb{C}^{n+1}$, each member endowed with the norm $\| - \|_p$. We now consider another such diagram where the norm is $\| - \|_q$. This is the natural habitat of the vectors $\{\Phi_n\}_{n \geq 0}$. A simple verification shows that the compatibility condition on the operators $\{F_n\}_{n \geq 0}$ is precisely the condition that the sequence $\{\iota_n(\Phi_n)\}_{n \geq 0}$ is advancing (Definition 5.8). The uniform boundedness of the operators corresponds exactly to the boundedness of this sequence. We now wish to invoke Proposition 5.10 on the advancing sequence but it requires that $q < \infty$, i.e., that $p > 1$. In that case the advancing sequence represents a unique element of ℓ_q. We just established the following.

Theorem 5.21 *The bounded linear functionals on ℓ_p, for $1 < p < \infty$, are in bijection with ℓ_q in a way that preserves the norms. The bounded linear functionals on c_0 are in bijection with ℓ_1.*

The proof above can be unpacked to produce a more direct expression for the sequence in ℓ_q that represents a given bounded functional F on ℓ_p. However, the diagrammatic approach is model-agnostic. As long as the same properties can be verified in a different diagram the same conclusion can be drawn. It can be shown that for $1 < p < \infty$ the diagrammatic properties leading to the ℓ_p and the L_p spaces agree, but not for $p = \infty$. This observation allows us to state the following result.

Theorem 5.22 *The bounded linear functionals on L_p, for $1 < p < \infty$, are in bijection with L_q in a way that preserves the norms. The bounded linear functional on $C([a, b], \mathbb{C})$ are more difficult to determine.*

In the measure-theoretic model of L_p this result can be established by an explicit expression. The approach above emphasizes that the result is not an artefact of a particular model but rather a consequence of the way the classical spaces are built out of the finite-dimensional building blocks.

This leaves us with the case $p = 1$. The reason why the same analysis above does not show that bounded functionals on ℓ_1 correspond to c_0 is that with respect to the norm $\| - \|_\infty$ the advancing sequence $\{\Phi_n\}_{n \geq 0}$ above, even if bounded, need not represent an element in the completion c_0; the space c_0 is not large enough for that to happen. Our analysis reveals how to correct the problem. To be able to represent all of the bounded functionals on ℓ_1 we must allow all bounded sequences, not just those that converge to 0.

Theorem 5.23 *The bounded functionals on ℓ_1 correspond to ℓ_∞ in a way that preserves the norm.*

For clarity we quote here the corresponding result for the function spaces.

Theorem 5.24 *The bounded functionals on L_1 correspond to L_∞ in a norm-preserving fashion.*

5.5.3   Non-separability

The common reason for the separability of ℓ_p and L_p for $1 \leq p < \infty$, as well as c_0 and $C([a, b], \mathbb{C})$, is the density of, respectively, c_{00} and $\mathbb{P}$. The space ℓ_∞ is larger than c_0 and similarly L_∞ is larger than $C([a, b], \mathbb{C})$. So much so in fact that they are no longer separable.

Theorem 5.25 *The spaces ℓ_∞ and L_∞ are not separable.*

Proof We treat the case of ℓ_∞. By Proposition 5.4 it suffices to exhibit an uncountable net in ℓ_∞. Consider the set $X = \{0, 1\}^\infty$ of all sequences composed of 0's and 1's. For any two such sequences x, y

$$\|x - y\|_\infty = \sup_{n \geq 1} |x_n - y_n| \in \{0, 1\}$$

since $x_n, y_n \in \{0, 1\}$. Thus, if $x \neq y$, then $\|x - y\|_\infty = 1$ and so X is a net. As for the uncountability of X, the set X is in bijection with $\mathcal{P}(\mathbb{N})$, the power set of the naturals, by means of the characteristic function. Cantor's Theorem tells us that the cardinality $|\mathcal{P}(\mathbb{N})|$ is strictly larger than that of $\mathbb{N}$, and thus uncountable. Using the measure-theoretic model of L_∞ a similar analysis will show its non-separability. $\square$

Exercises

Exercise 5.47 Prove that the set $(\mathbb{Q} + i\mathbb{Q})^n$ is dense in $\mathbb{C}^n$ for all $n \geq 1$. What happens if you replace n by an infinite cardinality?

Exercise 5.48 For the correspondence $F \mapsto \Phi$ described in Sect. 5.5.2 between linear functionals F on $(\mathbb{C}^n, \| - \|_p)$ and vectors Φ in $(\mathbb{C}^n, \| - \|_q)$, verify that $\|F\| = \|\Phi\|_q$. More thoroughly, provide the details to prove Proposition 5.16.

Exercise 5.49 Recall that $\text{Hom}(V, W)$ stands for the linear space of all linear operators $T: V \to W$. Let us write $\text{Hom}(V, W)_B$ for its subset consisting of the bounded operators once norms are involved. Provide all the details to show that with respect to the norm $\| - \|_p$ there are bijections

$$\text{Hom}(c_{00}, \mathbb{C})_B \cong \text{Hom}(\ell_p, \mathbb{C})_B \text{ and } \text{Hom}(\mathbb{P}, \mathbb{C})_B \cong \text{Hom}(L_p, \mathbb{C})_B$$

for all $1 < p < \infty$. Show the boundedness is critical by exhibiting unbounded linear operators that do not extend to the completions. Investigate the case $p = \infty$.

Exercise 5.50 Let $1 < p < q < \infty$ satisfy $1/p + 1/q = 1$. Given $x \in \ell_q$ define $F_x: \ell_p \to \mathbb{C}$ by

$$F_x(y) = \sum_{n=1}^{\infty} x_n y_n.$$

Show that F_x is a bounded linear functional on ℓ_p with $\|F_x\| = \|x\|_q$. Further, show that $x \mapsto F_x$ is a bijective correspondence between ℓ_q and all bounded linear operators on ℓ_p. Note that this particular construction is the model-specific version of Theorem 5.21 obtained by following its proof for the sequence model of ℓ_p.

Further Reading

The classical sequence spaces are part of a large family of sequence spaces in Banach spaces and we refer to the excellent book [4] for a thorough presentation. For the role of finite-dimensional spaces in quantum computation theory see [6]. An elementary approach to the Lebesgue integral, and thus to the classical L_p spaces, can be found in [5]. A gentle introduction to functional analysis that includes an introductory treatment of measure theory is [2]. Finally, the less orthodox categorical approach to the subject can be found in [3].

References

1. Aubrun, G., Nechita, I.: The multiplicative property characterizes ℓ_p and L_p norms. Confluentes Mathematici (2011)
2. Cannarsa, P., D'Aprile, T.: Introduction to Measure Theory and Functional Analysis, vol. 89. Springer (2015)
3. Castillo, J.M.F.: The Hitchhiker guide to categorical banach space theory. Part I. Extracta Mathematicae **25**(2), 103–149 (2010)

4. Diestel, J.: Sequences and Series in Banach Spaces, vol. 92. Springer Science & Business Media (2012)
5. Mikusiński, J., Mikusiński, P.: An Introduction to Analysis. World Scientific (2017)
6. Nielsen, M.A., Chuang I.L.: Quantum Computation and Quantum Information. Cambridge University Press, Cambridge (2010)

Chapter 6
Banach Spaces

Abstract normed spaces were introduced in Chap. 2 and the classical ℓ_p and L_p spaces were discussed in Chap. 5. The norm in a metric space endows the space with a metric structure that coexists particularly nicely with the linear space structure. We saw that in the finite-dimensional case the topology induced by the norm is an invariant of the dimension and plays no central role in the general theory. However, in the infinite-dimensional case not all norms are equivalent and topology starts exerting its influence. In the course of this chapter we shall see how to use the topology in order to gain more control over an infinite-dimensional normed space than would be possible only algebraically.

We shall promptly revisit the definition here and study the topology and metric structure of normed spaces.

Section 6.1 presents the definition of semi-normed, normed, and Banach spaces, establishes several fundamental continuity results in general normed spaces, including the Open Mapping Theorem, a result of great importance. Section 6.2 is an investigation of a powerful technique for solving a problem by recasting it in the form of a fixed point problem, and then resorting to Banach's Fixed Point Theorem. This technique is exemplified in detail for the solution of certain classes of differential and integral equations, namely to Volterra equations and to Fredholm equations. Section 6.3 starts off with a general study of inverse operators, motivated by an analysis of fixed point problems, and then proceeds to apply this more powerful technique to strengthen the results obtained in the previous section. Section 6.4 investigates dual spaces, presents the duals of the classical spaces, and the Hahn-Banach Theorem. The first four sections represent the core material of the theory of bounded linear operators between normed spaces. Section 6.5 is a short introduction to two fruitful directions in which the theory can be generalized, namely by considering unbounded operators between normed spaces and by introducing locally convex spaces.

© Springer Nature Switzerland AG 2021
C. Alabiso and I. Weiss, *A Primer on Hilbert Space Theory*, UNITEXT for Physics,
https://doi.org/10.1007/978-3-030-67417-5_6

6.1 Semi-norms, Norms, and Banach Spaces

Banach spaces are normed spaces in which the algebra and geometry mesh very well together, giving rise to powerful theorems. We introduce Banach spaces in a broad context, starting with general semi-normed and normed spaces, a basic investigation of the topology of normed spaces, followed by a study of bounded linear operators and the Open Mapping Theorem.

6.1.1 Semi-norms and Norms

The notion of a norm on a linear space is an association of lengths to vectors in such a way that the linear structure is taken into consideration. In practice, for various reasons, it is convenient to introduce a weaker structure, that of a semi-norm on a linear space.

Definition 6.1 Let V be a linear space over $K = \mathbb{R}$ or over $K = \mathbb{C}$. A *semi-norm* on V is a function associating with every vector $x \in V$ a non-negative real number $p(x)$, the *norm* of x, in such a way that

$$p(\alpha x) = |\alpha| p(x)$$
$$\textit{Triangle inequality:} \qquad p(x + y) \le p(x) + p(y)$$

for all $\alpha \in K$ and $x, y \in V$. If $p(x) = 0$ implies $x = 0$ for all $x \in V$, then the function $\|x\| = p(x)$ is exactly a norm on V. A linear space V together with a semi-norm on it is called a *semi-normed space*. We continue to refer to a (semi-)normed space V, leaving $p(x)$ or $\|x\|$, implicit.

Example 6.1 Numerous examples of normed spaces were already presented in Chap. 2 so we only consider here semi-norms. Firstly, and quite trivially, any linear space V supports the *trivial* semi-norm given by $p(x) = 0$ for all $x \in V$. Less trivially, consider the linear space $C(\mathbb{R}, \mathbb{R})$ of all continuous functions $x : \mathbb{R} \to \mathbb{R}$. The *evaluation* semi-norm is given by $\|x\| = |x(0)|$. The verification of the axioms is immediate:

$$\|\alpha x\| = |\alpha x(0)| = |\alpha||x(0)| = |\alpha|\|x\|$$

and

$$\|x + y\| = |x(0) + y(0)| \le |x(0)| + |y(0)| = \|x\| + \|y\|.$$

Note that it is obvious that $\|x\| = 0$ need not imply $x = 0$, only that $x(0) = 0$, so this semi-norm is not a norm. This construction is a typical one giving rise to semi-norms by ignoring some of the information embodied in the vectors. Naturally, one may evaluate at any point, not just at $x = 0$, and obtain a family of semi-norms.

There is another natural way to obtain semi-norms. The space $C([a, b], \mathbb{R})$ is a space of continuous functions. If one wishes to consider non-continuous functions and to utilize the integral in order to obtain a norm similar in spirit to the L_∞ or L_p norms on $C([a, b], \mathbb{R})$, then one encounters the following difficulty. It is well-known that for a non-continuous function x, it is possible that $\int dt \, |x(t)| = 0$ while $x \neq 0$. For instance, if the function is constantly 0 except at finitely many points. Thus, by allowing non-continuous functions into our linear spaces, the insensitivity of the integral to finite changes in the integrand implies that such norms are rare. Semi-norms on such spaces though are plentiful. We will also see below one more instance where semi-norms arise naturally from norms, namely when transferring a norm from a space to a quotient of it.

We now describe the process of turning a semi-normed space into a normed one. This, of course, cannot be achieved without paying a price. In this case all non-zero vectors of norm 0 must be quotiented out of existence. This is a small price to pay since vectors of norm 0 are, in a sense, negligible. This is a useful method for constructing normed spaces, since in practice it is common to have a natural choice of a semi-norm on a given space, and then one passes to the quotient we now describe.

Theorem 6.1 *Let V be a semi-normed space.*

1. *The set $U = \{x \in V \mid p(x) = 0\}$ is a linear subspace of V.*
2. *The function*

$$\|x + U\| = p(x)$$

is well defined on the quotient space V / U and with it V / U is a normed space.

Proof

1. In a semi-normed space we have $p(0) = p(0 \cdot 0) = |0| \cdot p(0) = 0$ and thus $0 \in U$, which is thus in particular non-empty. It remains to show that U is closed under addition and scalar products. Indeed, if $x, y \in U$, then

$$0 \leq p(x + y) \leq p(x) + p(y) = 0 + 0 = 0,$$

and thus $p(x + y) = 0$, so $x + y \in U$. Finally, if $x \in U$ and $\alpha \in K$ is an arbitrary scalar, then $p(\alpha x) = |\alpha| p(x) = |\alpha| \cdot 0 = 0$, so that $\alpha x \in U$.
2. To show that $x + U \mapsto p(x)$ is well defined suppose that $x' + U = x + U$. Then $x - x' \in U$, namely $p(x - x') = 0$, and thus

$$\|x + U\| = p(x) = p((x - x') + x') \leq p(x - x') + p(x') = p(x') = \|x' + U\|.$$

A similar argument shows that $\|x' + U\| \leq \|x + U\|$ and consequently $\|x + U\| = \|x' + U\|$ so that $x + U \mapsto p(x)$ is independent of the choice of representative. Next, clearly, $\|x + U\| \geq 0$ for all $x + U \in V/U$ and

$$\|\alpha(x + U)\| = \|\alpha x + U\| = p(\alpha x) = |\alpha| p(x) = |\alpha| \|x + U\|$$

for all $\alpha \in K$ and $x + U \in V/U$. A similar argument proves the triangle inequality. It remains to show that if $\|x + U\| = 0$, then $x + U = 0$ in the quotient space, namely that $x \in U$. Indeed, since $\|x + U\| = p(x)$, if $\|x + U\| = 0$, then $x \in U$ by definition of U. To conclude, $p(x)$ induces a norm on the quotient space V/U, as claimed.

□

We already proved that a normed space is automatically endowed with a distance function. We extend this simple (yet pivotal) observation to make a connection between semi-normed spaces and semi-metric spaces as well. Recall that a semi-metric space is a set together with a distance function $d : X \times X \to \mathbb{R}_+$ satisfying the axioms of a metric space, except that it is possible that $d(x, y) = 0$ for distinct points x and y. For the reader's convenience we repeat the statement and the proof for normed spaces.

Theorem 6.2 *Let V be a semi-normed space. Then defining $d(x, y) = p(x - y)$ for all vectors $x, y \in V$, endows V with the structure of a semi-metric space. If V is a normed space, then $d(x, y) = p(x - y) = \|x - y\|$ endows V with the structure of a metric space. These are, respectively, the* induced *semi-metric and metric structures.*

Proof Clearly $d(x, y) \geq 0$. Symmetry follows by

$$d(x, y) = p(x - y) = p((-1)(y - x)) = |-1|p(y - x) = d(y, x)$$

while the triangle inequality follows by

$$d(x, z) = p(x - z) = p(x - y + y - z) \leq p(x - y) + p(y - z) = d(x, y) + d(y, z).$$

Finally, if the semi-norm is in fact a norm, then

$$d(x, y) = 0 \implies \|x - y\| = 0 \implies x - y = 0 \implies x = y,$$

completing the proof.

□

Recall that there is a canonical construction for turning a semi-metric space into a metric space. In more detail, if (X, d) is a semi-metric space, then defining $x \sim y$ precisely when $d(x, y) = 0$ yields an equivalence relation on X and the function $d([x], [y]) = d(x, y)$ is well defined and turns the quotient set $X/\sim$ into a metric space. The next result shows that starting with a semi-normed space, turning it into a metric space by first constructing the associated semi-metric and then turning it into a metric space, or first turning it into a normed space and then associating a metric space with it yield the same end-result.

Theorem 6.3 *Let V be a semi-normed space. Consider the induced normed space $V/U = \{x + U \mid x \in V\}$, where $U = \{x \in V \mid p(x) = 0\}$, and the induced metric space $(V/U, d)$. Consider next the induced semi-metric space (V, d) and the induced metric space $(V/\sim, d)$ where $\sim$ is the equivalence relation on V given by $x \sim y$ precisely when $d(x, y) = 0$. Then the two resulting metric spaces are identical.*

Proof First we show that the two resulting metric spaces have identical underlying sets. Indeed, for the second described metric space we consider the quotient set $V/\sim$ where $x \sim y$ if, and only if, $d(x, y) = 0$. But $d(x, y) = p(x - y)$ and thus $x \sim y$ if, and only if, $x - y \in U$. It follows that $[x] = x + U$ and thus the quotient set $V/\sim$ coincides with the quotient set construction for V/U. Next we show that the metrics agree as well. The induced metric on V/U is given by

$$d(x + U, y + U) = \|x - y + U\| = p(x - y)$$

while the metric d on $V/\sim$ is given in terms of the semi-metric d on V as

$$d([x], [y]) = d(x, y) = p(x - y).$$

The two metrics are thus identical, as claimed. $\qquad\square$

A subspace U of a semi-normed space (V, p) is automatically a semi-normed space by defining $y \mapsto p(y)$ for every $y \in U$, using the semi-norm in the ambient space. Similarly, any subspace U of a normed space $(V, \|\cdot\|)$ is a normed space with norm given by $y \mapsto \|y\|$, for all $y \in U$. The proofs are immediate. The situation for quotient spaces is slightly more complicated since topological considerations enter the discussion. Recall that a metric induces a topology so the concepts of topology are available in any normed space. In particular, we may speak of closed subsets of a normed space V.

Theorem 6.4 *Let V be a normed space and U a linear subspace of V. Then the quotient space V/U, when endowed with the function*

$$p(x + U) = \inf_{u \in U} \|x + u\|,$$

is a semi-normed space. If U is a closed subspace of V, then V/U is a normed space.

Proof The quotient vector space V/U is the set $\{x + U \mid x \in V\}$ of translations of U with the linear operations determined by the representatives. We need to verify that the proposed formula

$$p(x + U) = \inf_{u \in U} \|x + u\|$$

satisfies the conditions for being a semi-norm. Clearly, $p(x + U) \geq 0$ since it is an infimum of non-negative real numbers. Given $\alpha \in K$ and $x + U \in V/U$ we need to show that

$$p(\alpha x + U) = |\alpha| p(x + U).$$

If $\alpha = 0$, then

$$p(\alpha x + U) = p(0 + U) = \inf_{u \in U} \|u\| = 0$$

(since $0 \in U$ and $\|0\| = 0$), and the equality holds. If $\alpha \neq 0$, then

$$p(\alpha x + U) = \inf_{u \in U} \|\alpha x + u\| = \inf_{u \in U} \|\alpha x + \alpha \frac{1}{\alpha} u\| = |\alpha| \inf_{u \in U} \|x + \frac{1}{\alpha} u\|$$

and noting that $\{(1/\alpha)u\}_{u \in U}$ and $\{u\}_{u \in U}$ are the same set of vectors yields the desired equality. To establish the triangle inequality we need to show that

$$\inf_{u \in U} \|x + y + u\| \leq \inf_{u' \in U} \|x + u'\| + \inf_{u'' \in U} \|y + u''\| = \inf_{u',u'' \in U} \|x + u'\| + \|y + u''\|.$$

It suffices to establish that, given $u', u'' \in U$,

$$\inf_{u \in U} \|x + y + u\| \leq \|x + u'\| + \|y + u''\|.$$

This is the case since

$$\inf_{u \in U} \|x + y + u\| \leq \|x + y + (u' + u'')\| = \|(x + u') + (y + u'')\| \leq \|x + u'\| + \|y + u''\|.$$

We thus far established that V/U, with the proposed function, is a semi-normed space. To complete the proof we show that if U is closed in V, then V/U is in fact a normed space, namely that for $\|x + U\| = p(x + U)$, if $\|x + U\| = 0$, then $x + U = 0$ in the quotient space, that is that $x \in U$. So suppose that

$$\|x + U\| = \inf_{u \in U} \|u - x\| = 0.$$

Now, by definition of infimum, there exists a sequence $u_n \in U$ with $\|u_n - x\| \to 0$. In other words, $d(u_n, x) \to 0$ and thus $u_n \to x$ in the induced topology. We thus exhibited x as a limit of a sequence in the closed set U, and thus x itself belongs to U, as required. $\square$

To see that if U is not closed then the quotient space may indeed only be a semi-normed space, consider any normed space V and a dense linear subspace U of it. The quotient space V/U with the induced semi-norm

$$p(x + U) = \inf_{u \in U} \|x + u\| = \inf_{u \in U} \|u - x\|$$

allows for $p(x + U) = 0$ even if $x \notin U$. In fact, $p(x + U) = 0$ for all $x \in V$, in other words the induced semi-norm on V/U degenerates to the trivial semi-norm: for $x \in V$ choose a sequence in U that converges to x and then

$$p(x + U) \leq \inf_{n} \|u_n - x\| = 0.$$

6.1.2 Banach Spaces

Any finite-dimensional normed space is complete. However, an infinite-dimensional normed space need not be complete, e.g., the space c_{00}. Non-complete normed spaces exhibit pathological behaviour. In some sense, such spaces are full of holes. In the rest of this chapter we explore some of the consequences of the powerful interaction between algebra and topology when a normed space is also a Banach space, namely a complete metric space.

Definition 6.2 A normed space that with the metric induced by the norm is complete is a *Banach space*, which we typically denote by $\mathcal{B}$.

Example 6.2 The discussion of the classical spaces in Chap. 5 provides us with important examples of Banach spaces. In particular, any finite-dimensional normed space is a Banach space. The sequence spaces ℓ_p for all $1 \leq p < \infty$ as well as c_0 of sequences that converge to 0 arise as completions of c_{00} with respect to suitable norms and are thus Banach spaces. Similarly, the function spaces L_p for all $1 \leq p < \infty$ as well as $C([a, b], \mathbb{R})$ and $C([a, b], \mathbb{C})$ of continuous functions are completions of the polynomial space $\mathbb{P}$ with respect to suitable norms so these too are Banach spaces. The spaces ℓ_∞ and L_∞ of bounded sequences and of essentially bounded functions, each with its own supremum norm $\| - \|_\infty$ are Banach spaces as well.

We emphasize that for all $1 \leq p < \infty$ the spaces $C([a, b], \mathbb{R})$ and $C([a, b], \mathbb{C})$ are not complete: convergence in the integral norm is not sufficiently strong to preserve the delicate notion of continuity.

Recall that a series $\sum_{k=1}^{\infty} a_k$ of real or complex numbers is said to converge to s if the sequence $\{s_m\}_{m \geq 1}$ of partial sums, i.e., $s_m = \sum_{k=1}^{m} a_k$, converges to s. We thus see that convergence for series is defined in terms of sequences. Moreover, one can recover the convergence of sequences in terms of that of series, as follows. Given a sequence $\{x_m\}_{m \geq 1}$ consider the series $\sum_{k=1}^{\infty} a_k$ where $a_k = x_k - x_{k-1}$ (and $a_1 = x_1$). The partial sum s_m is then nothing but x_m, and thus if the series converges, then so does the sequence. Some aspects of the interplay between sequences and series in $\mathbb{R}$ and $\mathbb{C}$ extend to normed spaces.

Definition 6.3 For elements $\{a_m\}_{m \geq 1}$ in a normed space V we say that the *series* $\sum_{k=1}^{\infty} a_k$ *converges* to $s \in V$ if the sequence of partial sums converges to s with respect to the norm in V, i.e., when

$$\lim_{m \to \infty} \left\| \left(\sum_{k=1}^{m} a_k \right) - s \right\| = 0.$$

We say that a series

$$\sum_{k=1}^{\infty} a_k$$

converges absolutely if the series of real numbers

$$\sum_{k=1}^{\infty} \|a_k\|$$

converges.

Recall that if a series $\sum_{k=1}^{\infty} a_k$ of real (or complex) numbers converges absolutely, that is if $\sum_{k=1}^{\infty} |a_k|$ converges, then the original series converges as well. Viewing $\mathbb{R}$ (or $\mathbb{C}$) as a normed space, this result is a special case of the following useful criterion for a normed space to be a Banach space.

Theorem 6.5 *A normed space V is a Banach space if, and only if, the absolute convergence of a series in V implies its convergence.*

The proof is left for the reader as we recount how the algebraic structure in a normed space interacts with the metric and topological structures induced by the norm, essentially repeating Theorem 5.2.

Proposition 6.1 *The following assertions hold for any normed space V.*

1. *The mapping $x \mapsto \|x\|$, as a function $V \to \mathbb{R}$, is non-expanding and thus the norm function is (uniformly) continuous.*
2. *For any fixed $x_0 \in V$ the translation mapping $x \mapsto x_0 + x$ is a global isometry and a homeomorphism.*
3. *For any $\alpha \in K$ with $\alpha \neq 0$ the scaling mapping $x \mapsto \alpha x$, as a function $V \to V$, is uniformly continuous and a homeomorphism.*
4. *The mapping $(x, y) \mapsto x + y$, as a function $V \times V \to V$, is continuous (and in fact uniformly continuous when $V \times V$ is given the additive product metric).*

Proof

1. Refering to Proposition 2.13 we have

$$d_{\mathbb{R}}(\|x\|, \|y\|) = |\|x\| - \|y\|| \leq \|x - y\| = d_V(x, y).$$

2. Since $\|(x_0 + x) - (x_0 + x')\| = \|x - x'\|$ any translation is an isometry. Further, the translation $x \mapsto x_0 + x$ is clearly invertible, its inverse being the translation $x \mapsto -x_0 + x$. An isometry is certainly continuous, and thus any translation is a continuous function with continuous inverse, and thus a homeomorphism.
3. Since $\|\alpha x - \alpha x'\| = |\alpha| \|x - x'\|$, it follows that any scaling mapping is uniformly continuous. The scaling mapping $x \mapsto \alpha x$ is clearly invertible, with inverse the scaling mapping $x \mapsto (1/\alpha)x$. Any non-zero scaling is thus a continuous function with a continuous inverse, thus a homeomorphism.
4. Left for the reader.

$\square$

6.1.3 Bounded Operators

Bounded linear operators were given a preliminary look in Sect. 5.1. The main result we establish now is that such operators provide a rich source of examples of Banach spaces: the collection of bounded linear operators between normed spaces is itself a normed space and is a Banach space if the codomain is a Banach space.

Recall that for an operator $A : V \to W$ between normed spaces the operator norm of A is

$$\|A\| = \sup_{\|x\|=1} \|Ax\|$$

and that A is bounded precisely when $\|A\| < \infty$. Moreover, boundedness for linear operators is equivalent to continuity with respect to the metrics induced by the norms (see Theorem 5.3).

We look at two familiar linear operators, only one of which is bounded.

Example 6.3 In $C([a, b], \mathbb{R})$, the linear space of continuous functions $x : [a, b] \to \mathbb{R}$ with the L_∞ norm, consider the integral operator:

$$(Ax)(s) = \int_a^b dt \, K(t, s) \, x(t)$$

with $K : [a, b] \times [a, b] \to \mathbb{R}$ a given continuous function. To see that the operator A is bounded note that if $M = \max_{t,s \in [a,b]} |K(t, s)|$, then

$$\|Ax\| = \max_{s \in [a,b]} \left| \int_a^b dt \, K(t, s) \, x(t) \right| \leq \max_{t,s \in [a,b]} |K(t, s)| \left| \int_a^b dt \, x(t) \right| \leq$$
$$\leq M(b - a) \max_{t \in [a,b]} |x(t)| = M(b - a)\|x\|.$$

In a sense, the operator owes its boundedness to the robustness of integration; small perturbations have small effects on the integral.

Example 6.4 Consider the operator $A : C^1([0, 1], \mathbb{R}) \to C([0, 1], \mathbb{R})$, where both the domain and the codomain are given the L_∞ norm, that maps a continuously differentiable function $x : [0, 1] \to \mathbb{R}$ to its derivative, i.e., $Ax = dx/dt$. The fact that A is not bounded is seen for example by considering $x_n(t) = t^n$ and noting that $\|x_n\| = 1$ but $Ax_n(t) = nt^{n-1}$, and thus $\|Ax_n\| = n$.

The unboundedness of differentiation is a reflection of its local nature, e.g., the function $t \mapsto t^n$ increases rapidly but over a very short interval near 1.

6.1.4 The Open Mapping Theorem

A mathematical structure typically has an underlying set and the relevant functions between the structures are the structure-preserving functions. If $f : A \to B$ is such a structure-preserving function and it is a bijection of the underlying sets, there is no a-priori reason for the inverse function $f^{-1} : B \to A$ to still preserve the structures. Whether this is the case depends delicately on the actual structure. Typically in algebraic situations, e.g., if A and B are linear spaces and f is a linear operator, then the inverse does preserve the structure. In more geometric scenarios, e.g., if A and B are topological spaces and f is a continuous function, then the inverse may very well fail to preserve the structures. A Banach space is a synergy of algebra and geometry and so it is not obvious whether a given bounded linear operator $A : \mathcal{A} \to \mathcal{B}$ between Banach spaces that is also a bijection has an inverse that is still a bounded linear operator. The inverse will certainly be a linear operator but will it be continuous? Since the question is primarily topological, we phrase it that way and in slightly greater generality. A continuous function between topological spaces need not be open even if it is surjective (i.e., it need not send open subsets to open subsets). We show that surjectivity is a sufficient condition for a bounded linear operator between Banach spaces to be open. The result is due to Stefan Banach and is our first example of a deep result in the theory of Banach spaces, one with profound consequences.

The technical heart of the argument is stated as a separate lemma. Its proof is omitted in favour of a more conceptual presentation of the the main result. We refer the reader to any text dedicated to Banach spaces in order to fill-in the details and obtain a fully rigorous proof.

It will be convenient to introduce the notation $0_V(s) = \{x \in V \mid \|x\| < s\}$, the open ball in a normed space V with centre the zero vector 0 and radius $s > 0$. We further write $n \cdot 0_V(s)$ for the point-wise scaling of $0_V(s)$ by n, namely

$$n \cdot 0_V(s) = \{nx \mid x \in 0_V(s)\} = 0_V(ns).$$

Lemma 6.1 *Let $A : \mathcal{B} \to V$ be a bounded linear operator from a Banach space $\mathcal{B}$ to a normed space V. It then follows that if $r, s > 0$ are such that*

$$0_V(s) \subseteq \overline{A(0_{\mathcal{B}}(r))},$$

then

$$0_V(s) \subseteq A(0_{\mathcal{B}}(r)).$$

Theorem 6.6 (Open Mapping Theorem) *A bounded surjective linear operator $A : \mathcal{B}_1 \to \mathcal{B}_2$ between Banach spaces is an open mapping.*

Proof Let $U \subseteq \mathcal{B}_1$ be an open set, that we may assume is non-empty. For some $x_0 \in U$ let us consider Ax_0, an arbitrary element in $A(U)$. As U is open, let $\varepsilon > 0$ be such that $B_\varepsilon(x_0) \subseteq U$, and notice that $B_\varepsilon(x_0) = 0_{\mathcal{B}_1}(\varepsilon) + x_0$, i.e., a ball with centre x_0 is the x_0 translation of a ball centered at the origin. By linearity of A we then have

$$A(U) \supseteq A(x_0 + 0_{\mathcal{B}_1}(\varepsilon)) = Ax_0 + A(0_{\mathcal{B}_1}(\varepsilon)).$$

Thus the proof is reduced to showing that for every $\varepsilon > 0$ there exists a $\delta > 0$ such that $0_{\mathcal{B}_2}(\delta) \subseteq A(0_{\mathcal{B}_1}(\varepsilon))$, since then

$$A(U) \supseteq Ax_0 + 0_{\mathcal{B}_2}(\delta) = B_\delta(Ax_0)$$

and thus Ax_0 is an interior point of $A(U)$. As Ax_0 was arbitrary, every point of $A(U)$ is interior, and thus $A(U)$ is open.

We thus fix an open ball $0_{\mathcal{B}_1}(\varepsilon)$, $\varepsilon > 0$, denote $X = A(0_{\mathcal{B}_1}(\varepsilon))$, and we seek out a $\delta > 0$ for which $0_{\mathcal{B}_2}(\delta) \subseteq X$. Noting that

$$\mathcal{B}_1 = \bigcup_{n \geq 1} n \cdot 0_{\mathcal{B}_1}(\varepsilon)$$

the surjectivity and homogeneity of A implies that

$$\mathcal{B}_2 = A(\mathcal{B}_1) = \bigcup_{n \geq 1} n \cdot X.$$

Taking closures and applying Baire's Theorem, there exists an $n \geq 1$ such that $\overline{n \cdot X}$ contains an open ball. In other words, there exists a $\delta > 0$ such that $B_\delta(y) \subseteq \overline{n \cdot X}$. We omit some details on translations and scalings that are used to conclude that in fact from $B_\delta(y) \subseteq \overline{n \cdot X}$ it follows that $0_{\mathcal{B}_2}(\delta/n) \subseteq \overline{X}$. Lemma 6.1 then concludes the proof. $\qquad \square$

A consequence of the Open Mapping Theorem illustrates some of the taming effect of the completeness requirement on Banach spaces.

Corollary 6.1 (Bounded Inverse Theorem) *If a bounded linear operator $A : \mathcal{B}_1 \rightarrow \mathcal{B}_2$ between Banach spaces is invertible, then its inverse A^{-1} is also a bounded linear operator.*

Proof By Proposition 2.8 the inverse of a linear operator is itself linear. It remains to show that A^{-1} is continuous, namely that $(A^{-1})^{-1}(U)$ is an open set whenever U is open. Since $(A^{-1})^{-1} = A$, we need to show that $A(U)$ is open whenever U is open, namely that A is an open mapping. But A is invertible, hence surjective, and thus A is an open mapping by the Open Mapping Theorem. $\qquad \square$

6.1.5 Banach Spaces of Linear and Bounded Operators

We conclude this section by studying the set of mappings $\mathbf{B}(V, W)$ of all bounded linear operators $A : V \to W$ between normed spaces. The main result is that when endowed with the operator norm it is a normed space. Furthermore, if the codomain is a Banach space, then so is the space of mappings into it.

Theorem 6.7 *Let* $A, B : V \to W$ *be bounded linear operators between normed spaces. Then*

1. $\|A\| = 0$ *if, and only if,* $A = 0$.
2. $\|\alpha A\| = |\alpha| \|A\|$ *for all scalars* α.
3. $\|A + B\| \leq \|A\| + \|B\|$.

Proof

1. If $A = 0$ then $\|Ax\| \leq 0 \cdot \|x\|$ and thus $\|A\| = 0$. Conversely, if $\|A\| = 0$, then $\|Ax\| \leq \|A\|\|x\| = 0$ and thus $Ax = 0$, for all $x \in V$.
2. $\|\alpha A\| = \sup_{\|x\|=1} \|\alpha Ax\| = |\alpha| \sup_{\|x\|=1} \|Ax\| = |\alpha| \|A\|$.
3. $\|A + B\| = \sup_{\|x\|=1} \|(A + B)x\| \leq \sup_{\|x\|=1} (\|Ax\| + \|Bx\|) = \|A\| + \|B\|$.

$\square$

Recall Theorem 2.3 stating that for linear spaces V and W the set $\mathrm{Hom}(V, W)$ of all linear operators $A : V \to W$ is a linear space.

Theorem 6.8 *For normed spaces* V *and* W *the set* $\mathbf{B}(V, W)$ *is a linear subspace of* $\mathrm{Hom}(V, W)$. *Endowed with the operator norm it is a normed space.*

Proof Theorem 6.7 states that $\mathbf{B}(V, W)$ contains 0 and is closed under addition and scalar multiplication, and is thus a linear subspace of $\mathrm{Hom}(V, W)$. Further, the same theorem establishes the axioms of a normed space for the operator norm on $\mathbf{B}(V, W)$ and the claim follows. $\square$

Since $\mathbf{B}(V, W)$ is a normed space it is also a metric space and thus a corresponding notion of convergence for operators is automatic.

Definition 6.4 (*Operator Norm Convergence or Strong Convergence*) Let V and W be normed spaces. Given a sequence $\{A_m\}_{m \geq 1}$ of operators in $\mathbf{B}(V, W)$ we say that $\{A_m\}_{m \geq 1}$ *converges strongly* or *converges in the operator norm* to an operator $A_0 \in \mathbf{B}(V, W)$ if $\|A_m - A_0\| \to 0$ (equivalently, when $A_m \to A_0$ with respect to the metric induced by the operator norm).

A weaker notion of convergence is also available in the space $\mathbf{B}(V, W)$.

Definition 6.5 Let V and W be normed spaces. We say that a sequence $\{A_m\}_{m \geq 1}$ of operators in the space $\mathbf{B}(V, W)$ *converges point-wise* to an operator $A_0 \in \mathbf{B}(V, W)$ if $A_m x \to A_0 x$ for all $x \in V$.

Since $\|A_m x - A_0 x\| \leq \|A_m - A_0\| \|x\|$ it follows that strong convergence implies point-wise convergence. We illustrate that the converse may fail.

Example 6.5 Consider the sequence of linear operators $F_m : \ell_2 \to \mathbb{C}$ given by $F_m x = x_m$, that is F_m is the projection of the m-th coordinate, and thus is clearly a linear operator. Since

$$\|F_m x\| = |x_m| \leq \|x\|$$

we see that $\|F_m\| \leq 1$ and thus F_m is bounded (and in fact $\|F_m\| = 1$ as there clearly exist $x \in \ell_2$ with $\|F_m x\| = \|x\|$).

We claim that $F_m \to 0$ point-wise, i.e., that $F_m x \to 0$ for all $x \in \ell_2$. Indeed, for $x \in \ell_2$ one has

$$\sum_{m=1}^{\infty} |x_m|^2 < \infty$$

and consequently, $x_m \to 0$. But $x_m = F_m x$, and the claim follows. However, as noted, $\|F_m\| = 1$ and thus $\{F_m\}_{m \geq 1}$ does not converge strongly to 0, and thus does not converge at all.

Finally, we prove the completeness of $\mathbf{B}(V, \mathcal{B})$ when the codomain is complete.

Theorem 6.9 *Let $\mathcal{B}$ be a Banach space and V an arbitrary normed space. Then the linear space $\mathbf{B}(V, \mathcal{B})$ with the operator norm is a Banach space.*

Proof We already know that with the operator norm $\mathbf{B}(V, \mathcal{B})$ is a normed space so it remains to show that it is complete. Consider a Cauchy sequence $\{A_m\}_{m \geq 1}$. Since $\|A_k x - A_m x\| \leq \|A_k - A_m\| \|x\| \longrightarrow 0$ for all $x \in V$, it follows at once that $\{A_m x\}_{m \geq 1}$ is also a Cauchy sequence, and, as $\mathcal{B}$ is complete, $A_m x \to y$. We thus define $A_0(x) = y$ and obtain the function $A_0 : V \to \mathcal{B}$ that we proceed to show is a bounded linear operator.

The linearity of A_0 follows by standard limit arguments. For instance, continuity of vector addition implies that

$$A_0(x + x') = \lim_{m \to \infty} A_m(x + x') = \lim_{m \to \infty} A_m x + \lim_{m \to \infty} A_m x' = Ax + Ax'$$

and thus A_0 is additive. A similar argument shows A_0 is homogenous. To show that A_0 is bounded we use the continuity of the norm to obtain that

$$\|Ax\| = \|\lim_{m \to \infty} A_m x\| = \lim_{m \to \infty} \|A_m x\| \leq \lim_{m \to \infty} \|A_m\| \|x\| \leq M \|x\|$$

where M is any upper bound of the sequence $\|A_m\|$ (which exists since any Cauchy sequence is bounded).

Lastly, we verify that $A_m \to A_0$ in the operator norm. Given $\varepsilon > 0$, using the Cauchy condition, there exists an $N \in \mathbb{N}$ such that $\|A_m - A_k\| < \varepsilon$ for all $m, k > N$. In particular, for all $x \in V$,

$$\|(A_m - A_k)x\| \le \|A_m - A_k\|\|x\| < \varepsilon\|x\|,$$

for all $m, k \ge N$. When k tends to ∞ we thus obtain

$$\|(A_m - A_0)x\| \le \varepsilon\|x\|$$

for all $x \in V$ and thus
$$\|A_m - A_0\| \le \varepsilon$$

for all $m > N$. In other words, $\|A_m - A_0\| \to 0$, as required. $\square$

Exercises

Exercise 6.1 The aim of this exercise is to show that an infinite-dimensional Banach space is very large: its dimension is uncountable. The items along the way are of separate interest. Assume to the contrary that $\mathcal{B}$ is an infinite-dimensional Banach space and that it has a countable Hamel basis $\{u_1, u_2, \dots\}$. Let X_n be the span of $\{u_1, \dots, u_n\}$.

1. Show that every finite-dimensional linear subspace of $\mathcal{B}$ is closed. Conclude that X_n is closed.
2. Show that every proper linear subspace of $\mathcal{B}$ has empty interior. Conclude that X_n has empty interior.
3. Show that

$$X = \bigcup_{n \ge 1} X_n$$

and use Baire's Theorem to arrive at a contradiction.

Exercise 6.2 Consider the Euclidean spaces $\mathbb{R}^n$ with the standard inner product and its associated norm. Prove that any linear operator $A : \mathbb{R}^n \to \mathbb{R}^m$ is bounded.

Exercise 6.3 Prove Proposition 5.6

Exercise 6.4 Prove Theorem 6.5.

Exercise 6.5 Let U be a subspace of a normed space V and let $A : U \to W$ be a linear operator to a normed space W. Suppose that $B : V \to W$ is a linear operator that extends A. Show that $\|A\| \le \|B\|$.

Exercise 6.6 Show that point-wise and strong convergence in $\mathbf{B}(\mathbb{R}^n, \mathbb{R}^m)$ coincide.

Exercise 6.7 Prove that every infinite-dimensional normed space V either over the field $K = \mathbb{R}$ or the field $K = \mathbb{C}$ admits a discontinuous linear operator $A : V \to K$. (Hint: Use a Hamel basis.)

Exercise 6.8 Show that Lemma 6.1 may fail if the domain of the linear operator is not a Banach space.

Exercise 6.9 Let $\{A_m\}_{m\geq 1}$ be a sequence of operators $A_m : V \to W$ between normed spaces. Prove that if $\{A_m\}_{m\geq 1}$ is Cauchy, then $\{\|A_m\|\}_{m\geq 1}$ is Cauchy. Does the converse implication hold as well?

Exercise 6.10 Let $A : V \to W$ be a linear operator between normed spaces. Prove that if A is open, then A is surjective.

6.2 Fixed Point Techniques in Banach Spaces

Since a Banach space $\mathcal{B}$ is in particular a complete and non-empty metric space the Banach Fixed Point Theorem (Theorem 4.9) guarantees that any contraction on $\mathcal{B}$ admits a unique fixed point. In more detail, a contraction on a Banach space $\mathcal{B}$ is a function (not necessarily linear) $F : \mathcal{B} \to \mathcal{B}$ such that there exists $0 < \alpha < 1$ with

$$\|Fx - Fy\| \leq \alpha\|x - y\|.$$

The guaranteed unique fixed point for F is an element $x_0 \in \mathcal{B}$ with $F(x_0) = x_0$. From the proof of the Banach Fixed Point Theorem, the fixed point x_0 is obtained as the limit $x_k \to x_0$ where $x_{k+1} = F(x_k)$, with $x_1 \in \mathcal{B}$ arbitrary.

It follows that any problem in a Banach space $\mathcal{B}$ whose solution can be expressed as a fixed point for a contraction on $\mathcal{B}$ can be solved iteratively. We explore this technique in three different scenarios: systems of linear equations, first order differential equations, and integral equations.

Remark 6.1 Notice that the more general problem

$$x_0 = F(x_0) + b$$

is subsumed by the technique described above since for $G(x) = F(x) + b$ one has that

$$\|G(x) - G(y)\| = \|F(x) - F(y)\|$$

and thus F is a contraction if, and only if, G is. Moreover, a fixed point for G is precisely a solution of the equation $x_0 = F(x_0) + b$. It is customary to simplify the notation and write $x = Fx + b$. Notice that if F is a linear operator, then F is a contraction if, and only if, $\|F\| < 1$.

6.2.1 Systems of Linear Equations

Consider a linear system of n equations in n unknowns:

$$\sum_{i=1}^{n} a_{ki}x_i = b_k, \qquad k = 1, 2, \ldots, n.$$

In terms of operators this system of equations can be written as $Ax = b$ where A is the corresponding $n \times n$ matrix of coefficients viewed as a linear operator $\mathbb{R}^n \to \mathbb{R}^n$, with $x = (x_1, \cdots, x_n)$ representing the unknown vector and $b = (b_1, b_2, \ldots, b_n)$ the vector of free coefficients. Theoretically precise methods for solving such systems of linear equations are computationally demanding. For instance, Cramer's rule requires the computation of n^2 determinants and thus, for large n, is unfeasible. Moreover, numerical difficulties resulting from rounding may render some computation techniques unreliable. We now investigate the applicability of an iterative fixed point technique in this case. Such a solution is often more robust than naive applications of theoretical exact solutions (though we will not delve into a numerical analysis of the solution here).

Since $Ax = b$ if, and only if, $x = (-A + I_n)x + b$, where I_n is the $n \times n$ identity matrix, we are led to consider the equation

$$x = Fx + b$$

where $F = -A + I_n$. The original problem is equivalent to the fixed point problem

$$x = Gx$$

where $Gx = Fx + b = (-A + I_n)x + b$ is viewed as an operator $G : \mathbb{R}^n \to \mathbb{R}^n$. Let us assume some norm on $\mathbb{R}^n$ is chosen. For the Banach Fixed Point Theorem to be applicable G must be a contraction. Since F is a linear operator this is determined by the norm $\|F\|$: if $\|F\| < 1$, then F, and thus G, is a contraction and the original problem is amenable to the iterative solution $x_{k+1} = Gx_k$ with an arbitrary $x_1 \in \mathbb{R}^n$.

Whether F is a contraction is typically strongly dependent on the chosen norm for the ambient space $\mathbb{R}^n$. We consider two cases below, recalling that the Kronecker δ function is given by

$$\delta_{k,i} = \begin{cases} 1 & \text{if } k = i, \\ 0 & \text{if } k \neq i. \end{cases}$$

Example 6.6 Consider $\mathbb{R}^n$ with the ℓ_∞ norm: $\|x\| = \max_{1 \leq k \leq n} |x_k|$. Then, for any $n \times n$ matrix C

$$\|Cx - Cy\| = \max_{1 \leq k \leq n} \left| \sum_{i=1}^{n} c_{k,i}(x_i - y_i) \right| \leq \max_{1 \leq k \leq n} \left[\sum_{i=1}^{n} |c_{k,i}| \, |x_i - y_i| \right] \leq$$

$$\leq \max_{1 \leq k \leq n} \left[\sum_{i=1}^{n} |c_{k,i}| \right] \max_{1 \leq i \leq n} |x_i - x_i| = \max_{1 \leq k \leq n} \left[\sum_{i=1}^{n} |c_{k,i}| \right] \|x - y\|.$$

Thus, if the inequality

$$\max_{1 \leq k \leq n} \left[\sum_{i=1}^{n} |c_{k,i}| \right] < 1$$

holds, then the operator C is a contraction. Returning to the system $Ax = b$ and the related operator $Gx = (-A + I_n)x + b$, we see that G is guaranteed to be a contraction if

$$\max_{1 \leq k \leq n} \sum_{i=1}^{n} |-a_{k,i} + \delta_{k,i}| < 1,$$

in which case the original system can be solved iteratively.

Example 6.7 Let us now consider $\mathbb{R}^n$ with the ℓ_2 norm $\|x\|^2 = \sum_{k=1}^{n} x_k^2$, giving rise to the Euclidean metric. For an $n \times n$ matrix C viewed as an operator on $\mathbb{R}^n$ we have

$$\|Cx\|^2 = \sum_{k=1}^{n} (c_k \cdot x)^2$$

where c_k is the k-th row of the matrix C and $c_k \cdot x$ is the scalar product in $\mathbb{R}^n$. Applying the Cauchy-Schwarz Inequality we obtain

$$\|Cx\|^2 \leq \sum_{k=1}^{n} \|c_k\|^2 \|x\|^2$$

and thus we see that

$$\|C\| \leq \sqrt{\sum_{k=1}^{n} \|c_k\|^2} = \sqrt{\sum_{k,i=1}^{n} c_{k,i}^2}$$

Going back to the linear system $Ax = b$, we conclude that it is iteratively solvable if

$$\sum_{k,i=1}^{n} (-a_{k,i} + \delta_{k,i})^2 < 1.$$

6.2.2 Cauchy's Problem and the Volterra Equation

Consider Cauchy's problem for the first order differential equation:

$$\frac{dx}{dt}(t) = f[t, x(t)] \quad \text{with} \quad x(t_0) = x_0.$$

The associated integral equation

$$x(t) = \int_{t_0}^{t} ds \ f[s, x(s)] + x_0$$

is called a *Volterra integral equation*. Suppose that, for $|t - t_0| \le \varepsilon$ and $|x - x_0| \le \eta$, the function $f(t, x)$ is continuous and $|f(t, x)| < M$. Noting that a solution of the Volterra equation is also a solution to the original Cauchy problem, we introduce the operator

$$(Bx)(t) = \int_{t_0}^{t} ds \ f[s, x(s)] + x_0$$

since its fixed points are precisely solutions to the Volterra equation. If B can be seen to be an operator

$$B : C(I, \mathbb{R}) \to C(I, \mathbb{R})$$

for a suitable interval $I \subseteq [a, b]$, and if B is moreover a contraction with respect to some norm, then locally Cauchy's problem has a unique solution that, starting with an arbitrary x_0, may be computed iteratively:

$$x_1(t) = \int_{t_0}^{t} ds \ f(s, x_0) + x_0, \quad x_2(t) = \int_{t_0}^{t} ds \ f[s, x_1(s)] + x_0, \ldots, \quad x_n(t) = \ldots$$

Since the norm we choose must turn the space $C(I, \mathbb{R})$ into a Banach space (since otherwise Banach's Fixed Point Theorem does not apply) the only norm we can consider is the L_∞ norm (and now the importance of the L_p spaces becomes apparent as they are complete and thus they allow one to consider other norms).

Let us examine under which assumptions is it guaranteed that B is a contraction for a suitable interval. Consider $C(I, \mathbb{R})$ where $I = [t_0 - \varepsilon, t_0 + \varepsilon]$ and choose some interval $J = [x_0 - \eta, x_0 + \eta]$ contained in I. We seek to identify conditions on ε, η, and M that will assure the existence of a solution, at least locally. We note the following.

1. The function Bx is certainly a continuous function, given the assumptions on $f[t, s]$.
2.

$$\|Bx - x_0\| = \max_{t \in I} \left| \int_{t_0}^{t} ds \ f[s, x(s)] \right| \le$$
$$\le \max_{t \in I} |t - t_0| \ \max_{t \in I, \, x \in J} |f(t, x)| \le \varepsilon M.$$

We thus see that the operator B transforms the restriction of x to J into a function taking values in J provided that $\varepsilon \le \eta/M$, a condition we now assume is met.
3. For all $x, y \in J$ we have that

$$\|Bx - By\| \leq \max_{t \in I} \int_{t_0}^{t} ds \, \big|f[s, x(s)] - f[s, y(s)]\big|.$$

4. Suppose the function $f(t, x)$ is not just continuous and bounded but also satisfies the *Lipschitz* condition:

$$\big|f(t, x) - f(t, y)\big| \leq K|x - y|, \quad \forall \, x, y \in J,$$

uniformly for all $t \in I$. Then

$$\|Bx - By\| \leq \max_{t \in I} \int_{t_0}^{t} ds \, \big|f[s, x(s)] - f[s, y(s)]\big| \leq K \max_{t \in I} \int_{t_0}^{t} ds \, \big|x(s) - y(s)\big| \leq$$
$$\leq K\varepsilon \max_{s \in I} \big|x(s) - y(s)\big| = K\varepsilon \|x - y\|, \quad \forall \, x, y.$$

Collecting these observations together we conclude that if $\varepsilon \leq \eta/M$ and $\varepsilon \leq 1/K$, then the operator B is a contraction and Cauchy's problem has one, and only one, solution inside the *reduced interval* $I = [t_0 - \varepsilon, t_0 + \varepsilon]$, with $\varepsilon \leq \min(\eta/M, 1/K)$. Furthermore, the solution can be obtained iteratively as a fixed point solution of the associated Volterra equation.

6.2.3 Fredholm Equations

The Volterra equations, considered above, are examples of integral equations with variable limits. We turn to consider *Fredholm equations*, belonging to the family of integral equations with fixed limits.

First consider the general form of a non-linear Fredholm equation

$$x(t) = \lambda \int_{a}^{b} ds \, K\big[t, s, x(s)\big] = [Ax](t),$$

where we consider A as an operator and we seek to identify conditions under which the iterative fixed point method is applicable. Let $K(t, s, x)$ be a continuous function in $[a, b] \times [a, b] \times [-\eta, \eta]$ and bounded by $M > 0$. Let $J = [-\eta, \eta]$ and consider $C(J, \mathbb{R})$ with the L_∞ norm. Noticing that

$$\|Ax\| \leq \max_{t \in [a,b]} \bigg|\lambda \int_{a}^{b} ds \, K\big[t, s, x(s)\big]\bigg| \leq |\lambda| M(b - a)$$

we see that if $|\lambda| \leq \eta/\big[M(b - a)\big]$, then $\|Ax\| \leq \eta$ and so the operator A is well defined when restricting to the interval J. Furthermore,

$$\|Ax - Ay\| = \max_{t \in [a,b]} \left| \lambda \int_a^b ds \left\{ K[t, s, x(s)] - K[t, s, y(s)] \right\} \right| \le$$

$$\le |\lambda| \max_{t \in [a,b]} \int_a^b ds \left| K[t, s, x(s)] - K[t, s, y(s)] \right|.$$

If we also make the assumption that the function $K(t, s, x)$ is Lipschitz with respect to x, uniformly in the square $[a, b] \times [a, b]$, with Lipschitz constant K, then:

$$\|Ax - Ay\| \le |\lambda|(b - a)K \max_{s \in [a,b]} |x(s) - y(s)| \le |\lambda|(b - a)K \|x - y\|.$$

It thus follows that if $|\lambda|(b - a)K < 1$, then the operator A is a contraction, and thus the original non-linear Fredholm equation has a unique solution obtained by the iterative scheme

$$x_{n+1}(t) = \int_a^b ds\, K[t, s, x_n(s)]$$

with $x_0(t)$ any continuous function on $[a, b]$ satisfying $\max_t |x_0(t)| \le \eta$. The condition

$$|\lambda|(b - a) \le \min\{\eta/M, 1/K\}$$

guarantees the applicability of the iterative solution.

Consider now the linear Fredholm equation:

$$x(t) = \lambda \int_a^b ds\, K(t, s)x(s) + f(t)$$

with λ a real parameter, $f(t)$ a continuous function in $[a, b]$, and $K(t, s)$ continuous in the square $[a, b] \times [a, b]$ and bounded by $M > 0$. Still considering $C([a, b], \mathbb{R})$ with the L_∞ norm, consider the operator

$$(Bx)(t) = \lambda \int_a^b ds\, K(t, s)x(s) + f(t).$$

We now have that,

$$\|Bx - By\| = \max_{t \in [a,b]} \left| \lambda \int_a^b ds\, K(t, s)[x(s) - y(s)] \right| \le$$

$$\le |\lambda|M(b - a) \max_{s \in [a,b]} |x(s) - y(s)| = |\lambda|M(b - a)\|x - y\|$$

and thus, for $|\lambda| < 1/[M(b - a)]$, the operator B is a contraction and the uniform limit (i.e., with respect to the L_∞ norm) of the iterative process

$$x_{n+1}(t) = \lambda \int_a^b ds\, K(t, s)x_n(s) + f(t) = \ldots$$

exists and gives the solution to the given Fredholm equation. A possible choice of
the initial function is given by $x_0(t) = f(t)$.

Example 6.8 Consider in $C([0, 1], \mathbb{R})$ the Fredholm equation:

$$x(t) = \lambda \int_0^1 ds \, ts x(s) + \alpha t$$

with α a real parameter. Since $\max_{t,s \in [0,1]} K(t, s) = 1$ and $(b - a) = 1$, it follows
that for $|\lambda| < 1/[M(b - a)] = 1$ the operator is a contraction and one can solve the
problem by successive approximations:

$$x_0 = \alpha \, t = \alpha_0 \, t, \quad x_1 = \lambda \int_0^1 ds \, (t \, s) \, (\alpha_0 \, s) + \alpha \, t = (\alpha + \alpha_0 \, \lambda/3) \, t = \alpha_1 \, t, \, , \ldots$$

$$x_{n+1} = \lambda \int_0^1 ds \, (t \, s) \, (\alpha_n \, s) = (\alpha + \alpha_n \, \lambda/3) \, t = \alpha_{n+1} \, t, \, , \ldots$$

For $|\lambda| < 1$ the limit $x = \lim_{n \to \infty} x_n$ exists and thus also $\beta = \lim_{n \to \infty} \alpha_n$ exists.
When passing to the limit in the last equality we obtain that $\alpha + \beta \lambda/3 = \beta$, from
which we conclude that $\beta = 3\alpha/(3 - \lambda)$ and, finally, that $x(t) = 3\alpha/(3 - \lambda)t$.

We got this solution by assuming $|\lambda| < 1$, but one can easily see that the limit of α_n
as $n \to \infty$ also exists if $|\lambda| < 3$, as suggested by the last relation. In fact:

$$\alpha_{n+1} = \alpha + \alpha_n \, \lambda/3 = \cdots = \alpha \left[1 + \lambda/3 + (\lambda/3)^2 + \cdots + (\lambda/3)^{n+1} \right] \implies$$

$$\implies \quad \alpha_n(t) \to 3\alpha/(3 - \lambda) \quad \text{for} \quad |\lambda| < 3.$$

The condition $|\lambda| < 1$ is a sufficient condition for the convergence of the iterative
process which, as just seen, has a broader domain of convergence. More generally,
the function $x(t)$ found above is actually a solution of the original problem for every
$\lambda \neq 3$, as is easily checked with a test function $x(t) = \gamma t$, with γ unknown.

Exercises

Exercise 6.11 Show that a contraction on an arbitrary normed space need not have
a fixed point.

Exercise 6.12 Give an example of an isometry $F : \mathbb{R} \to \mathbb{R}$ without any fixed points
(here $\mathbb{R}$ is given the Euclidean metric structure).

Exercise 6.13 A function $f : X \to Y$ between metric spaces is a *Lipschitz* function
if there exists a number $K \in \mathbb{R}$, called a *Lipschitz constant* such that

$$d(f(x), f(x')) \leq K d(x, x')$$

for all $x, x' \in X$. Show that, with respect to the Euclidean metric on $\mathbb{R}$, a function $f : \mathbb{R} \to \mathbb{R}$ with a bounded derivative is Lipschitz.

Exercise 6.14 Prove that a Lipschitz function $f : X \to Y$ between metric spaces is uniformly continuous, but the converse may fail.

Exercise 6.15 Give an example of a function $F : X \to Y$ between metric spaces which satisfies $d(Fx, Fy) < d(x, y)$ for all $x, y \in X$, but is not a contraction.

Exercise 6.16 Perform a similar analysis as done in Example 6.7 but consider the ℓ_1 norm on $\mathbb{R}^n$ and attempt to obtain a reasonable criterion for the applicability of the fixed point technique for solving a system of linear equations.

Exercise 6.17 Referring to Example 6.4, prove, once more, that the differentiation operator d/dt on $C([a, b], \mathbb{R})$ is unbounded by considering the sequence

$$x_n(t) = \sin[n\pi(b - t)/(b - a)].$$

Exercise 6.18 With successive substitutions, solve the Volterra equation:

$$x(t) = t - \int_0^t ds(t - s)\, x(s).$$

Exercise 6.19 Following Example 6.8, solve the equation:

$$x(t) = \frac{5}{6}t + \frac{1}{2} \int_0^1 ds\, ts\, x(s).$$

Exercise 6.20 Following Example 6.8, solve the equation:

$$x(t) = e^t - 1 + \int_0^1 ds\, t\, x(s).$$

6.3 Inverse Operators

From Chap. 2 we know that if a linear operator $A : V \to W$ admits an inverse function $A^{-1} : W \to V$, then that function is automatically a linear operator. The Open Mapping Theorem implies that if the linear spaces are Banach spaces and the operator is bounded, then its inverse is also bounded. In this section we will establish a condition on A that guarantees the existence of a bounded inverse without assuming the spaces are Banach. We will then obtain a simple yet powerful result that presents the inverse of an operator in terms of a convergent series of operators. The analysis will reveal the close connection between invertibility and fixed point solutions. We will then have a second, deeper, look at the Volterra and Fredholm equations in light of the results below.

6.3.1 Existence of Bounded Inverses

We present conditions that guarantee the existence of inverse operators, providing estimates for their norm, and, under suitable conditions, we obtain the inverse as the sum of a strongly convergent series.

Theorem 6.10 *Let $A : V \to W$ be a bounded surjective linear operator between normed spaces. Then A is invertible with bounded inverse A^{-1} if, and only if, there exists a constant $C > 0$ such that $\|Ax\| \geq C\|x\|$, for all $x \in V$. Moreover, with such a constant C the operator norm of the inverse satisfies $\|A^{-1}\| \leq 1/C$.*

Proof Suppose that $\|Ax\| \geq C\|x\|$ holds. Then if $Ax = 0$, then $0 = \|Ax\| \geq C\|x\|$, and thus $x = 0$. In other words, $\mathrm{Ker}(A) = \{0\}$ and thus A is injective. Since it is given that A is surjective we conclude that A is invertible and that its inverse A^{-1} is automatically a linear operator. Moreover, for all $y \in W$ let $x \in V$ with $Ax = y$, and thus $x = A^{-1}y$. We then have

$$\|y\| = \|Ax\| \geq C\|x\| = C\|A^{-1}y\|,$$

showing that $\|A^{-1}\| \leq 1/C$. In the other direction, suppose that A^{-1} exists and is bounded. Then, for all $x \in V$

$$\|x\| = \|A^{-1}Ax\| \leq \|A^{-1}\|\|Ax\|.$$

Noticing that the norm of an invertible operator is strictly positive, we conclude that

$$\|Ax\| \geq \frac{1}{\|A^{-1}\|}\|x\|$$

and thus we may take $C = 1/\|A^{-1}\|$. $\qquad\square$

Consider now an inhomogenous fixed point problem $x = Ax + b$, where $A : \mathcal{B} \to \mathcal{B}$ is a linear operator on a Banach space $\mathcal{B}$. As we know, if A is a contraction, namely if $\|A\| < 1$, then a unique solution exists. Notice though that the problem may be re-written as $(I - A)x = b$, where we write I for the identity operator on $\mathcal{B}$. Now, if $(I - A)$ is invertible, then we obtain the unique solution $x = (I - A)^{-1}b$. Since the latter does not rely on the norm it actually holds in any linear space. We thus have two criteria for the solution of the equation $x = Ax + b$: one is that A be a contraction and the other is that $I - A$ be invertible. The next result shows that the former implies the latter, provided the ambient space is a Banach space.

Proposition 6.2 *Let $A : \mathcal{B} \to \mathcal{B}$ be a linear operator on a Banach space $\mathcal{B}$. If $\|A\| < 1$, then $I - A$ is invertible. Moreover, $\|(I - A)^{-1}\| \leq 1/(1 - \|A\|)$.*

Proof Since $\|Ax\| < \|A\|\|x\|$, if follows that

$$\|(I - A)x\| = \|x - Ax\| \geq \|x\| - \|Ax\| \geq \|x\| - \|A\|\|x\| = (1 - \|A\|)\|x\|.$$

The claim would follow from Theorem 6.10 if we can show that $I - A$ is surjective. In other words, given $y \in W$, we wish to solve the equation

$$(I - A)x = y$$

or, equivalently, the equation

$$x = Ax + y.$$

But this is precisely an inhomogenous fixed point problem, and since A is assumed to be a contraction, a solution exists.   □

We see that the invertibility of $I - A$ is important for the solution of fixed point problems, and thus also to the solution of systems of linear equations, differential equations, and integral equations. While the condition $\|A\| < 1$ gives a sufficient condition for the existence of $(I - A)^{-1}$ it does not provide us with any feasible means of obtaining the inverse. The following result remedies the situation.

Definition 6.6 Let $A : V \to V$ be a linear operator on a normed space V. The series $\sum_{k=0}^{\infty} A^k$, where A^0 is interpreted as the identity operator, is called the *Neumann series* of A.

Recall the notation $\mathbf{B}(V, W)$ for the set of all bounded linear operators $A : V \to W$. When $V = W$ we will write $\mathbf{B}(V)$ instead of $\mathbf{B}(V, W)$. Similarly, for a Banach space $\mathcal{B}$, we write $\mathbf{B}(\mathcal{B})$ instead of $\mathbf{B}(\mathcal{B}, \mathcal{B})$.

Theorem 6.11 *Let $A : V \to V$ be a linear operator on a normed space V. If the Neumann series for A converges in the operator norm, then the Neumann series is the inverse of $I - A$.*

Proof Suppose that the Neumann series converges to the operator S, that is

$$S = \lim_{m \to \infty} \sum_{k=0}^{m} A^k$$

with respect to the operator norm in the space $\mathbf{B}(V)$. Since

$$(I - A) \sum_{k=0}^{m} A^k = I - A^{m+1},$$

when m tends to ∞ we obtain

$$(I - A) \sum_{k=0}^{\infty} A^k = I - \lim_{m \to \infty} A^{m+1}.$$

We leave it as an exercise to the reader to verify that if the Neumann series of A converges, then $A^m \to 0$ in the operator norm, and thus we see that the Neumann series is a right inverse of $I - A$. A similar agument shows it is also a left inverse, and the proof is complete. □

The expected sufficient condition on an operator A on a Banach space guaranteeing the convergence of the Neumann series is given in the next result.

Corollary 6.2 *Let* $A : \mathcal{B} \to \mathcal{B}$ *be a linear operator on a Banach space* $\mathcal{B}$ *with* $\|A\| < 1$. *Then the Neumann series of* A *converges, and in particular*

$$(I - A)^{-1} = \sum_{k=0}^{\infty} A^k,$$

where the series converges in the operator norm.

Proof All we have to do is establish the convergence in $\mathbf{B}(\mathcal{B})$ of the Neumann series. Since $\mathcal{B}$ is a Banach space, so is the space $\mathbf{B}(\mathcal{B})$. By Theorem 6.5 it suffices to show the convergence of the series $\sum_{k=0}^{\infty} \|A^k\|$. But $\|A^k\| \le \|A\|^k$ and $\sum_{k=0}^{\infty} \|A\|^k$ is a convergent geometric series. □

Corollary 6.3 *If* A *is as above and* $\lambda \in \mathbb{R}$ *is such that* $\|\lambda A\| = |\lambda| \|A\| < 1$, *then* $I - \lambda A$ *is invertible and its inverse is given by the Neumann series*

$$(I - \lambda A)^{-1} = \sum_{k=0}^{\infty} \lambda^k A^k.$$

6.3.2 Fixed Point Techniques Revisited

The fixed point techniques in Sect. 6.2 relied on point-wise convergence. The convergence of the Neumann series though is strong convergence (i.e., in the operator norm), a form of convergence much stronger than point-wise convergence. We thus expect to have greater control over the situation when the Neumann series can be brought into the playground when solving a fixed point problem. We now look again at some of the differential and integral equations we solved above and witness the enhancement due to the strong convergence.

Example 6.9 Consider the linear Fredholm equation

$$x(t) = \lambda \int_a^b ds\, K(t,s)x(s) + f(t) = \lambda Ax + f$$

with $K(t, s)$, known as the *kernel*, continuous in the square $[a, b] \times [a, b]$, and $f(t)$ continuous in the interval $[a, b]$. Since

$$|K(t,s)| \le M \quad \Longrightarrow \quad \|A\| \le |\lambda| M(b-a) = q$$

we see that for those values of λ such that $q < 1$, the Neumann series

$$\sum_{k=0}^{\infty} A^k$$

of A converges (in the operator norm). Here $A^0 = \mathrm{id} = I$ and

$$A^1 = \int_a^b ds\, K_1(t,s)(*), \quad A^2 = \int_a^b ds\, K_2(t,s)(*), \ldots, \quad A^n = \int_a^b ds\, K_n(t,s)(*), \ldots,$$

where

$$K_1(t,s) = K(t,s), \quad K_2(t,s) = \int_a^b du\, K(t,u)\, K(u,s), \ldots,$$

$$K_n(t,s) = \int_a^b du\, K_{n-1}(t,u)\, K(u,s), \ldots$$

Notice further that for all $t, s \in [a, b]$

$$|\lambda^i K_i(t,s)| \le |\lambda|^i M^i (b-a)^i \le q^i,$$

and, since $0 \le q < 1$ by assumption, the series

$$K(t,s;\lambda) = \sum_{i=1}^{\infty} \lambda^i K_i(t,s),$$

converges uniformly in $[a,b] \times [a,b]$. The function $K(t,s;\lambda)$ is called the *resolvent kernel* and is a continuous function in the square $[a,b] \times [a,b]$ (as a uniform limit of continuous functions). From here it follows that

$$x(t) = f(t) + \int_a^b ds\, K(t,s;\lambda) f(s).$$

Example 6.10 Consider the Fredholm integral equation

$$x(t) = \sin t - \frac{t}{4} + \frac{1}{4} \int_0^{\pi/2} ds\, t\, s\, x(s).$$

With reference to the previous example, we have the estimates

$$|K(t,s)| = |t\, s| \le \frac{\pi^2}{4}, \quad (b-a) = \frac{\pi}{2}, \quad \lambda = \frac{1}{4} < \frac{1}{M(b-a)} = \frac{8}{\pi^3}$$

and thus we can perform the same series expansion as above. The resolvent kernel is now given by

$$K(t, s; \lambda) = \lambda K_1(t, s) + \lambda^2 K_2(t, s) + \lambda^3 K_3(t, s) + \cdots,$$

where

$$K_1(t, s) = t s, \quad K_2(t, s) = \int_0^{\pi/2} du\ (t u)\,(u s) = \alpha\, t s,$$

$$K_3(t, s) = \int_0^{\pi/2} du\ K_2(t, u) K_1(u, s) = \alpha \int_0^{\pi/2} du\ (t u)\,(u s) = \alpha^2 t s, \quad \ldots$$

$$K_n(t, s) = \alpha^{n-1} t s + \cdots \quad \text{with} \quad \alpha = \int_0^{\pi/2} du\ u^2 = \pi^3/24.$$

The expansion of all nuclei is given by:

$$K(t, s; \lambda) = \lambda\left[t s + \lambda \alpha\, t s + (\lambda \alpha)^2 t s + \cdots\right] = \lambda\, t s\, (1 - \lambda \alpha)^{-1},$$

since $\lambda \alpha = \pi^3/(4 \cdot 24) < 1$. The solution of our problem is then

$$x(t) = \sin t - \frac{t}{4} + \frac{1}{4} \int_0^{\pi/2} ds\ \frac{1}{1 - \alpha/4}\, t s \left(\sin s - \frac{s}{4}\right) =$$

$$\sin t - \frac{t}{4} + \frac{t}{4 - \alpha} \int_0^{\pi/2} ds\ \left(s \sin s - \frac{s^2}{4}\right).$$

Since

$$\int_0^{\pi/2} s \sin s = -s \cos s \Big|_0^{\pi/2} + \int_0^{\pi/2} ds\ \cos s = \sin s \Big|_0^{\pi/2} = 1$$

we finally obtain

$$x(t) = \sin t - \frac{t}{4} + \frac{t}{4 - \alpha} \left(1 - \frac{\alpha}{4}\right) = \sin t - \frac{t}{4} + \frac{t}{4} = \sin t.$$

Let us now reconsider the Volterra equation, in light of the more powerful technique of the Neumann series.

Example 6.11 Consider in $C([a, b], \mathbb{R})$ the linear Volterra equation

$$x(t) = \lambda \int_a^t ds\ K(t, s)\, x(s) + f(t) = \lambda A x + f,$$

with the usual assumption of continuity and boundedness on $K(t, s)$ and $f(t)$, and let

$$M = \max_{\{t,s\}\in[a,b]} |K(t,s)|, \quad F = \max_{t\in[a,b]} |f(t)|.$$

Compared with the previous treatment of the Fredholm equation by means of point-wise fixed point solutions, here the integral does not extend over the entire interval $[a, b]$ but only up to t, which in turn becomes the integration variable of the next iteration. Since usually integration makes functions *smoother*, we expect to obtain better conditions on the values of λ for which the iterative solution is converging. Then, consider the solution

$$x(t) = \sum_{i=0}^{\infty} \lambda^i A^i f = \sum_{i=0}^{\infty} \lambda^i \phi_i,$$

which certainly exists at least for those values of λ found for the linear Fredholm equation. The following estimates hold

$$|\phi_1(t)| \le \int_a^t ds\, |K(t,s)\, f(s)| \le M\, F\, (t-a),$$

$$|\phi_2(t)| \le \int_a^t ds\, |K(t,s)\, \phi_1(s)| \le M^2\, F\, \int_a^t ds\, (s-a) = M^2 F (t-a)^2/2, \dots,$$

$$|\phi_n| \le M^n\, F\, \frac{(t-a)^n}{n!}, \dots$$

It follows that

$$\left| \sum_{i=0}^{\infty} \lambda^i \phi_i \right| \le \sum_{i=0}^{\infty} |\lambda^i|\, |\phi_i| \le F \sum_{i=0}^{\infty} |\lambda|^i\, M^i\, \frac{(t-a)^i}{i!} = F\, \exp\left[|\lambda|\, M\, (t-a) \right],$$

is convergent for all values of λ. Therefore, the function $x(t)$ is continuous and is a solution of the Volterra equation for every λ.

Example 6.12 Let us see how the linear homogeneous Fredholm equation admits solutions other than the zero solution. Given the equation

$$x(t) = \lambda \int_0^1 ds\, (t+s)\, x(s),$$

we introduce the constants c_1 and c_2 given by

$$x(t) = \lambda t \int_0^1 ds\, x(s) + \lambda \int_0^1 ds\, s\, x(s) = \lambda t\, c_1 + \lambda\, c_2.$$

By inserting this form of $x(t)$ into the first equation, one obtains

$$\lambda t\, c_1 + \lambda c_2 = \lambda \int_0^1 ds\ (t+s)(\lambda s\, c_1 + \lambda c_2) = \lambda^2 \left(\frac{c_1}{2} t + c_2 t + \frac{c_1}{3} + \frac{c_2}{2} \right).$$

The coefficients of similar terms in t must be equal, so that

$$\begin{cases} c_1 = \lambda c_1/2 + \lambda c_2 \\ c_2 = \lambda c_1/3 + \lambda c_2/2. \end{cases}$$

The two equations are compatible for $\lambda = 6 \pm 4\sqrt{3}$ and the system, since it is homogeneous, admits infinitely many solutions, namely

$$c_1 = \frac{2\lambda}{2-\lambda}\, c, \qquad c_2 = c,$$

with c an arbitrary parameter. The non-trivial solution of our equation is then given by

$$x(t) = \lambda \left(\frac{2\lambda t}{2-\lambda} + 1 \right),$$

up to an arbitrary proportionality factor. Obviously, $|\lambda| = \left| 6 \pm 4\sqrt{3} \right| > 1/2$, below which value the solution is unique.

Remark 6.2 In the previous examples the L_∞ norm has always been assumed. This was done in order to allow for the machinery developed thus far. It should be noted though that this does not exclude the existence of non-continuous solutions, i.e., ones that we have excluded a priori.

Example 6.13 Consider the homogeneous Volterra equation

$$x(t) = \int_0^t ds\ s^{t-s}\, x(s), \quad 0 \le t \le 1.$$

The kernel $K(t,s) = s^{t-s}$ is continuous in the integration domain with $s \le t$, so that the equation admits the unique continuous solution $x = 0$. However, there exist infinitely many solutions not belonging to $C([a,b], \mathbb{R})$, given by $x(t) = c\, t^{t-1}$ with c an arbitrary constant, as can be checked directly

$$\int_0^t ds\ s^{t-s}\, x(s) = \int_0^t ds\ s^{t-s} s^{s-1} = \int_0^t ds\ s^{t-1} = \frac{s^t}{t} \Big|_0^t = t^{t-1} = x(t).$$

As expected, the function $x(t)$ is singular, since $x(t) \to \infty$ as $t \to 0$.

Exercises

Exercise 6.21 Prove that if the Neumann series of an operator $A : V \to V$, where V is a normed space, converges, then $A^m \to 0$ in the operator norm.

Exercise 6.22 Show that linear operators can be as wild as one desires by showing that given any function $f : X \to Y$, where X is an arbitrary set and Y is a linear space over K, there exists a linear space $\hat{X}$ and an injection $i : X \to \hat{X}$ together with a linear operator $F : \hat{X} \to Y$ such that $f = F \circ i$.

Exercise 6.23 Repeat the previous exercise with $\mathbb{C}$ instead of $\mathbb{R}$.

Exercise 6.24 Let $\mathcal{B}$ be a Banach space and I the identity operator on $\mathcal{B}$. Prove that any linear operator $A : \mathcal{B} \to \mathcal{B}$ satisfying $\|A - I\| < 1$ is invertible.

Exercise 6.25 Let $\mathcal{B}$ be a Banach space and I the identity operator on $\mathcal{B}$. Suppose $A : \mathcal{B} \to \mathcal{B}$ is a linear operator with $\|A - I\| = 1$, is A necessarily not invertible?

Exercise 6.26 Let $B : \mathcal{B} \to \mathcal{B}$ be an invertible linear operator on a Banach space $\mathcal{B}$. Prove that any linear operator $A : \mathcal{B} \to \mathcal{B}$ satisfying $\|A - B\| < \|B\|$ is invertible.

Exercise 6.27 Let $B : \mathcal{B} \to \mathcal{B}$ be an invertible linear operator on a Banach space $\mathcal{B}$. If $A : \mathcal{B} \to \mathcal{B}$ is a linear operator satisfying $\|A - B\| > \|B\|$, is A necessarily not invertible?

Exercise 6.28 Let $A : \mathcal{B} \to \mathcal{B}$ be an invertible linear operator on a Banach space $\mathcal{B}$. Prove that given $\varepsilon > 0$ there exists a $\delta > 0$ such that $\|B^{-1} - A^{-1}\| < \varepsilon$ for all linear operators $B : \mathcal{B} \to \mathcal{B}$ satisfying $\|B - A\| < \delta$.

6.4 Dual Spaces

A good understanding of an object is intimately coupled with a sound understanding of how it relates, via structure preserving functions, to other objects. For a normed space V the simplest non-trivial case is understanding the linear operators $V \to K$ to the ground field. If V is finite dimensional, then all such operators are bounded and the situation is reduced to elementary linear algebra. Let us quickly recall the details where, for simplicity, we take the ground field to be $\mathbb{R}$. The *dual space* of a finite-dimensional linear space V is the linear space $V^* = \text{Hom}(V, \mathbb{R})$ of all linear operators $A : V \to \mathbb{R}$ where $\mathbb{R}$ is viewed as a linear space over itself in the usual way. Given a basis for V one can associate with it its dual basis, forming a basis for V^*. In particular, showing that V and V^* have the same dimension, and thus are isomorphic.

In the infinite-dimensional case the boundedness constraint is significant. A bounded operator is rather tame yet at the same time it is significantly harder to construct one since not only the linear structure but also the topology of V must be taken into consideration. A Hamel basis for V can still be used to produce linear operators but, even if one actually had a Hamel basis written down, ensuring boundedness is hopeless.

With that in mind we tend to define the dual space in the infinite-dimensional context, look at some dual spaces of classical spaces, and have a brief look at the Hahn-Banach Theorem. The latter is the main theoretical tool that addresses the difficulty of constructing exemplars in the dual space. As said, Hamel bases are blind to the topology and thus unhelpful for constructing bounded operators. The Hahn-Banach Theorem is the surrogate tool. Its proof uses the Axiom of Choice in a significant way and so one should not expect it to produce explicit operators.

6.4.1 Duality

A first look at linear functionals was given in Sect. 5.5 for the classical spaces ℓ_p and L_p where we looked at the bounded functionals on these spaces. More generally, the bounded linear operators to the ground field on any normed space is an important object of study.

Definition 6.7 A *linear functional* on a normed space V is a linear operator $V \to K$ to the ground field K (either $K = \mathbb{R}$ or $K = \mathbb{C}$). The set of all bounded linear functionals $F : V \to K$ is denoted by V^* and is called the *dual space* of V.

Of course, the general theory of linear operators specializes to linear functionals so, in particular, V^* is a Banach space. In Chap. 5 we discussed evaluation functionals as well as summation and integration functionals. Taking the limit of a sequence is a linear functional on the space c of convergent sequences, and there are many other functionals of practical interest.

Recall from Definition 5.6 that a linear isometry $A: V \to W$ between normed spaces is a linear isomorphism that preserves the norms. Real normed spaces V and W are said to be *congruent* if there exists a linear isometry between them. In the complex case we say that V and W are congruent if there exists a function $A: V \to W$ satisfying the same properties as a linear isometry except for conjugating the homogeneity condition, i.e., $A(\alpha x) = \overline{\alpha} A(x)$. For some spaces one can, with varying amounts of effort, compute the dual space and understand it well up to congruency. For example, write $(\mathbb{C}^n, \| - \|_p)$ for the linear space $\mathbb{C}^n$ equipped with the specific norm $\| - \|_p$ for $1 \le p \le \infty$ and let q satisfy $1/p + 1/q = 1$. Then from the results of Sect. 5.5 we have

$$(\mathbb{C}^n, \| - \|_p)^* \cong (\mathbb{C}^n, \| - \|_q)$$

and, with similar notation, also

$$(\mathbb{P}_n, \| - \|_p)^* \cong (\mathbb{P}_n, \| - \|_q).$$

More substantially, the results of Sect. 5.5 constitute almost all the details for a proof of the following classical result due to Riesz.

Theorem 6.12 *For the classical sequence and function spaces the following congruences hold.*

1. $c_0^* \cong \ell_1$ *and* $(C([a, b], \mathbb{C}), \| - \|_\infty)^* \cong L_1$.
2. $\ell_1^* \cong \ell_\infty$ *and* $L_1^* \cong L_\infty$.
3. $\ell_p^* \cong \ell_q$ *and* $L_p^* \cong L_q$ *for all* $1 < p, q < \infty$ *such that* $1/p + 1/q = 1$.

These explicit expressions for the dual space reveal a special property of the spaces ℓ_p and L_p for all $1 < p < \infty$, namely that the dual of the dual is the original space. To formalize this observation, we note that generally, given a Banach space $\mathcal{B}$, there is a canonical function $\Psi \colon \mathcal{B} \to \mathcal{B}^{**}$ to the *double dual* of $\mathcal{B}$, namely the dual space of $\mathcal{B}^*$. The function Ψ must map a vector $x \in \mathcal{B}$ to a bounded linear operator $\Psi(x) \colon \mathcal{B}^* \to K$. It is defined by

$$\Psi(x)(F) = F(x)$$

and it is bounded since

$$\|\Psi(x)(F)\| = \|F(x)\| \leq \|F\| \cdot \|x\|$$

that in fact shows that

$$\|\Psi(x)\| \leq \|x\|$$

and it is natural to expect Ψ to be an isometry. To show that $\|\Psi(x)\| = \|x\|$ it would be very helpful to have a bounded functional F_x of norm 1 with the property that $F_x(x) = \|x\|$, since then the norm of $\Psi(x)$ would be attained at F_x and give us the desired equality. Constructing a functional F_x with $F_x(x) = \|x\|$ is a triviality: if $x = 0$ take $F = 0$, otherwise extend x to a Hamel basis and then obtain F_x by linear extension. However, there is no guarantee that such a functional will be bounded. In the next subsection we present the Hahn-Banach Theorem that, among other things, will enable us to construct the bounded functional we are after. Therefore, we will be able to show that the function Ψ is in fact an isometry. However, in general it need not be surjective and thus not an isometric isomorphism. Spaces for which this is the case form an important class of Banach spaces.

Definition 6.8 If the canonical mapping $\mathcal{B} \to \mathcal{B}^{**}$ is an isometric isomorphism, then the Banach space $\mathcal{B}$ is *reflexive*.

We can now restate the previous observations as follows. In the families of Banach spaces ℓ_p and L_p the reflexive spaces are those where $1 < p < \infty$.

6.4.2 The Hahn-Banach Theorem

We saw that it is possible to exhibit elements in the dual space for some infinite-dimensional Banach spaces. Often, though, one has a bounded linear operator defined only on a portion of the space and one wants to extend it to a bounded linear operator on the whole space. Here the linear structure and the topology pull in different directions. Preserving linearity is a rigid requirement that is difficult to achieve while maintaining boundedness. The solution is the Hahn-Banach Theorem. It does not rely on the completeness of the norm and is thus valid for general normed spaces. After establishing it we present some of its consequences that further relate a space and its dual.

Theorem 6.13 (Hahn-Banach) *Let V be a normed space over K, with either $K = \mathbb{R}$ or $K = \mathbb{C}$. Suppose that $U \subseteq V$ is a linear subspace and $f : U \to K$ a bounded linear functional on the subspace U. There exists then a linear functional $F : V \to K$ such that $F|_U = f$ and $\|F\| = \|f\|$.*

Proof For simplicity, we only give the proof for real normed spaces. Thus we assume that a bounded linear functional $f : U \to \mathbb{R}$ is given. Clearly, if $f = 0$, then $F = 0$ is the desired extension to all of V. We may thus proceed under the assumption that $f \neq 0$, and thus, normalizing if needed, that $\|f\| = 1$. Suppose that $x_0 \in V \setminus U$ and let U_1 be the span of $U \cup \{x_0\}$. It is easy to see that U_1 consists of all vectors of the form $x + \beta x_0$, where $x \in U$ and $\beta \in \mathbb{R}$. Notice that, given any $\alpha \in \mathbb{R}$, the function

$$f_1(x + \beta x_0) = f(x) + \beta \alpha$$

is a linear functional on U_1 and it extends U. We will now show that an $\alpha \in \mathbb{R}$ can be chosen so as to assure that $\|f_1\| = 1$. By the definition of the norm $\|f_1\|$ it suffices to guarantee that

$$|f(x) + \beta \alpha| \le \|x + \beta x_0\|,$$

for all $x \in U$ and $\beta \in \mathbb{R}$. Stated differently (replace x by $-\beta x$ and divide both sides by $|\beta|$), we seek an α such that

$$|f(x) - \alpha| \le \|x - x_0\|$$

for all $x \in U$. We thus wish to find an $\alpha \in \mathbb{R}$ with

$$f(x) - \|x - x_0\| \le \alpha \le f(x) + \|x - x_0\|$$

for all $x \in U$. In other words, α is common to all the intervals $[A_x, B_x]$, for all $x \in U$, where $A_x = f(x) - \|x - x_0\|$ and $B_x = f(x) + \|x - x_0\|$. The existence of such an α will follow by showing that $A_x \le B_y$ for all $x, y \in U$. And indeed,

$$B_y - A_x = f(y) - f(x) + \|y - x_0\| + \|x - x_0\|$$

and since

$$|f(y) - f(x)| = |f(y - x)| \le \|y - x\| \le \|y - x_0\| + \|x - x_0\|$$

it follows that

$$B_y - A_x \ge 0.$$

We thus established that f, and in fact any linear operator defined on a proper subspace, can be extended to the span of its current domain with one additional vector added, without altering the norm.

The stage is now clear for an application of Zorn's Lemma to further extend f_1, one dimension at a time, until we arrive at a linear operator whose domain is the entire space V. Let P be the collection of all pairs (U', f') where $U' \supseteq U$ is a linear subspace of V and $f' : U' \to \mathbb{R}$ is a linear functional extending f and satisfying $\|f'\| = 1$. Introduce a partial order on P by means of

$$(U', f') \le (U'', f'')$$

precisely when $U' \subseteq U''$ and when f'' extends f'. The fact that P is then a poset, is immediate. We must now establish the conditions of Zorn's Lemma. Firstly, P is non-empty since (U, f) is a member of P. Next, given a chain $\{(U_i, f_i)\}_{i \in I}$ in P, consider $U' = \bigcup_{i \in I} U_i$ and define $f' : U' \to \mathbb{R}$ as follows. Given $u \in U'$ there exists an index $i \in I$ with $u \in U_i$. If $j \in I$ is another index with $u \in U_j$, then, since we are given a chain, either f_i extends f_j or f_j extends f_i. In either case setting $f'(u) = f_i(u) = f_j(u)$ produces a well-defined function. The function f' is a linear functional. Indeed, if $u_1, u_2 \in U'$, then $u_1 \in U_{i_1}$ and $u_2 \in U_{i_2}$ for some $i_1, i_2 \in I$. But, by the chain condition, we may assume without loss of generality that $U_{i_1} \subseteq U_{i_2}$. Thus

$$f'(u_1 + u_2) = f_{i_2}(u_1 + u_2) = f_{i_2}(u_1) + f_{i_2}(u_2) = f'(u_1) + f'(u_2)$$

and thus f' is additive. A similar argument shows that f is also homogenous. We leave it to the reader to similarly show that $\|f'\| = 1$ and thus it is now clear that (U', f') is in P and is an upper bound for the given chain. Note however that there is no guarantee that f' is the desired linear functional F, since U' may be properly contained in V.

With the conditions of Zorn's Lemma now verified, the existence of a maximal element $(U_M, F) \in P$ is guaranteed. If $U_M = V$, then F is a linear functional on all of V, it extends f, and its norm is 1, as required. Assume thus that U_M is properly contained in V, and let $x_0 \in V - U_M$. But, at the beginning of the proof we saw that in such a situation F can be extended to the linear subspace U_1, the span of $U_M \cup \{x_0\}$, without altering the norm. That would give rise to an element $(U_1, f_1) \in P$ with $(U_M F) < (U_1, f_1)$, an impossibility. The proof is thus complete.   □

We explore a handful of the applications of the Hahn-Banach Theorem.

Theorem 6.14 *Given a normed space V and $x_0 \in V$ with $x_0 \neq 0$ there exists a linear functional $F \in V^*$ such that $\|F\| = 1$ and $F(x_0) = \|x_0\|$.*

Proof Let

$$V_0 = \{\alpha x_0 \mid \alpha \in K\}$$

be the subspace of V spanned by x_0 and let $F_0 : V_0 \to K$ be given by

$$F_0(\alpha x_0) = \alpha \|x_0\|.$$

Clearly, F_0 is a linear functional, $\|F_0\| = 1$, and $F_0(x_0) = \|x_0\|$. By the Hahn-Banach Theorem, an extension $F \in V^*$ exists, establishing the claim. $\square$

Corollary 6.4 *For all $x \in V$*

$$\|x\| = \sup_{F \in V^*, \|F\|=1} |F(x)|.$$

Corollary 6.5 *For all vectors x in a normed space V, if $F(x) = 0$ for all $F \in V^*$, then $x = 0$.*

We may now return to the natural embedding $\Psi : \mathcal{B} \to \mathcal{B}^{**}$ of $\mathcal{B}$ in its double dual, given by $\Psi(x)(F) = F(x)$.

Theorem 6.15 *For all Banach spaces $\mathcal{B}$ the embedding Ψ is an isometry.*

Proof We already noted that $\|\Psi(x)\| \leq \|x\|$ by direct computation, but now we can do better:

$$\|\Psi(x)\| = \sup_{F \in V^*, \|F\|=1} |\Psi(x)(F)| = \sup_{F \in V^*, \|F\|=1} |F(x)| = \|x\|.$$

$\square$

Another result uses the Hahn-Banach Theorem to show that there are sufficiently many bounded functionals to separate a point from the closure of a subspace.

Theorem 6.16 *Let V be normed space and U a linear subspace of it. Given $x_0 \in U$ it holds that x_0 is in the closure of U if, and only if, there exists no bounded linear functional f on V such that $f(x) = 0$ for all $x \in U$ but $f(x_0) \neq 0$.*

Proof Suppose that x_0 is in the closure of U. There exists then a sequence $\{u_n\}_{n \geq 1}$ of element in U converging to x_0. Remembering that a bounded linear functional $f : V \to K$ is in particular continuous, it follows that $f(u_n) \to f(x_0)$. But then, if $f(x) = 0$ for all $x \in U$, it follows that $f(x_0) = 0$. In the other direction, suppose x_0 is not in the closure of U. Then there exists a positive δ such that $\|x - x_0\| > \delta$, for all $x \in U$. Consider the span U' of the set $U \cup \{x_0\}$. Define now $f_1(x + \beta x_0) = \beta$ (remember that the elements of the span are of the form $x + \beta x_0$, with $\beta \in K$ and $x \in U$). This is clearly a linear functional on U_1 and since

$$\delta|\beta| \leq |\beta|\|x_0 + \frac{1}{\beta}x\| = \|\beta x_0 + x\|$$

we have that $\|f_1\| \leq 1/\delta$. Since, by definition, $f_1(x) = 0$ for all $x \in U$ and $f_1(x_0) = 1$, applying the Hahn-Banach Theorem to extend f' to all of V establishes the claim. $\square$

Exercises

Exercise 6.29 Let $\mathcal{B}$ be a Banach space whose norm is induced by an inner product. Given $x \in \mathcal{B}$ show that $y \mapsto \langle x, y \rangle$ defines a bounded linear functional F_x on $\mathcal{B}$ and that $\|F_x\| = \|x\|$.

Exercise 6.30 With notation as in the previous exercise show that for the inner product spaces $\mathbb{R}^n$ and $\mathbb{C}^n$ every linear functional F is of the form F_y for a unique y. In Chap. 8 we will see that this is a special case of a broader phenomenon.

Exercise 6.31 Consider the space $X = C([a, b], \mathbb{R})$ of continuous functions with the $\| - \|_\infty$ norm.

1. Prove that

$$F(x) = \int_0^1 dt \, tx(t)$$

   defines a bounded linear functional and that the norm $\|F\|$ is attained.
2. Show that $\{x \in X \mid x(1) = 0\}$ is a closed linear subspace of V.
3. Prove that the restriction of F to the subspace from the previous item is still bounded but its norm is no longer attained.

Exercise 6.32 Let V be a normed space. Prove that V^* is bounded if, and only if, $\dim(V) = 0$.

Exercise 6.33 Let V be a normed space and $f \in V^*$. Prove that $p_f : V \to \mathbb{R}$ given by $p_f(x) = |f(x)|$ is a semi-norm on V. Under which conditions on V can it be a norm?

Exercise 6.34 Let $\{x_1, \cdots, x_n\}$ be n linearly independent vectors in a normed space V and $z_1, \cdots, z_n \in \mathbb{C}$. Prove the existence of a linear functional $F \in V^*$ with

$$F(x_k) = z_k$$

for all $1 \leq k \leq n$.

Exercise 6.35 Consider the space $C([a, b], \mathbb{R})$ with the L_∞ norm. Given an integrable function $y : [a, b] \to \mathbb{R}$, prove that the function $F_y : C([a, b], \mathbb{R}) \to \mathbb{R}$ given by

$$F_y(x) = \int_a^b dt \, y(t)x(t)$$

is a linear functional and show that

$$\|F\| = \int_a^b dt \, |y(t)|.$$

Exercise 6.36 Show that the extension guaranteed by the Hahn-Banach Theorem is, generally speaking, far from unique by presenting a linear operator with infinitely many extensions. Proceed in two ways, one by analyzing the proof of the Hahn-Banach Theorem, the other by giving explicit extensions of a well-chosen linear operator. Hint: consider finite-dimensional spaces.

Exercise 6.37 In the previous exercise you constructed explicit extensions of a linear functional. Convince yourself that for infinite-dimensional linear spaces it may be very hard, if at all possible, to obtain an explicit extension of an arbitrary functional.

Exercise 6.38 Use the Hahn-Banach Theorem to prove that it is possible to assign to every bounded sequence $\{x_m\}_{m\geq 1}$ of real numbers a real number $\lim_{m\to\infty} x_m$ such that

1. If $\{x_m\}$ is a convergent sequence, then $\lim_{m\to\infty} x_m$ is the limit of the sequence in the usual sense.
2. $\lim_{m\to\infty}(x_m + y_m) = \lim_{m\to\infty} x_m + \lim_{m\to\infty} y_m$ for all bounded sequences $\{x_m\}_{m\geq 1}$ and $\{y_m\}_{m\geq 1}$.
3. $\lim_{m\to\infty} \alpha x_m = \alpha \lim_{m\to\infty} x_m$ for all bounded sequences $\{x_m\}_{m\geq 1}$ and all scalars $\alpha \in \mathbb{R}$.

Can you determine $\lim_{m\to\infty}(-1)^m$?

6.5 Unbounded Operators and Locally Convex Spaces

The theory presented so far is that of bounded linear operators between normed spaces. In this section we present two directions for generalization of the theory. The first one stems from the observation that while bounded operators encompass many examples of interest, some important operators, the differentiation operator for instance, are not bounded. One is thus compelled to consider unbounded operators. The second generalization we consider originates from the fact that at times the space one is considering does not support a norm. It turns out that if instead of a norm one has a suitable family of semi-norms, then one can still recover much of the general theory. Spaces with such a family of semi-norms are called locally convex spaces. We will only investigate the definition and some illustrative examples without delving any deeper than that.

6.5.1 Closed Operators

The development of the theory of unbounded linear operators was motivated by the unbounded nature of the differentiation operator, as well as the development of a mathematical framework for quantum mechanics (e.g., to account for unbounded observables). Arbitrary unbounded operators may be too wild to tame, and so some restrictions must be placed in order to identify a suitable class of operators for which a reasonable theory emerges. Closed operators form a class of operators broader than the bounded ones admitting a strong general theory that allows one to analyze situations that lie out of reach of the theory of bounded operators. The results we present below are only the tip of the unbounded iceberg.

In the context of unbounded linear operators it is common to re-interpret the definition of an operator between linear spaces to be one that is not necessarily defined on all of the specified domain.

Definition 6.9 Let $\mathcal{B}_1$ and $\mathcal{B}_2$ be Banach spaces. An *(unbounded) linear operator* $A : \mathcal{B}_1 \to \mathcal{B}_2$ consists of a linear subspace $\mathcal{D}(A) \subseteq \mathcal{B}_1$ and a function $A : \mathcal{D}(A) \to \mathcal{B}_2$ that is additive and homogenous. Further, A is a *closed linear operator* when the following condition holds. For every sequence $\{x_m\}_{m \geq 1}$ in $\mathcal{B}_1$ if $x_m \to x_0$ for some $x_0 \in \mathcal{B}_1$ and $Ax_m \to y_0$ for some $y_0 \in \mathcal{B}_2$, then $x_0 \in \mathcal{D}(A)$ and $Ax_0 = y_0$.

Example 6.14 In Example 6.4 we saw that the differentiation operator d/dt : $C^1([0, 1], \mathbb{R}) \to C([0, 1], \mathbb{R})$, with the L_∞ norm, is an unbounded linear operator. In the context of unbounded operators we may thus consider

$$d/dt : C([0, 1], \mathbb{R}) \to C([0, 1], \mathbb{R}),$$

with the L_∞ norm, where we may specify various different subspaces as the domain of definition. Different choices for the actual domain may lead to completely different properties of the operator. For instance, if we take the domain to be $C^1([a, b], \mathbb{R})$, then we obtain an unbounded closed operator. Indeed, this claim follows from the elementary result that if $\{x_m\}_{m \geq 1}$ is a sequence of functions that converges (pointwise) to x_0, and their derivatives x'_m converge uniformly to y_0, then x_0 is differentiable and $x'_0 = y_0$. In contrast, taking the domain of d/dt to be $C^\infty([0, 1], \mathbb{R})$ yields a non-closed operator.

The defining condition of a closed operator $A : \mathcal{B}_1 \to \mathcal{B}_2$ can be restated in terms of its graph $\Gamma(A) = \{(x, Ax) \in \mathcal{B}_1 \times \mathcal{B}_2 \mid x \in \mathcal{D}(A)\}$, as follows.

Proposition 6.3 *An unbounded linear operator $A : \mathcal{B}_1 \to \mathcal{B}_2$ between Banach spaces is closed if, and only if, the graph $\Gamma(A)$ is closed in $\mathcal{B}_1 \times \mathcal{B}_2$.*

The proof is left for the reader. We prove that a closed linear operator is automatically bounded, provided its domain is closed.

Theorem 6.17 (Banach's Closed Graph Theorem) *Let $A : \mathcal{B}_1 \to \mathcal{B}_2$ be a closed linear operator between Banach spaces. If the domain $\mathcal{D}(A)$ is closed in $\mathcal{B}_1$, then A is bounded.*

Proof First, it is straightforward to verify that the space $\mathcal{B}_1 \times \mathcal{B}_2$ with norm given by

$$\|(x, y)\| = \|x\| + \|y\|$$

is a Banach space. Since A is closed, its graph $\Gamma(A) \subseteq \mathcal{B}_1 \times \mathcal{B}_2$ is thus a closed subset of a complete space and thus is itself a Banach space. Similarly, by assumption, $\mathcal{D}(A)$ is closed in $\mathcal{B}_1$ and thus is a Banach space too.

We now consider the operator $P : \Gamma(A) \to \mathcal{D}(A)$ given by

$$P(x, Ax) = x,$$

clearly a linear operator. Since

$$\|P(x, Ax)\| = \|x\| \leq \|x\| + \|Ax\| = \|(x, Ax)\|$$

we see that P is a bounded linear operator. Moreover, P is invertible (its inverse P^{-1} obviously given by $x \mapsto (x, Ax)$) and thus, by Corollary 6.1, P^{-1} is bounded, say by M. Therefore,

$$\|P^{-1}x\| = \|x\| + \|Ax\| \leq M\|x\|$$

for all $x \in \mathcal{D}(A)$, and thus

$$\|Ax\| \leq (M - 1)\|x\|$$

showing A is bounded. $\qquad\qquad\square$

This is yet another deep result in the theory of Banach spaces; this short proof uses a corollary of the Open Mapping Theorem. In fact, each of the open mapping theorem, the bounded inverse theorem, and the closed graph theorem can be used to prove the other two.

6.5.2 Locally Convex Spaces

As motivation for the next concept consider the set $\mathbb{R}^\infty$ of all sequences of real numbers (with obvious modifications one may also consider $\mathbb{C}^\infty$). Of course the linear structure of $\mathbb{R}$ carries over to $\mathbb{R}^\infty$ by point-wise operations and thus $\mathbb{R}^\infty$ is a linear space. However, there is no natural way to turn $\mathbb{R}^\infty$ into a normed space. In particular, each ℓ_p norm is only defined on a proper subset of $\mathbb{R}^\infty$ and cannot be extended to a norm on $\mathbb{R}^\infty$. There is however a natural choice of a family of semi-norms, defined as follows. For each $i \in \mathbb{N}$ let $p_k : \mathbb{R}^\infty \to \mathbb{R}$ be given by $p_k(x) = |x_k|$, the absolute value of the k-th component of x. It is easily seen that p_k is indeed a semi-norm.

We thus see that in the absence of a norm a space may still admit a family of semi-norms and it is the case that, under suitable conditions, significant portions of the theory developed above (in particular the Hahn-Banach Theorem) admit suitable analogues. We present the definition and explore some of the most basic aspects of these spaces.

Definition 6.10 A *locally convex linear space* is a linear space V together with a family $\{P_i\}_{i \in I}$ of semi-norms on V. The family of semi-norms is said to *separate points* if from $p_i(x) = 0$ for all $i \in I$ it follows that $x = 0$.

Example 6.15 Of course any normed space and any semi-normed space is a locally convex space. It is not hard to show that any locally convex space given by a finite family of semi-norms is in fact semi-normed. Thus the concept of locally convex spaces becomes substantial only when once considers families of infinitely many semi-norms, as in the case of $\mathbb{R}^\infty$ above.

We now consider the topology induced by the family of semi-norms in a locally convex space.

Definition 6.11 Let V be a locally convex space given by the family $\{p_i\}_{i \in I}$ of semi-norms. For each $i \in I$ and $y \in V$ let $f_{i,y} : V \to \mathbb{R}$ be the function

$$f_{i,y}(x) = p_i(x - y).$$

The *induced topology* on V is the smallest topology such that each of $f_{i,y} : V \to \mathbb{R}$ is continuous.

The existence of such a smallest topology is guaranteed by Proposition 3.11. In fact, from the proof of that proposition, a local base at y for the induced topology is given by the collection $U_{B,\varepsilon}$, where $B \subseteq I$ is a finite subset of indices, $\varepsilon > 0$, and

$$U_{B,\varepsilon} = \{x \in V \mid p_i(x - y) < \varepsilon \;\; \forall i \in B\}.$$

The following result follows easily and its proof is left for the reader.

Proposition 6.4 *Let V be a locally convex space given by the family $\{p_i\}_{i \in I}$ of semi-norms.*

1. *A sequence $\{x_m\}_{m \geq 1}$ in V converges to $x_0 \in V$ in the induced topology if, and only if, $p_i(x_m - x_0) \to 0$ for all $i \in I$.*
2. *The induced topology is Hausdorff if, and only if, the family of semi-norms separates points.*
3. *The linear space operations are continuous.*

With these observations made, the similarities between the theory of normed spaces we developed above and the theory of locally convex spaces with separated points is only starting to become visible. Without a doubt, it is more convenient to have a single norm and appeal to the rich theory of normed spaces whenever possible. However, often enough a norm is simply not available. The theory of locally convex spaces is rich enough so as to justify the somewhat cumbersome management of a family of semi-norms instead of a single norm. As mentioned above, significant portions of the theory of (semi-)normed spaces transfers quite smoothly to the theory of locally convex spaces, however we do not explore this any further.

Exercises

Exercise 6.39 Let $F: \mathcal{B} \to K$ be a linear functional on a Banach space $\mathcal{B}$ over the field K. Show that $\mathrm{Ker}(F)$ is a closed subspace of $\mathcal{B}$ if, and only if, F is bounded. Show further that if F is unbounded, then its kernel is a dense linear subspace of $\mathcal{B}$.

Exercise 6.40 Prove that if $\mathcal{B}_1$ and $\mathcal{B}_2$ are Banach spaces, then so is $\mathcal{B}_1 \times \mathcal{B}_2$ with norm given by $\|(x, y)\| = \|x\| + \|y\|$.

Exercise 6.41 Prove that if $A : \mathcal{B}_1 \to \mathcal{B}_2$ is a closed linear operator, then its kernel $\mathrm{Ker}(A)$ is a closed subspace of $\mathcal{B}_1$.

Exercise 6.42 Prove that any bounded linear operator is a closed operator.

Exercise 6.43 Prove Proposition 6.3.

Exercise 6.44 Let V be a normed space. We may now consider an induced topology on it in two ways. Namely, we may view it as a locally convex space where the family of semi-norms consists of just the given norm on V, and obtain the induced topology, or we may consider the induced metric $d(x, y) = \|x - y\|$ and consider its induced topology. Show that the two topologies coincide.

Exercise 6.45 Prove that every locally convex linear space V defined using a finite family $\{p_i\}_{i \in I}$ of semi-norms is semi-normed. In more detail, construct a single semi-norm on V which induces the same topology as the finite family of semi-norms does.

Exercise 6.46 Let V be a locally convex space given by a countable family $\{p_i\}_{i \geq 1}$ of semi-norms. Consider the function $d : V \times V \to \mathbb{R}$ given by

$$d(x, y) = \sum_{i=1}^{\infty} \frac{1}{2^n} \frac{p_i(x - y)}{p_i(x - y) + 1}.$$

Prove that (V, d) is a semi-metric space, and that it is a metric space if, and only if, the family of norms separates points.

Exercise 6.47 Continuing the previous exercise, show that the topologies induced by the semi-metric and by the family of semi-norms coincide.

Exercise 6.48 Consider $\mathbb{R}$ with the standard structure of an inner product space and let $V = C(\mathbb{R}, \mathbb{R})$ be the linear space of all continuous functions $x : \mathbb{R} \to \mathbb{R}$. For each $t \in \mathbb{R}$ let

$$p_t(x) = |x(t)|.$$

For each $n \in \mathbb{N}$ let

$$q_n(x) = \max_{-n \le t \le n} |x(t)|.$$

For each compact subset $C \subseteq \mathbb{R}$ let

$$r_C(x) = \max_{t \in C} |x(t)|.$$

For each of the families $\{p_t\}_{t \in \mathbb{R}}$, $\{q_n\}_{n \in \mathbb{N}}$, and $\{r_C\}_C \subseteq \mathbb{R}$, where C ranges over the compact subsets of $\mathbb{R}$, decide whether it endows V with the structure of a locally convex space, whether the family of semi-norms separates points, and compare the induced topologies.

Exercise 6.49 Prove Proposition 6.4.

Further Reading

The hitchhiker's guide to infinite-dimensional analysis [1] offers a comprehensive source on all the topics covered in this chapter and far beyond. For a more concise introduction to Banach spaces, complete with a detailed preliminaries chapter, see [2]. For a classic introduction to functional analysis see [5], and for a text somewhat different in style, including an entire chapter devoted to unbounded operators, see [4]. To read more about locally convex spaces the reader may consult [3].

References

1. Aliprantis, C.D., Border, K.C.: Infinite Dimensional Analysis (A Hitchhiker's Guide), 3rd edn., p. xxii+703. Springer, Berlin (2006)
2. Carothers, N.L.: A Short Course on Banach Space Theory. London Mathematical Society Student Texts, vol. 64, p. xii+184. Cambridge University Press, Cambridge (2005)
3. Osborne, M.S.: Locally Convex Spaces. Graduate Texts in Mathematics, vol. 269, p. viii+213. Springer, Cham (2014)
4. Pedersen, G.K.: Analysis Now. Graduate Texts in Mathematics, vol. 118, p. xiv+277. Springer, New York (1989)
5. Yosida, K.: Functional Analysis. Classics in Mathematics, p. xii+501. Springer, Berlin (1995)

Chapter 7
Topological Groups

Banach spaces admit a particularly rich theory since the algebraic structure (i.e., the norm) induces a particularly nice metric (i.e., a complete one) resulting in a powerful interaction between algebra and geometry. Quite often, when an algebraic structure and a topological structure are allowed to interact, the fusion of the two theories results in an intricate and interesting new theory. One such fusion is the focus of this chapter.

There are at least two motivating reasons for considering topological groups. The first one is that just as groups model symmetry, topological groups model continuous symmetry. The second reason for studying topological groups is that quite often a portion of, say, a Banach space may fail to be a Banach space on its own but it may still retain enough of the algebra and the topology of the ambient space to form a topological group. This is often the case since a group is a much weaker algebraic structure compared to a linear space, and a topology is much weaker than a metric (whether or not induced by a norm). Topological groups thus are much more common-place since one only requires a group structure and a topology, rather than a full linear space and a metric structure, yet the theory of topological groups still presents a powerful fusion between algebra and geometry.

Section 7.1 introduces groups and group homomorphisms, without assuming any prior knowledge of groups. The main objects of study, namely topological groups, are then presented in Sect. 7.2. Section 7.3 presents topological subgroups and a first encounter with the interesting interaction between algebra and topology, namely that the closure of a subgroup is a subgroup. The quotient construction for topological groups is the topic of Sect. 7.4, including a rather detailed account of normal subgroups. Finally, Sect. 7.5 discusses another, deeper, consequence of the interaction between algebra and topology, namely the uniformizability of the topology of a topological group, thus enabling one to import significant portions of the uniform machinery of metric spaces into the realm of topological groups.

© Springer Nature Switzerland AG 2021

C. Alabiso and I. Weiss, *A Primer on Hilbert Space Theory*, UNITEXT for Physics,
https://doi.org/10.1007/978-3-030-67417-5_7

7.1 Groups and Homomorphisms

The axioms defining a group are an abstraction of structure present in many familiar algebraic systems, namely an associative and unital binary operation with respect to which every element is invertible. The historical development of group theory is rooted in the study of roots of polynomials, famously with the work of Evariste Galois and Niels Henrik Abel on finite groups. There is a distinct difference between the theory of finite groups and the theory of infinite groups. In finite group theory various counting arguments play an important role in determining the structure of groups. The recent classification of all finite simple groups, the result of an almost unfathomable collaboration spanning thousands of articles, is a milestone of finite group theory. Infinite groups though can be so wild that a full classification of all groups seems beyond reasonable expectations. Some limitations on size must be placed, and one way to impose such size constraints is through the introduction of a topology, and demanding that the group, as a topological space, be compact. The classification of compact Hausdorff topological groups is a much more manageable project.

With that in mind we present groups with an eye towards their common applications in Physics (as a formalism of symmetry) and as a step towards defining topological groups (and thus finite groups are only glanced at).

Definition 7.1 A *group* is a triple $(G, \cdot, e)$ where G is a set, $\cdot$ is a binary operation $G \times G \to G$, usually denoted by $(g, h) \mapsto g \cdot h$, or even just $(g, h) \mapsto gh$, and $e \in G$ is a chosen element, such that the following axioms hold.

1. Associativity, i.e., $(g_1 g_2) g_3 = g_1 (g_2 g_3)$ for all $g_1, g_2, g_3 \in G$.
2. The element e is an *identity* element, i.e., $eg = g = ge$ for all $g \in G$.
3. Existence of *inverses*, i.e., for all $g \in G$ there exists an element, denoted by $g^{-1} \in G$ and called the *inverse* of g, satisfying $gg^{-1} = e = g^{-1}g$.

If $gh = hg$ holds for all $g, h \in G$, then the group is said to be *commutative* or *abelian*. In an abelian group it is customary to write $g + h$ instead of gh, and to denote e, the neutral element, by 0. The cardinality of G is called the *order* of the group.

When there is no danger of confusion it is common to refer to a group G, leaving the operation $\cdot$ and the identity e implicit.

Proposition 7.1 *In a group G:*

1. *The identity is unique, i.e., if $e' \in G$ has the property that $e'g = g = g'e$ for all $g \in G$, then $e' = e$.*
2. *Inverses are unique, i.e., for every element $g \in G$, if $h_1, h_2 \in G$ satisfy*

$$gh_2 = gh_1 = e = h_1 g = h_2 g,$$

then $h_1 = h_2$.

Proof The arguments are very similar to those given in the proof of Proposition 2.1. The adaptation is left for the reader.   □

Remark 7.1 An inspection of the definition of linear space (Definition 2.1) reveals that we have already encountered abelian groups. Indeed, in that definition there are three sets of axioms, the first of which states, precisely, that $(V, +, 0)$ is an abelian group for any linear space V. Thus, had we chosen to present groups prior to linear spaces, we could have defined a linear space as an abelian group together with a scalar product, satisfying the other two sets of axioms in the definition of linear space.

Many examples of groups arise as the set of symmetries of an object. For instance, fixing an equilateral triangle in the plane, its symmetries correspond to the various ways in which rigid motions of the plane map the triangle onto itself. The set of all rigid motions that map the triangle to itself (of which there are precisely six: the identity, three reflections, and two rotations) is a group under the operation of composition. The verification is straightforward: the composition of rigid motions that fix the triangle is again a rigid motion that fixes the triangle, the identity function is a rigid motion that fixes the triangle (and thus serves as the identity element of the group), and the inverse function of a rigid motion that fixes the triangle is again a rigid motion that fixes the triangle. The resulting group is called the *dihedral group* of order 6, one member in the following family of groups

Theorem 7.1 *Let P be a regular n-gon in the plane (i.e., a regular polygon with n vertices), $n \geq 2$. The set D_n of all rigid motions (i.e., functions of the plane to itself that preserve lengths and angles) together with the operation of composition of functions is a group.*

Proof The reader is invited to turn the argument for D_3 given above into a proof of the general case.   □

Remark 7.2 The groups $\{D_n\}_{n \geq 2}$ are known as the *dihedral groups*. It can be shown that D_n has order $2n$. For that reason, the dihedral group D_n is also commonly (and confusingly) denoted by D_{2n}.

Of course, one may vary the shape being fixed, as well as the degree to which the mappings preserve the ambient geometry, or change the ambient geometry itself. As an extreme example, given a set S, the set of all bijections $\sigma : S \to S$, together with composition of functions, is easily seen to be a group. This is the group of symmetries of the set S, clearly essentially determined only by the cardinality of S. This group is denoted by $\text{Sym}(S)$, or, when $S = \{1, 2, 3, \ldots, n\}$, by S_n.

Following the same general idea, the following is an important example in the theory of topological groups.

Theorem 7.2 *For $n \geq 1$ let $\text{GL}_n(\mathbb{R})$ be the set of all invertible linear operators $A : \mathbb{R}^n \to \mathbb{R}^n$. With the operation of composition of functions, $\text{GL}_n(\mathbb{R})$ is a group.*

Proof The straightforward verification of the axioms, using established results from Chap. 2, is left for the reader.   □

The group $GL_n(\mathbb{R})$ is known as the *general linear group*. It is the group of linear symmetries of the linear space $\mathbb{R}^n$. By specifying a basis for $\mathbb{R}^n$ one may identify $GL_n(\mathbb{R})$ with the set of all invertible $n \times n$ matrices with real entries. The group $GL_n(\mathbb{R})$ is essentially the same as the group of invertible $n \times n$ matrices over $\mathbb{R}$ with the operation of matrix multiplication. Other important groups are defined in terms of certain linear symmetries, or, equivalently, as spaces of matrices satisfying various conditions. See the solved problems section for this chapter for detailed examples.

As a matter of fact, the claim above is a special case of a much more general source of groups, as we now show.

Theorem 7.3 *Let $\mathcal{B}$ be a Banach space. The subset G of $\mathbf{B}(\mathcal{B})$ consisting of all invertible bounded linear operators on $\mathcal{B}$, with the operation of composition of functions, is a group.*

Proof The implicit claim that the composition of invertible bounded linear operators is again an invertible bounded linear operator follows from the fact that

$$\|A_1 A_2\| \le \|A_1\| \|A_2\|$$

for all operators $A_1, A_2 \in \mathbf{B}(\mathcal{B})$, as is easily established. Next, clearly, the identity $\mathrm{id}_{\mathcal{B}}$ is a member of $\mathbf{B}(\mathcal{B})$, and it serves as the identity element of the group. The fact that composition of functions is associative is trivially verified. The fact that the inverse operator of a bounded operator is bounded is Corollary 6.1 of the Open Mapping Theorem.   □

We now turn to the structure preserving mappings between groups.

Definition 7.2 A function $\psi : G \to H$ between groups is a *group homomorphism* (or simply a *homomorphism*) if

$$\psi(g_1 g_2) = \psi(g_1)\psi(g_2)$$

for all $g_1, g_2 \in G$. If ψ is also bijective, then it is called an *isomorphism*, and then the groups G and H are said to be *isomorphic*, denoted by $G \cong H$.

Example 7.1 As said above, in the presence of a basis for $\mathbb{R}^n$ there is a bijection between $GL_n(\mathbb{R})$ and $GL(n; \mathbb{R})$, given by mapping an invertible linear operator to its representing matrix. In fact, this correspondence is a group isomorphism between the two groups. The determinant function $\det : GL_n(\mathbb{R}) \to \mathbb{R}$, that assigns to a linear operator $A : \mathbb{R}^n \to \mathbb{R}^n$ its determinant (i.e., the determinant of its representing matrix in any basis) is a group homomorphism $\det : GL_n(\mathbb{R}) \to \mathbb{R}^*$. Here $\mathbb{R}^*$ is the set of non-zero real numbers and the group operation is ordinary multiplication. When $n = 1$ the determinant $\det : GL_1(\mathbb{R}) \to \mathbb{R}^*$ is a group isomorphism.

A group homomorphism is only required to preserve the group product in the domain and codomain. It follows immediately though that the rest of the group structure is also preserved. For the proof it is helpful to note that the existence of inverses in a group immediately implies left and right cancelation laws: if $gh = gh'$, then $h = h'$ and if $hg = h'g$, then $h = h'$.

Proposition 7.2 *Let $\psi : G \to H$ be a group homomorphism. Then $\psi(e) = e$ and $\psi(g^{-1}) = \psi(g)^{-1}$, for all $g \in G$.*

Proof We only establish that $\psi(e) = e$ (notice that the e on the left is the identity in G while on the right it is the identity in H, and thus they may be different). Notice that $\psi(e)\psi(e) = \psi(ee) = \psi(e) = e\psi(e)$. The cancelation law now implies that $\psi(e) = e$. □

The verification of the following properties of homomorphisms and isomorphisms is very similar to the analogous results established for linear operators, for instance in Proposition 2.8. We thus omit the details of the proof.

Proposition 7.3 *If $G_1 \xrightarrow{\varphi} G_2 \xrightarrow{\psi} G_3$ are group homomorphisms, then*

1. *The composition $\psi \circ \varphi$ is a homomorphism.*
2. *If both φ and ψ are isomorphisms, then so is $\psi \circ \varphi$.*
3. *If ψ is an isomorphism, then so is ψ^{-1}.*
4. *The identity function $\mathrm{id}_G : G \to G$ is a group isomorphism.*

7.2 Topological Groups and Homomorphism

A topological group is a group together with a topology on it that is required to be compatible with the group structure in the sense that the group operations are to be continuous. Some of the far-reaching consequences resulting from this fusion between algebra and topology will be explored in the subsequent sections. In particular, it will be shown that the topology is then sufficiently metric-like so as to render such concepts as uniform continuity and Cauchy sequences meaningful.

Definition 7.3 A *topological group* is a group G together with a topology on the set G such that the group operation $\cdot : G \times G \to G$ is continuous (with respect to the given topology on the codomain and the product topology on the domain), and such that the function $G \to G$ given by $g \mapsto g^{-1}$ is continuous.

Example 7.2 The following examples of topological groups are all subsets of $\mathbb{R}$ or $\mathbb{C}$, and we consider each of them with the subspace topology induced, in each case, by the metric $d(x, y) = |x - y|$.

1. The circle $S^1 = \{z \in \mathbb{C} \mid |z| = 1\}$ with multiplication.
2. The set $\mathbb{R}$ of all real numbers with addition.
3. The set $\mathbb{R}^*$ of all non-zero real numbers with multiplication.
4. The set $\mathbb{C}$ of all complex numbers with addition.
5. The set $\mathbb{C}^*$ of all non-zero complex numbers with multiplication.

The next result provides a large repository of examples.

Lemma 7.1 *Let V be a normed space. Then $(V, +, 0)$, i.e., the additive structure of the space, is a topological group.*

Proof We already noted that the additive part of a linear space is an abelian group. The continuity requirements are satisfied by Proposition 6.1.   □

Another significant source of topological groups arises from groups of bounded linear operators on a Banach space.

Theorem 7.4 *Let $\mathcal{B}$ be a Banach space. The subset G of $\mathbf{B}(\mathcal{B})$ consisting of the invertible bounded linear operators on $\mathcal{B}$ is a topological group when endowed with the operator norm topology.*

Proof Having already observed that G is a group under composition of functions it remains to verify the continuity of the composition and the inverse mapping with respect to the operator norm. The continuity of the composition follows at once from the inequality

$$\|A_1 A_2\| \leq \|A_1\| \|A_2\|$$

while the continuity of the inverse mapping is Exercise 6.28.   □

Corollary 7.1 *The general linear group $\mathrm{GL}_n(\mathbb{R})$, when endowed with the operator norm topology, is a topological group (which is not abelian if $n > 1$).*

Proof $\mathrm{GL}_n(\mathbb{R})$ is precisely the group of invertible operators on the Banach space $\mathbb{R}^n$.   □

Since a topological group is a fusion between a group and a topological space, so are homomorphisms of topological groups a fusion between group homomorphisms and continuous mappings.

Definition 7.4 A function $\psi : G \to H$ between topological groups is said to be a *homomorphism* if ψ is both continuous and a group homomorphism. We say that ψ is an *isomorphism* if ψ is both a group isomorphism and a homeomorphism. If an isomorphism between topological groups G and H exists, then the groups are said to be *isomorphic*, denoted by $G \cong H$.

Example 7.3 In a topological group G, the mapping $g \mapsto g^{-1}$ is a homeomorphism but not usually a group homomorphism. It is a group homomorphism if G is abelian, since then

$$(gh)^{-1} = g^{-1}h^{-1}$$

while in general one only has the equality

$$(gh)^{-1} = h^{-1}g^{-1}.$$

For any fixed element $a \in G$, the *left translation* mapping $g \mapsto ag$ and the *right translation* mapping $g \mapsto ga$ are both homeomorphisms, but are generally not group homomorphisms. The determinant function det : $GL_n(\mathbb{R}) \to \mathbb{R}^*$ is continuous (since the determinant is polynomial in the entries of the matrix) and, as we have seen, is a group homomorphism, and thus is a homomorphism of topological groups. det : $GL_1(\mathbb{R}) \to \mathbb{R}^*$ is an isomorphism.

The reader is invited to state and prove the analogous result of Proposition 7.3 for topological homomorphisms and isomorphisms.

7.3 Topological Subgroups

A topological subgroup of a topological group is a subset that is on its own right a topological group with the induced algebraic and topological structures from the ambient topological group. In other words, a topological subgroup is a fusion of the concepts subgroup and topological subspace. Fortunately, it turns out that the subspace topology on a subgroup is automatically compatible with the group structure, and thus it is only the algebraic structure that dictates what the topological subgroups are. Consequently, there is no difference between the topological subgroups of a topological group and its subgroups when the topology is forgotten.

In light of the above, we focus on the notion of subgroup and, anticipating the contents of the next section, already discuss normal subgroups. We then observe the first non-trivial consequence of the interaction of the algebra and the topology, namely that the closure of a subgroup is a subgroup.

Definition 7.5 A subset H of a group G is a *subgroup* of G if the group operation in G restricts to a group operation on H. Equivalently, H is a subgroup of G if $e \in H$ and for all $h_1, h_2 \in H$

$$h_1 h_2^{-1} \in H.$$

The motivation for the following definition is presented in the next section.

Definition 7.6 A subgroup H of a group G is said to be *normal* if $ghg^{-1} \in H$ for all $h \in H$ and $g \in G$. The element ghg^{-1} is called the *conjugate* of h by g. Thus, a subgroup is normal when it is closed under conjugation.

Remark 7.3 Note that the condition for normality of H in G is equivalent to: for all $g \in G$ and $h \in H$, there exists $h' \in G$ such that

$$hg = gh'.$$

Thus, the condition of normality can be seen as a weak form of commutativity; elements in G commute with elements in H, up to a replacement by another element in H. If G is abelian, then every subgroup H of G is normal.

Example 7.4 Given a group homomorphism $\psi : G \to H$ its *kernel* is the set $\mathrm{Ker}(\psi) = \{g \in G \mid \psi(g) = e\}$. The kernel of a homomorphism is easily seen to be a normal subgroup of G. One may also easily verify (compare with Theorem 2.8) that $\mathrm{Ker}(\psi) = \{e\}$, i.e., the kernel is the *trivial subgroup* of G, if, and only if, ψ is injective.

Regarding subgroups of a topological group G, the presence of a topology entails no complications, at least in the following sense.

Lemma 7.2 *Let G be a topological group and H a subgroup of G. When endowed with the subspace topology, H is a topological group.*

The proof is immediate, relying on the simple fact that the restriction of a continuous mapping to a subspace remains continuous.

We may now demonstrate one consequence of the fruitful interaction between the group structure and the topology.

Theorem 7.5 *If H is a (normal) subgroup of a topological group G, then $\overline{H}$, the closure of H, is also a (normal) subgroup of G.*

Proof To show that $\overline{H}$ is a subgroup, we need to show that $gh^{-1} \in \overline{H}$ for all $g, h \in \overline{H}$. In other words, for the function $f : G \times G \to G$ given by $f(g, h) = gh^{-1}$ we need to establish that $f(\overline{H} \times \overline{H}) \subseteq \overline{H}$. Since f is clearly continuous and $\overline{H}$ is closed, it follows that $f^{-1}(\overline{H})$ is closed. Next, $H \times H \subseteq f^{-1}(H) \subseteq f^{-1}(\overline{H})$, where the first inclusion follows from the fact that H is a subgroup and the second inclusion since $H \subseteq \overline{H}$. Taking closures leads to $\overline{H \times H} \subseteq \overline{f^{-1}(\overline{H})} = f^{-1}(\overline{H})$. The result follows by noting that $\overline{H} \times \overline{H} = \overline{H \times H}$ (this latter fact is a general property of closures that has nothing to do with the group structure on H). We leave it to the reader to verify that if H is a normal subgroup, then so is $\overline{H}$.   □

7.4 Quotient Groups

Whereas the notion of subgroup of a topological group was quite straightforward, with a seamless interaction between the algebraic and the topological demands, the situation for the quotient of a topological group by a subgroup is more intricate. The main difficulty lies with the algebraic structure, namely that for an arbitrary subgroup H it is not always possible to obtain a group structure on the relevant quotient set. Topologically however, there are no difficulties since the quotient topology is always available.

Given a group G and a subgroup H defining $g_1 \sim g_2$ precisely when $g_1^{-1} g_2 \in H$ is easily seen to be an equivalence relation on G. The equivalence class of $g \in G$ is the set

$$gH = \{gh \mid h \in H\}$$

and is called the *left coset of g*. In particular, two left cosets $g_1 H$ and $g_2 H$ are equal if, and only if, $g_1^{-1} g_2 \in H$. Similarly, defining $g_1 \sim g_2$ precisely when $g_1 g_2^{-1} \in H$ gives rise to an equivalence relation whose equivalence classes are all of the form

$$Hg = \{hg \mid h \in H\}$$

and are called *right cosets* (of H in G). Of course, if G is abelian, then the left and right cosets coincide, but in general they may be different. However, any general result concerning left cosets in a group has a corresponding dual result about right cosets. In fact the translation between results on left and right cosets is achieved by means of the correspondence $gH \leftrightarrow Hg^{-1}$. For this reason, we may safely concentrate on the left cosets.

Any choice of a subgroup H of G gives rise to an equivalence relation $\sim$ with corresponding quotient set

$$G/H = \{gH \mid g \in G\},$$

the set of all left cosets of H in G, with the corresponding canonical projection $\pi : G \to G/H$ given by $\pi(g) = gH$. We address the question whether the group structure on G induces a group structure on G/H.

Theorem 7.6 *Let H be a subgroup of a group G. If the operation on G/H given by $(g_1 H) \cdot (g_2 H) = (g_1 g_2)H$ is well defined, then it defines a group structure on G/H and $\pi : G \to G/H$ is a group homomorphism.*

Proof The verification of the group axioms is trivial. For instance, noticing that the coset eH is precisely the set H we show that H is an identity element for the given operation. Let $gH \in G/H$ be an arbitrary element. Then

$$H \cdot gH = eH \cdot gH = (eg)H = gH$$

and similarly $gH \cdot H = gH$. We leave the verification of the other group axioms to the reader. The claim that π is a group homomorphism is nothing but the observation that $\pi(g_1 g_2) = (g_1 g_2)H = g_1 H \cdot g_2 H = \pi(g_1)\pi(g_2)$. $\square$

We thus see that as soon as the operation $(g_1 H) \cdot (g_2 H) = (g_1 g_2)H$ is well defined the quotient set G/H acquires a group structure. For the operation to be well defined one needs to verify that if $g_1 H = g_3 H$ and $g_2 H = g_4 H$, then necessarily $g_1 g_2 H = g_3 g_4 H$. In other words, given that $g_1^{-1} g_3 \in H$ and $g_2^{-1} g_4 \in H$, it must follow that $(g_1 g_2)^{-1} g_3 g_4 \in H$. Let us denote $g_1^{-1} g_3 = h_1$ and $g_2^{-1} g_4 = h_2$. Noting that

$$(g_1 g_2)^{-1} g_3 g_4 = g_2^{-1} g_1^{-1} g_3 g_4 = g_2^{-1} h_1 g_4,$$

if G were abelian, then we would obtain that

$$(g_1 g_2)^{-1} g_3 g_4 = g_2^{-1} g_4 h_1 = h_2 h_1 \in H,$$

as desired. However, G need not be abelian. A closer look though reveals that a commutativity demand is far too strong. Indeed, if H is a normal subgroup of G, then for g_4 and h_1 as above, there exists $h' \in H$ such that $h_1 g_4 = g_4 h'$, and then

$$(g_1 g_2)^{-1} g_3 g_4 = g_2^{-1} h_1 g_4 = g_2^{-1} g_4 h' = h_1 h' \in H.$$

We summarize this discussion as follows, remarking first that for a normal subgroup H in G it follows at once that the left and right cosets of H in G coincide, and thus one may simply speak of the cosets of H in G.

Theorem 7.7 *If H is a normal subgroup of G, then the quotient set G/H of all cosets of H in G is a group when endowed with the operation*

$$(g_1 H) \cdot (g_2 H) = (g_1 g_2) H.$$

The canonical projection $\pi : G \to G/H$ is then a group homomorphism. The group G/H is called the quotient group *of G by H.*

Corollary 7.2 *A subgroup H of a group G is normal if, and only if, H is the kernel of some group homomorphism whose domain is G.*

Proof It was already noted that the kernel of a homomorphism is a normal subgroup. For the converse, show that

$$\mathrm{Ker}(\pi) = H$$

for the canonical projection $\pi : G \to G/H$.   □

The situation now for a topological group G and a subgroup H of it is summarized in the following result whose proof is left for the reader.

Theorem 7.8 *Let G be a topological group and H a subgroup.*

1. The set $\{gH \mid g \in G\}$ of all left cosets of H in G is the quotient set for the equivalence relation $g_1 \sim g_2 \iff g_1^{-1} g_2 \in H$. When endowed with the quotient topology it is a topological space but it need not be a group.
2. The set $\{Hg \mid g \in G\}$ of all right cosets of H in G is the quotient set for the equivalence relation $g_1 \sim g_2 \iff g_1 g_2^{-1} \in H$. When endowed with the quotient topology it is a topological space but it need not be a group.
3. If H is normal in G, then left and right cosets coincide, as do the two quotient sets above, which are then denoted by G/H. The quotient set G/H acquires a group structure from G and with the quotient topology it is a topological group.

7.5 Uniformities

In the presence of a topology notions such as convergence and continuity become available. However, a topology is too weak to speak of Cauchy sequences (and thus

of completeness) or of uniform continuity, notions that do exist in the presence
of a metric. However, there exists a structure, known as a uniformity, that is in
between a topology and a metric, that does allow one to speak of Cauchy sequences,
completeness, completions, and uniform continuity. Moreover, every topological
group is automatically endowed with two (often distinct) such uniformities.

We first define the concept of uniformity in general and then proceed to describe
the canonical uniformities present in any topological group. As motivation, recall
that the axiomatization of a topology can be seen as the result of purifying a distance
function, eliminating any trace of details irrelevant for continuity. Similarly, the
axiomatization of a uniformity can be seen to arise from a similar purification of a
distance function from anything irrelevant to uniform continuity.

Definition 7.7 Let X be a set. A non-empty collection $\mathcal{E} = \{E_i\}_{i \in I}$ of relations
$E_i \subseteq X \times X$ is said to be a *uniform structure* or a *uniformity* on X if the following
conditions hold (where we write $x \sim_E y$ to indicate that $(x, y) \in E$).

1. $x \sim_E x$ for all $E \in \mathcal{E}$ and $x \in X$.
2. For all $E \in \mathcal{E}$ there exists an $F \in \mathcal{E}$ such that $x \sim_F y$ implies $y \sim_E x$.
3. For all $E \in \mathcal{E}$ there exists an $F \in \mathcal{E}$ such that $x \sim_F y$ and $y \sim_F z$ together imply
   that $x \sim_E y$.
4. If $E, F \in \mathcal{E}$, then so is $E \cap F$.
5. If $E \subseteq F \subseteq X \times X$ and $E \in \mathcal{E}$, then $F \in \mathcal{E}$.

The pair $(X, \mathcal{E})$ is called a *uniform space* and the members of $\mathcal{E}$ are its *entourages*. If
context clarifies any ambiguity it is common to refer to a uniform space X, leaving
the collection $\mathcal{E}$ of entourages implicit.

Example 7.5 Given a metric space (X, d), a uniformity on the set X is induced by
the distance function as follows. For every $\varepsilon > 0$ consider the set

$$E_\varepsilon = \{(x, y) \in X \times X \mid d(x, y) < \varepsilon\}.$$

It then follows that the collection

$$\mathcal{E} = \{E \subseteq X \times X \mid \exists \varepsilon > 0 \ E \supseteq E_\varepsilon\}$$

is a uniformity. The verification of the axioms is straightforward. For instance, con-
dition 3 follows from the triangle inequality: given $E \supseteq E_\varepsilon$, let $F = E_{\varepsilon/2}$, then
if $x \sim_F y$ and $y \sim_F z$, namely $d(x, y), d(y, z) < \varepsilon/2$, then $d(x, z) < \varepsilon$, so that
$x \sim_E z$. The resulting uniformity is called the *induced uniformity*.

Definition 7.8 A function $f : X \to Y$ between uniform spaces is said to be *uniformly continuous* if $f^{-1}(E)$ is an entourage in X for every entourage E in Y (here
$f^{-1}(E)$ stands for $\{(x, x') \in X \times X \mid f(x) \sim_E f(x')\}$).

Topology captures the notion of continuity in the sense that a function $f : X \to$
Y between metric spaces is continuous in the metric sense if, and only if, f is

continuous with respect to the induced topologies. We establish the analogous result
for uniformities and uniformly continuous functions.

Theorem 7.9 *A function $f : X \to Y$ between metric spaces is uniformly continuous
in the metric sense if, and only if, f is uniformly continuous with respect to the induced
uniformities.*

Proof Suppose that f is uniformly continuous in the sense that for every entourage
F in Y, $f^{-1}(F)$ is an entourage in X. Given $\varepsilon > 0$ consider the entourage

$$F_\varepsilon = \{(y, y') \in Y \times Y \mid d(y, y') < \varepsilon\}.$$

By definition of the induced uniformity $f^{-1}(F_\varepsilon)$ contains an entourage of the form
$E_\delta = \{(x, x') \in X \times X \mid d(x, x') < \delta\}$, for some $\delta > 0$. In particular, if $d(x, x') <$
δ, then $(x, x') \in E_\delta$ and thus $(f(x), f(x')) \in F_\varepsilon$, namely $d(f(x), f(x')) < \varepsilon$. In
other words, f is metrically uniformly continuous.

   In the other direction, suppose that f is uniformly continuous in the sense that for
all $\varepsilon > 0$ there exists a corresponding $\delta > 0$ such that $d(f(x), f(x')) < \varepsilon$ provided
that $d(x, x') < \delta$. To show that f is uniformly continuous with respect to the induced
uniformities, let F be an entourage in Y, namely there exists an $\varepsilon > 0$ such that
$F \supseteq F_\varepsilon$ where

$$F_\varepsilon = \{(y, y') \in Y \times Y \mid d(y, y') < \varepsilon\}.$$

With $\delta > 0$ corresponding to ε, we claim that $f^{-1}(F) \supseteq E_\delta$, where

$$E_\delta = \{(x, x') \in X \times X \mid d(x, x') < \delta\}.$$

Indeed, if $(x, x') \in E_\delta$, then $d(x, x') < \delta$ and thus $d(f(x), f(x')) < \varepsilon$. It follows
that $(x, x') \in f^{-1}(F_\varepsilon)$ and thus $E_\delta \subseteq f^{-1}(F_\varepsilon) \subseteq f^{-1}(F)$. This shows that $f^{-1}(E)$
contains an entourage in X, and thus is itself an entourage, which is what we needed
to show.   □

Clearly, the distance function cannot be reconstructed from the induced uniformity.
It is a simple matter to find different metrics that induce the same uniformity (for
instance, by scaling all distances by a positive constant). A uniform space $(X, \mathcal{E})$
whose uniform structure $\mathcal{E}$ is the induced uniformity for some distance function d
on X is said to be a *metrizable* uniform space. Thus, so far, we established that
the concept uniform space is weaker than that of metric space, and that it correctly
captures the notion of uniformly continuous functions. Next, we show that every
uniform space induces a topology.

Theorem 7.10 *Let $(X, \mathcal{E})$ be a uniform space. The collection τ of sets $U \subseteq X$ such
that for all $x \in U$ there exists an entourage $E \in \mathcal{E}$ such that $\{y \in X \mid x \sim_E y\} \subseteq U$
is a topology on X called the* induced topology.

Proof The proof is left as an exercise for the reader.   □

The passage from a distance function to the induced uniformity loses information and is thus not a reversible process. Similarly the passage from a uniformity to the induced topology is not reversible. A topological space (X, τ) whose topology τ is the induced topology for some uniformity on X is called a *uniformizable* topological space.

We are now ready to describe the two uniformities present on any topological group. Recall that every element $a \in G$ gives rise to a left translation map $g \mapsto ag$ and to a right translation map $g \mapsto ga$, both homeomorphisms. We extend the translation mappings so that they operate on subsets $S \subseteq G$, i.e., we define $aS = \{as \mid s \in S\}$, and similarly $Sa = \{sa \mid s \in S\}$.

Theorem 7.11 *Let G be a topological group and $\mathcal{B}$ a local basis at e. For every open set $U \in \mathcal{B}$ let $R_U = \{(x, y) \in X \times X \mid xy^{-1} \in U\}$. The collection*

$$\mathcal{E}_R = \{E \subseteq X \times X \mid \exists U \in \mathcal{B} \ \ E \supseteq R_U\}$$

is a uniformity on X called the induced right uniformity. *Similarly, for each $U \in \mathcal{B}$ let $L_U = \{(x, y) \in X \times X \mid x^{-1}y \in U\}$. The collection*

$$\mathcal{E}_L = \{E \subseteq X \times X \mid \exists U \in \mathcal{B} \ \ E \supseteq L_U\}$$

is a uniformity on X called the induced left uniformity.

Proof This is a routine verification of the axioms of a uniform space. □

The following properties of the induced uniformities justify their names.

Theorem 7.12 *Let G be a topological group and let $\mathcal{E}_L$ and $\mathcal{E}_R$ be the left and right induced uniformities, respectively. Then*

1. *The induced topology by either $\mathcal{E}_L$ or $\mathcal{E}_R$ is the original topology on G.*
2. *Every left translation $g \mapsto ag$ is uniformly continuous with respect to $\mathcal{E}_L$.*
3. *Every right translation $g \mapsto ga$ is uniformly continuous with respect to $\mathcal{E}_R$.*

Proof The proof is straightforward. □

Remark 7.4 We conclude this short account of topological groups with the following remark. A sequence $\{x_m\}_{m \geq 1}$ in a uniform space X is said to be a *Cauchy sequence* if for every entourage E there exists an $N \in \mathbb{N}$ such that $x_k \sim_E x_n$ for all $k, n > N$. A uniform space X is said to be *complete* if every Cauchy sequence in X converges (under the topology induced by the uniformity). The process of completion of a metric space (as given in Chap. 4 by means of Cauchy sequences) extends to a completion process for every uniform space (however not by means of Cauchy sequences, as they are too weak for general uniform spaces). Similarly, other uniform concepts of metric spaces have analogous results valid for all uniform spaces.

The result above shows that the topology of a topological group is always uniformizable (and the uniformity can be chosen to further be compatible with all right

or all left translations). Thus, in a sense, the topology of topological groups is rather
tame, behaving more like metric spaces do than the more wild topological spaces
out there do.

Exercises

Exercise 7.1 Let G be a group and Aut(G) the set of all isomorphisms $\psi : G \to G$.
Prove that, under composition of functions, Aut(G) is a group. (This is the group of
symmetries of the group G)

Exercise 7.2 Given a group G and an element $g \in G$, prove that $h \mapsto ghg^{-1}$ is an
isomorphism $G \to G$. Such an isomorphism is called an *inner* isomorphism. Prove
that the set of all inner isomorphisms of G is a normal subgroup of Aut(G).

Exercise 7.3 Prove that if H is a normal subgroup of G, then $gH = Hg$ for all
$g \in G$. Use a suitable subgroup H of the dihedral group D_3, the group of symmetries
of an equilateral triangle, to refute the converse.

Exercise 7.4 Prove that if G is a topological group and if, for some $g \in G$, the set
$\{g\}$ is closed, then G is Hausdorff.

Exercise 7.5 Prove that for all topological groups G, the quotient group $G/\overline{\{e\}}$ exists
and is a Hausdorff topological group.

Exercise 7.6 Prove that an open subgroup H of a topological group G is also closed.
(Hint: The complement of H is a union of translates of H).

Exercise 7.7 Prove that the topology induced by the uniformity induced by a metric
is identical to the topology induced by the metric.

Exercise 7.8 For subsets S and T of a topological group G consider the point-wise
product $ST = \{st \mid s \in S, t \in T\}$. Prove that if S and T are compact, then so is ST.

Exercise 7.9 Prove Theorems 7.10–7.12.

Exercise 7.10 The following is a well-known theorem in the theory of uniform
spaces: A uniform space $(X, \mathcal{E})$ is metrizable if, and only if, the uniformity $\mathcal{E}$ is
generated by countably many entourages, where $\mathcal{E}$ is *generated* by the collection
$\mathcal{F} \subseteq \mathcal{E}$ if
$$\mathcal{E} = \{E \subseteq X \times X \mid \exists F \in \mathcal{F} \;\; E \supseteq F\}.$$

Prove that a first countable Hausdorff topological group is metrizable.

Further Reading

For a broad physics-oriented introduction to groups see [4] or, for an introduction to groups in the context of Quantum Mechanics, see [5]. For a friendly introduction to topological groups see Chap. 3 of [1] (and the rest of the book for a delightful deeper study of the subject). For an elementary introduction to uniform spaces alongside topological spaces see [2]. Finally, for a comprehensive introduction to topological groups, starting with an independent treatment of topological spaces and metric spaces, see [3].

References

1. Diestel, J., Spalsbury, A.: The Joys of Haar Measure. Graduate Studies in Mathematics, vol. 150, p. xiv+320. American Mathematical Society, Providence, RI (2014)
2. James, I.M.: Topological and Uniform Spaces. Undergraduate Texts in Mathematics, 164 pp. Springer (1987)
3. Markley, N.G.: Topological Groups: An Introduction, 367 pp. Wiley (2010)
4. Sternberg, S.: Group Theory and Physics, 444 pp. Cambridge University Press, Cambridge (1995)
5. Tinkham, M.: Group Theory and Quantum Mechanics, 352 pp. Dover Publications (2003)

Chapter 8
Hilbert Spaces

This chapter is different in style than the previous ones. Presumably, the reader will soon venture off to consult books that delve much deeper into Hilbert space territory. This chapter offers a glimpse of the shores. With the background knowledge developed thus far each of the following sections concentrates on, roughly, a single aspect of Hilbert space theory that is markedly different to the situation in Banach space theory. This is done in the form of a discussion starting off with a consideration of general interest, contemplation of the complexity of the situation in the realm of Banach spaces, and culminating with tremendous improvement in Hilbert space theory.

Before summarizing the contents of the following sections, here is a preliminary discussion. A Banach space is a linear space together with a norm that induces a complete metric space. In that case the geometry borne by the norm meshes very well with the underlying algebra of the vectors. This was witnessed in the chapter on Banach spaces. Since a normed space does not carry a notion of angles Banach space theory, by necessity, is confined to focus solely on angle-free geometry. Hilbert space theory incorporates angles into the analysis.

Let us recall some facts from elementary geometry of the plane. By Pythagoras' Theorem the length (i.e., the norm) of a vector $x = (x_1, x_2) \in \mathbb{R}^2$ is given by $\sqrt{x_1^2 + x_2^2}$. Consider now two vectors $x, y \in \mathbb{R}^2$. These vectors then determine a triangle with side lengths $\|x\|$, $\|y\|$, and $\|x - y\|$. Let θ be the internal triangle angle formed by x and y. From elementary trigonometry, namely the law of cosines (a generalization of Pythagoras' Theorem), the lengths of the sides of the triangle and the angle θ are related by the formula

$$\|x - y\|^2 = \|x\|^2 + \|y\|^2 - 2\|x\|\|y\|\cos\theta.$$

Using the above formula for the length of a vector, this equality becomes

© Springer Nature Switzerland AG 2021
C. Alabiso and I. Weiss, *A Primer on Hilbert Space Theory*, UNITEXT for Physics,
https://doi.org/10.1007/978-3-030-67417-5_8

$$(x_1 - y_1)^2 + (x_2 - y_2)^2 = x_1^2 + x_2^2 + y_1^2 + y_2^2 - 2\|x\|\|y\|\cos\theta,$$

which simplifies to

$$x_1 y_1 + x_2 y_2 = \|x\|\|y\|\cos\theta.$$

Recalling the standard inner product in $\mathbb{R}^2$, given by $\langle x, y \rangle = x_1 y_1 + x_2 y_2$, we obtain

$$\langle x, y \rangle = \|x\|\|y\|\cos\theta$$

and thus we have discovered the standard inner product and the Cauchy-Schwarz Inequality in $\mathbb{R}^2$ by elementary geometry.

We thus see that with a sufficiently strong geometry (i.e., the presence of angles), a notion of a norm gives rise to an inner product. Conversely, an inner product gives rise to a norm by means of $\|x\| = \sqrt{(x, x)}$, and, via the Cauchy-Schwarz Inequality, also gives rise to angles.

In light of this discussion a Hilbert space is essentially a Banach space fortified with angles.

Definition 8.1 A *Hilbert space* is an inner product space that with the induced norm is a Banach space.

Prominent examples are the sequence space ℓ_2 and the function space L_2 (Theorem 5.17). Indeed, among all the Banach spaces ℓ_p, $1 \le p \le \infty$, only ℓ_2 is a Hilbert space and similarly L_2 is exceptional in its family.

Let us recall that a norm $\| - \|$ is induced by an inner product if, and only if, the parallelogram identity

$$2\left(\|x\|^2 + \|y\|^2\right) = \|x + y\|^2 + \|x - y\|^2$$

holds for all vectors x and y. In other words, a Banach space is a Hilbert space if, and only if, the parallelogram identity (which only mentions the norm) is valid. This chapter investigates some of this identity's far-reaching implications.

Section 8.1 is concerned with the geometric aspect known as the closest point property, chosen as a starting point due to its intuitive quality and pivotal role in establishing the results of the following sections. Section 8.2 continues with the strong geometric flavor and utilizes the closest point property to establish the existence of orthogonal complements. Section 8.3 is a brief view of bases, from Hamel, through Schauder, to Hilbert. Section 8.4 presents classical Fourier series through the Hilbert basis lens and Sect. 8.5 establishes the Riesz Representation Theorem. Each section concludes with a brief pointer to a famous result directly related to the core of the section, thus providing a somewhat richer historical context to the journey.

8.1 The Closest Point Property

Recall the point-to-set distance

$$d(x, S) = \inf\{d(x, s) \mid s \in S\},$$

meaningful in any metric space X, that measures the distance from a point x to a subset $S \subseteq X$. Generally speaking there is no reason for the infimum to be attained. In other words, there need not be, for the given point x, any single point in S that realizes the distance from x to S and if such a point exists it need not be unique. For instance, in $\mathbb{R}$ with the Euclidean metric consider the point $x = 0$ and the open interval $S = (1, 2)$; then $d(0, S) = 1$ while $d(0, s) > 1$ for all $s \in S$. If, instead, one takes $S = [-2, -1] \cup [1, 2]$, then $d(0, S) = 1$ and is realized at both $s = 1$ and $s = -1$. From this discussion we distill the following definition.

Definition 8.2 Let X be a metric space and $S \subseteq X$ a subset. Then S has the *closest point property* if for all $x \in X$ there is a unique point $s \in S$ with

$$d(x, s) = d(x, S).$$

Further examples in Euclidean spaces of subsets that fail to have the closest point property can easily be constructed, e.g., in $\mathbb{R}^n$, by choosing S either not closed or not convex (recall that a subset S in a linear space is convex if $tx + (1 - t)y \in S$ for all $x, y \in S$ and $0 \leq t \leq 1$, i.e., when the line joining any two points in S is contained in S). It turns out that closedness and convexity are precisely the obstructions to the closest point property not just in $\mathbb{R}^n$ but in any Hilbert space. We will shortly show how the completeness and the parallelogram law play crucial roles in the proof that any non-empty closed and convex subset of a Hilbert space has the closest point property.

First, though, we pause to demonstrate the necessity of the Hilbert structure; in a Banach space a non-empty closed and convex subset need not satisfy the closest point property. Consider $\mathbb{R}^2$ with the ℓ_∞ norm. Let $x = 0$ be the origin and $S = [1, 3] \times [-1, 1]$ a square. Then $d(0, S) = 1$ and this distance is attained at each point on the left side $\{1\} \times [-1, 1]$ of the square S. There are thus uncountably many points in S realizing the distance.

More dramatic is the observation that in a Banach space the distance from x to S need not be realized even by a single point. First, let $F: \mathcal{B} \to \mathbb{R}$ be a bounded linear functional on a real Banach space $\mathcal{B}$ and consider the set $S = \{x \in \mathcal{B} \mid F(x) = 1\}$. The set S is closed since it is the inverse image of the closed set $\{1\} \subseteq \mathbb{R}$ under the continuous function F. It is empty only if F is trivial (since F is homogenous) and it is convex (since F is also additive). The point-to-set distance $d(0, S)$ is then

$$\inf_{x \in S} \|x\|$$

which, we show, is precisely $1/\|F\|$, where $\|F\|$ is the operator norm of F (we are now assuming $F \neq 0$). Indeed, given an arbitrary $x \in S$, namely $F(x) = 1$, we have $1 = |F(x)| \leq \|F\|\|x\|$ and so

$$1/\|F\| \leq \inf_{x \in S} \|x\|.$$

For the reverse inequality let $\varepsilon > 0$ be given. By the definition of the operator norm there exists $y \in \mathcal{B}$ with $|F(y)| > (\|F\| - \varepsilon)\|y\|$. If such a y can be found in S, then we are done. But that is easily arranged by considering $(1/F(y))y$.

We thus see that the distance from 0 to $F^{-1}(1)$ computes the reciprocal of the operator norm of F. If that distance is attained, say at $x \in S$, then we have

$$1/\|F\| = \|x\| \quad \text{and} \quad F(x) = 1$$

and then for $y = \|F\|x$ we obtain

$$\|y\| = 1 \quad \text{and} \quad F(y) = \|F\| \cdot F(x) = \|F\|$$

so the norm of F is attained at y. All that is left now is to recall that not all bounded linear functionals on a Banach space (of infinite dimension) attain their norms. For any such non-trivial bounded linear functional F the point-to-set distance $d(0, S)$ to the non-empty, closed, and convex set $S = F^{-1}(1)$ is not attained. Bounded functionals that fail to attain their norm arise naturally (see Exercise 6.31).

Now that we have established the appropriate depth of the situation we return to the main question. Let $S \subseteq \mathcal{H}$ be an arbitrary non-empty, closed, and convex subset in a Hilbert space $\mathcal{H}$ and $x \in \mathcal{H}$ a point. Before showing the existence of a point $s \in S$ realizing the distance $d(x, S)$ we remark that its uniqueness follows easily, and without recourse to the completeness of the space. Indeed, suppose that two distinct points $s, t \in S$ realize the distance $d(x, S)$. The points x, s, t are then the vertices of a triangle with equal side lengths $d(x, s) = d(x, t) = \alpha$. The point $w = 1/2(s + t)$ lies in S (by convexity) and so $d(x, w) \geq d(x, S) = \alpha$. To summarize, we have an isosceles triangle in which the side length is no greater than the height. Let us formalize this situation in an arbitrary inner product space and obtain a contradiction. Let the sides of the triangle be given by vectors a, b. The assumptions are then that $a \neq b$ and $\|a\| = \|b\| = \alpha$ while $\|1/2(a + b)\| \geq \alpha$. Applying the parallelogram law yields

$$2(\|a\|^2 + \|b\|^2) = \|a + b\|^2 + \|a - b\|^2$$

and the left-hand side is simply $4\alpha^2$. Using the inequality

$$\|a + b\|^2 = 4\|1/2(a + b)\|^2 \geq 4\alpha^2$$

we see that $\|a - b\|^2 \leq 0$, forcing $a = b$ and thus a contradiction.

Having seen that the distance $d(x, S)$ is realized at no more than one point it remains to establish the existence of a point $s \in S$ with $d(x, S) = d(x, s)$. The strategy is to obtain the desired point s as the limit of a sequence that approximates the distance $d(x, S)$. Indeed, suppose that $\{s_n\}_{n \geq 1}$ is a sequence in S with

$$\lim_{n \to \infty} d(x, s_n) = d(x, S)$$

and further that

$$\lim_{n \to \infty} s_n = s$$

where the convergence is in $\mathcal{H}$. Since the distance function is continuous (even uniformly so) it follows that

$$d(x, S) = \lim_{n \to \infty} d(x, s_n) = d(x, \lim_{n \to \infty} s_n) = d(x, s)$$

showing that the distance is attained at s. It remains to argue that $s \in S$, but S is closed and s is the limit of a sequence in it.

To finish the proof we must show the existence of a convergent sequence with

$$\lim_{n \to \infty} d(x, s_n) = d(x, S).$$

The strategy is clear: by definition of $d(x, S)$ one can choose points in S with $d(x, s)$ lying arbitrarily close to $d(x, S)$. If chosen to form a Cauchy sequence, then completeness of $\mathcal{H}$ would guarantee the convergence of the sequence. The parallelogram law will tell us how to choose s_n so as to fulfil the Cauchy condition. Let us look at two points $u, v \in S$. The idea is that if u and v are close to realizing the distance $d(x, S)$, then they must be close to each other. We thus apply the parallelogram law to the vectors $x - u$ and $x - v$:

$$\|(x - u) + (x - v)\|^2 + \|(x - u) - (x - v)\|^2 = 2 \left(\|x - u\|^2 + \|x - v\|^2 \right)$$

giving us

$$\|u - v\|^2 = 2 \left(\|x - u\|^2 + \|x - v\|^2 \right) - \|2x - (u + v)\|^2 =$$

$$2 \left(\|x - u\|^2 + \|x - v\|^2 \right) - 4 \left\| x - \frac{u + v}{2} \right\|^2.$$

Noting that $1/2(u + v) \in S$ by convexity we conclude that

$$\|u - v\|^2 \leq 2 \left(\|x - u\|^2 + \|x - v\|^2 \right) - 4d(x, S).$$

We now see how to arrange for the Cauchy condition, e.g., choose $s_n \in S$ (remember that $S \neq \emptyset$) to satisfy

$$d(x, s_n)^2 < d(x, S)^2 + \frac{1}{n}$$

for all $n \geq 1$. For such a choice for the sequence $\{s_n\}_{n \geq 1}$ we clearly have

$$\lim_{n \to \infty} d(x, s_n) = d(x, S)$$

and

$$d(s_n, s_m)^2 = \|s_n - s_m\|^2 < 4d(x, S)^2 + 2\left(\frac{1}{n} + \frac{1}{m}\right) - 4d(x, S)^2 = 2\left(\frac{1}{n} + \frac{1}{m}\right)$$

and thus $\{s_n\}_{n \geq 1}$ is a Cauchy sequence.

We summarize the discussion with the statement of what we achieved.

Theorem 8.1 *In a Hilbert space $\mathcal{H}$ any non-empty closed and convex subset $S \subseteq \mathcal{H}$ has the closest point property.*

In a Hilbert space the closest point property settles the discussion. In a Banach space the set of closest points can range from empty to infinitely large. The study of its geometric properties is of much interest. We shall only mention here a theorem essentially due to Robert Clarke James from 1964 to the effect that in a Banach space $\mathcal{B}$ the condition that every non-empty closed and convex subset admits at least one closest point is equivalent to the reflexivity (Definition 6.8) of $\mathcal{B}$.

8.2 Orthogonal Complements

A linear subspace V can be a large place (particularly if it is infinite dimensional). Divide and conquer is a powerful paradigm whose incarnation in this context is the decomposition of V as the direct sum of two complementary subspaces. Recall that U, W, two subspaces of V, are complementary if every $x \in V$ can be written uniquely as

$$x = u + w$$

with $u \in U$ and $w \in W$, in which case we write

$$V = U \oplus W.$$

For instance, the familiar decomposition of a function f as

$$f(x) = \frac{f(x) + f(-x)}{2} + \frac{f(x) - f(-x)}{2}$$

can (with some verification) be phrased as

$$C(I, \mathbb{R}) = C_e(I, \mathbb{R}) \oplus C_o(I, \mathbb{R})$$

expressing the Banach space of continuous functions on an interval $I = [-a, a]$ as the direct sum of its subspaces of even and odd functions, respectively. A simple observation makes this statements stronger: each of these subspaces is closed in the topology of the ambient space induced by the L_∞ norm and thus the Banach space is decomposed as the direct sum of two complementary Banach spaces. This is a most desirable situation. After all, if only one, or none, of the two subspaces was Banach, then the divide and conquer method would force us out of the convenience of Banach space theory.

It is often the case that the ambient space of interest is rather complicated and one is not fortunate enough to easily recognize a representation of the space as the direct sum of two convenient subspaces in it. Instead, one singles out just one convenient subspace $U \subseteq \mathcal{B}$ of a Banach space $\mathcal{B}$ and the need to identify a complementary subspace W of it must be resolved. Therefore, a pertinent question arises: given a closed subspace U in a Banach space $\mathcal{B}$ is there a *closed* subspace W complementary to U?

En route to an answer let us first observe that if $\mathcal{B}$ is finite dimensional, then the answer is affirmative. Indeed, it is commonplace to use a basis argument in order to show that any subspace of a finite-dimensional linear space admits a complementary subspace (and almost always infinitely many such subspaces). Since in finite dimensions all subspaces are closed, the issue is resolved. The basis argument essentially still applies in infinite dimensions. Let us outline how. We fix a Banach space $\mathcal{B}$ and a subspace U. Let S be the set of all subspaces $W \subseteq \mathcal{B}$ with the property that $U \cap W = \{0\}$ (equivalently, so that no $x \in V$ decomposes in more than one way as a sum $x = u + w$ with $u \in U, w \in W$). Ordering S by inclusion yields a poset that is non-empty (justify!) and satisfies the condition of Zorn's Lemma. There is thus a maximal subspace $W \in S$ and it can be shown to necessarily complement U.

An immediate issue with this result is its reliance on Zorn's Lemma; it is not computationally tractable and so cannot be used to divide and conquer in a practical situation. In some sense, through the use of the Axiom of Choice we have lost control over the resulting complementary subspace. But matters are even more severe. If our initial choice of the subspace U in $\mathcal{B}$ is closed there is no guarantee that the complementary subspace produced by Zorn's Lemma is closed. Worse still, the space c_0 is a closed subspace of the Banach space ℓ_∞ of bounded sequences with the ℓ_∞ norm yet none of the infinitely many complementary subspaces of c_0 in ℓ_∞ is closed. An accessible proof of this fact is given in [3].

Let us now focus on the situation in a Hilbert space $\mathcal{H}$, assuming a closed subspace $M \subseteq \mathcal{H}$ is given. We seek a closed complementary subspace N, namely to write a vector $x \in \mathcal{H}$ as a sum $x = y + z$ with $y \in M$ and $z \in N$. There is a natural candidate for the point $y \in M$: the closest point to x. A unique $y \in M$ with $d(x, M) = d(x, y)$ exists by the closest point property in $\mathcal{H}$ (from the previous section) noting that M is non-empty (since $0 \in M$), convex (since it is a linear subspace) and closed (by assumption). We are thus well under way: $x = y + z$ where $y \in M$ is the unique point satisfying $d(x, M) = d(x, y)$, and $z = x - y$.

Intuitively, the linear subspace M in $\mathcal{H}$ is a flat portion of the ambient space. The shortest distance from x to M should be realized by dropping a perpendicular from x to M. In other words, we expect the vector $z = x - y$ to be orthogonal to every vector in M. We thus define

$$M^{\perp} = \bigcap_{m \in M} m^{\perp}$$

where $m^{\perp} = \{w \in \mathcal{H} \mid \langle w, m \rangle = 0\}$, the set of all vectors perpendicular to m. Each $m^{\perp}$ is easily seen to be a linear subspace and so $M^{\perp}$, as the intersection of linear subspaces is itself a linear subspace. Moreover, since $m^{\perp}$ is the inverse image of the closed set $\{0\}$ under the continuous function $w \mapsto \langle w, m \rangle$ to the ground field it follows that $m^{\perp}$ is closed. Thus, as the intersection of closed subsets, the space $M^{\perp}$ is closed. By our geometric intuition we expect that $z \in M^{\perp}$. If that is the case, then $N = M^{\perp}$ would indeed be a closed subspace complementing M: the decomposition $x = y + z$ is already given and the fact that $M \cap M^{\perp} = \{0\}$ is immediate from the general fact that $\langle w, w \rangle = 0 \iff w = 0$.

It remains to show that $z = x - y$ belongs to $M^{\perp}$. We note that z satisfies

$$\|z\| \leq \|z - m\|$$

for all $m \in M$ since

$$\|z\| = \|x - y\| = d(x, M)$$

while the fact that $y + m \in M$ tells us that

$$\|z - m\| = \|x - (y + m)\| \geq d(x, M).$$

The argument will be completed by showing that this property of z implies that $z \in M^{\perp}$ (incidentally, the converse is true as well and completeness is never used). So, fix $m \in M$, which we may assume to be non-zero. We work in the real case. By assumption $\|z\|^2 \leq \|z - \alpha m\|^2$ for all scalars α and after expansion and immediate algebra we get

$$|\langle z, m \rangle| \leq \frac{|\alpha|}{2} \|m\|^2$$

and allowing α to tend to 0 shows $\langle z, m \rangle = 0$. The complex case is handled by multiplying α by a suitable complex number of modulus 1.

We summarize the above as the following theorem.

Theorem 8.2 *In a Hilbert space any closed subspace M has its orthogonal complement $N = M^{\perp}$ as a closed complementary subspace.*

As noted above, the Banach space ℓ_{∞} does not have this property. It is a deep and celebrated result of Yoram Lindenstrauss and Lior Tzafriri from 1971 that a Banach space in which every closed subspace admits a closed complementary subspace is in fact a Hilbert space, provided one is allowed to lightly tweak the norm in a way that does not affect the topology.

8.3 Bases: Hamel, Schauder, and Hilbert

In a finite-dimensional linear space V a choice of a basis endows the space with a notion of coordinates. If the basis vectors are $u_1, \ldots, u_n$, then each vector x can be written uniquely as

$$x = \sum_{i=1}^{n} \alpha_i u_i$$

and we may treat the coefficients $\alpha_1, \ldots, a_n$, in this order, as the the coordinates of x in the given basis. For concreteness, let us assume the ground field is $\mathbb{C}$. What we have then is a process $V \to \mathbb{C}^n$. Let us now consider arbitrary n vectors $u_1, \ldots, u_n$ in V and attempt to use them for the purpose of obtaining coordinates. This is a simple matter: if $a = (\alpha_1, \ldots, \alpha_n)$ is a vector in $\mathbb{C}^n$, then let

$$\psi(a) = \sum_{i=1}^{n} \alpha_i u_i$$

thus defining a function $\psi \colon \mathbb{C}^n \to V$. Clearly, ψ is a linear operator where $\mathbb{C}^n$ is given the usual linear space structure.

When the vectors $u_1, \ldots, u_n$ are thought of as a specification of directions in the space V the requirements for them to form a basis are the following. The directions may not be redundant, i.e., they are linearly independent. This is equivalent to the injectivity of ψ. The directions must also provide sufficient coverage to reach any vector, i.e., they are spanning. This is equivalent to the surjectivity of ψ. Thus, the vectors $u_1, \ldots, u_n$ are a basis if, and only if, ψ is a bijection and so $V \cong \mathbb{C}^n$. Obviously, a different choice of basis produces a different coordinates function ψ. Therefore, a desire to conclude that V and $\mathbb{C}^n$ should, or even can, be identified should immediately be scrutinized. An identification of this sort depends on ψ — of which there are many. Carelessly identifying V with $\mathbb{C}^n$ throws away the vital information present in the function ψ.

Let us thus remain precise and appreciate the utility of the coordinates function $\psi \colon \mathbb{C}^n \to V$ and its dependency on the basis. An important aspect of ψ is its computability. After all, in practical situations one naturally constructs the linear space V, decides on some basis in it, and then wishes to find, for a vector $x \in V$, its coordinates. This amounts to solving the inverse problem

$$\psi(a) = x$$

for the associated coordinates function $\psi \colon \mathbb{C}^n \to V$. Typically, such inverse problems are hard to solve but the linearity of ψ and finite dimensionality lead to efficient algorithms for doing just that (this sentence summarizes quite a bit of elementary linear algebra).

When we turn to infinite dimensions not only is the tractability of ψ compromised but Hamel bases, the infinite-dimensional algebraic analogues of bases in finite dimensions, can rarely be exhibited. And if a basis cannot be written down there is certainly no hope of computing coordinates according to it. For this reason Hamel bases are of very limited use in Banach space theory. For instance, in the Banach space ℓ_∞ the linear subspace c_{00} of the eventually 0 sequences is an infinite-dimensional linear space with a Hamel basis given by the standard vectors e_n. However, c_{00}, with the ℓ_∞ norm, is not a Banach space as it is not closed in ℓ_∞. Its closure in ℓ_∞ is the Banach space c_0, the space of sequences converging to 0. In a sense, c_0 is the simplest non-trivial example of a Banach space, obtained from the very manageable space c_{00}, yet already here exhibiting a Hamel basis is hopeless.

Algebraically this is a pessimistic state of affairs. Computationally, our preference is to work with finite lists of numbers. Inductively approximating would allow us to even work with countable lists of numbers, but we cannot move beyond the countable. Applications drive us to infinite-dimensional spaces. To tame those and get a useful theory completeness is highly desirable. But then Exercise 6.1 shows the inevitable problem: as soon as a Banach space is infinite dimensional no countable Hamel basis exists. So, for precisely the linear spaces of interest the jump from finite to infinite dimensions skips over the countable option.

To the rescue comes the presence of a topology in a Banach space. Suppose that we have a sequence $\{u_n\}_{n\geq 1}$ of vectors in a Banach space $\mathcal{B}$. These vectors form a *Schauder basis* of $\mathcal{B}$ if for every $x \in \mathcal{B}$ there exists a unique sequence $a = \{a_i\}_{i\geq 1}$ of complex numbers with

$$x = \sum_{i=1}^{\infty} \alpha_i u_i$$

in the sense that

$$\lim_{n\to 0} \left\| x - \sum_{i=1}^{n} \alpha_i u_i \right\| = 0.$$

For example, it is rather straightforward that in each of the Banach spaces c_0 and ℓ_p, $1 \leq p < \infty$, the sequence $\{e_n\}_{n\geq 1}$ is a Schauder basis. Now, given a Schauder basis in $\mathcal{B}$ let C be the set of all sequences $\{\alpha_i\}_{i\geq 1}$ as above for which the series converges. Due to the delicate issue of convergence in $\mathcal{B}$ this set can be hard to understand. Regardless, a function $\psi \colon C \to B$ can then be defined by

$$\psi(a) = \sum_{i=1}^{\infty} \alpha_i u_i$$

and so we are in a similar situation to the finite-dimensional one: given a vector x we seek to solve $\psi(a) = x$ to find the coordinates of x. We would also like to retain some of the computational tractability of this inverse problem. In general Banach spaces this is still very challenging even if a relatively simple Schauder basis is given.

Before making the final step toward an effective notion of basis let us note that the existence of a Schauder basis is not guaranteed. Indeed, if one exists, then $\mathcal{B}$ must be separable. This is no surprise; a Schauder basis is a countable set that topologically generates the entire space so it is expected that the entire space is topologically not too big. For a proof suppose $\{u_n\}_{n \geq 1}$ is a Schauder basis. Note first that the set of all finite linear combinations of these vectors is dense in $\mathcal{B}$ since it contains all partial sums of all series that suffice to generate $\mathcal{B}$. This set though is not countable but restricting to its subset in which the coefficients have rational real and imaginary parts is countable. It is still dense since $\mathbb{Q}$ is dense in $\mathbb{R}$.

In a finite-dimensional inner product space V orthonormal bases $u_1, \ldots, u_n$ are phenomenally pleasant to work with. In this case solving the inverse problem $\psi(a) = x$ for any given $x \in V$ trivializes: the coefficients are given by the inner product $\alpha_i = \langle u_i, x \rangle$. We thus consider a Hilbert space $\mathcal{H}$ together with an orthonormal sequence $\{u_i\}_{i \geq 1}$ of vectors in it, namely the vectors are pair-wise orthogonal and each is of norm 1: $\langle u_i, u_j \rangle = \delta_{i,j}$, the Kronecker delta. We seek to use these vectors to obtain coordinates for all vectors $x \in \mathcal{H}$.

Suppose

$$x = \sum_{i=1}^{\infty} \alpha_i u_i$$

for some sequence $a = \{\alpha_i\}_{i \geq 1}$. Then

$\|x\|^2 =$

$\left\| \sum_{i=1}^{\infty} \alpha_i u_i \right\|^2 =$ meaning of the infinite series

$\left\| \lim_{n \to \infty} \sum_{i=1}^{n} \alpha_i u_i \right\|^2 =$ continuity of the norm

$\left(\lim_{n \to \infty} \left\| \sum_{i=1}^{n} \alpha_i u_i \right\| \right)^2 =$ generalized Pythagoras' Theorem (Exercise 2.74)

$\lim_{n \to \infty} \sum_{i=1}^{n} \|\alpha_i u_i\|^2 =$ norm properties and $\|u_i\| = 1$

$\lim_{n \to \infty} \sum_{i=1}^{n} |\alpha_i|^2 =$ meaning of infinite sum in $\mathbb{R}$

$$\sum_{i=1}^{\infty} |\alpha_i|^2 = \qquad\qquad \text{definition of norm in } \ell_2$$

$$\|a\|_2^2$$

and so $\|x\| = \|a\|_2$ and in particular $a \in \ell_2$. In other words, coordinates with respect to an orthonormal sequence must come from ℓ_2. The converse is also true: if $a \in \ell_2$, then the series

$$\sum_{i=1}^{\infty} \alpha_i u_i$$

converges in $\mathcal{H}$. This is so since the sequence of partial sums

$$x_n = \sum_{i=1}^{n} \alpha_i u_i$$

satisfies, by the generalized Pythagoras' Theorem,

$$\|x_{n+p} - x_n\|^2 = \left\| \sum_{i=n+1}^{n+p} \alpha_i u_i \right\|^2 = \sum_{i=n+1}^{n+p} |\alpha_i|^2$$

which together with the assumed convergence of

$$\|a\|^2 = \sum_{i=1}^{\infty} |\alpha_i|^2$$

implies that $\{x_n\}_n \geq 1$ is Cauchy and thus converges, and this is precisely the convergence we are after. So, starkly differently than the case with a Schauder basis, ℓ_2 precisely parametrizes all possible coordinates once an orthonormal sequence is chosen. We may thus define the coordinates function

$$\psi : \ell_2 \to \mathcal{H}$$

analogously to the finite-dimensional case. On top of being a linear operator the function ψ, as we just computed, preserves norms, i.e., $\|\psi(a)\| = \|a\|_2$. But more is true. The function ψ respects the inner products; the given one on $\mathcal{B}$ and the standard one on ℓ_2. In more detail,

$$\langle \psi(a), \psi(b) \rangle = \langle a, b \rangle$$

for all $a, b \in \ell_2$. This is shown by a computation very similar to the one above where instead of the continuity of the norm one uses the continuity of the inner product.

Ideally, we would like ψ to be bijective and since it is an isometry, and thus injective, the only obstruction is surjectivity. In other words, we would like each and every vector $x \in \mathcal{H}$ to have coordinates $a \in \ell_2$, namely to be able to solve $\psi(a) = x$. When this is the case we say that $\{u_i\}_{i \geq 1}$ is a *Hilbert basis*. If ψ fails to be surjective, then it is because $\{u_i\}_{i \geq 1}$ consists of too few vectors; not enough to provide sufficient coverage to reach all vectors in $\mathcal{H}$. Intuitively, if that happens, we would simply add another orthogonal vector to the list, make sure it is normalized, and thus increase the coverage. In this manner we can improve the sequence forever and, hopefully, end up with a Hilbert basis. The problem is that if $\mathcal{H}$ is very large the sequence may become so long that it is no longer countable. Let us look into that matter.

We say that a set $S \subseteq \mathcal{H}$ is an *orthonormal system* if $\langle s, t \rangle = \delta_{s,t}$ for all $s, t \in S$. Using Zorn's Lemma we show that $\mathcal{H}$ contains a maximal complete orthonormal system. The setup is standard: let $\mathcal{A}$ be the set of all orthonormal systems in $\mathcal{H}$ ordered by inclusion. It is clear that $\mathcal{A}$ is non-empty and that every chain in it has an upper bound given by its union. There is thus a maximal orthonormal system $S_{\max}$ in $\mathcal{H}$. Without further conditions $S_{\max}$ may not be countable. But, if $\mathcal{H}$ is separable, then $S_{\max}$ must be countable by Proposition 5.4. This is so since for all $x, y \in S_{\max}$ one has by Pythagoras' Theorem that $\|x - y\| = \sqrt{2}$ and so $S_{\max}$ is a net in $\mathcal{H}$.

We now know that a separable Hilbert space has a countable maximal orthonormal system $S_{\max}$ and proceed to show that it is a Hilbert basis. Enumerating $S_{\max}$ arbitrarily we have $S_{\max} = \{e_1, e_2, \dots\}$ without repetition. Fixing $x \in \mathcal{H}$ we must show that

$$x = \sum_{i=1}^{\infty} \alpha_i e_i$$

for suitable coordinates. Firstly, we discover what these coordinates must be; assuming the equality, we would obtain through a continuity argument that for any $1 \leq j < \infty$

$$\langle e_j, x \rangle = \langle e_j, \lim_{n \to \infty} \sum_{i=1}^{n} \alpha_i e_i \rangle = \lim_{n \to \infty} \langle e_j, \alpha_i e_i \rangle = \alpha_j$$

so in other words we must show

$$x = \sum_{i=1}^{\infty} \langle e_i, x \rangle e_i.$$

Let us for the moment assume that the series converges and write $\hat{x}$ for its sum. Our goal is to show that $x = \hat{x}$. If that was not the case, then for $y = x - \hat{x}$ another simple limit argument shows that $\langle e_i, y \rangle = 0$, namely y is orthogonal to $S_{\max}$ and thus also orthogonal to its linear span and consequently orthogonal to its closure $M = \overline{\text{span}(S_{\max})}$. In other words, we may write $\mathcal{H} = M \oplus M^{\perp}$ where M is the closure of the span of the maximal orthonormal system $S_{\max}$ and $y \in M^{\perp}$. But then if $y \neq 0$, then $S_{max} \cup \{y/\|y\|\}$ would be an orthonormal system strictly larger than S_{max}. This impossibility forces $y = 0$ and so $x = \hat{x}$.

It remains to see that the series giving rise to $\hat{x}$ converges. We have seen that for such a series convergence is equivalent to the coefficients belonging to ℓ_2. The coefficients in our case do belong to ℓ_2 as a result of Bessel's Inequality

$$\sum_{i=1}^{\infty} |\langle e_i, x \rangle|^2 \leq \|x\|^2$$

that we now prove. Let

$$y_n = \sum_{i=1}^{n} \langle e_i, x \rangle e_i$$

and note that $x - y_n$ is orthogonal to $e_1, \ldots, e_n$ and thus to y_n. Therefore,

$$\|x\|^2 = \|(x - y_n) + y_n\|^2 = \|x - y_n\|^2 + \|y_n\|^2 \geq \|y_n\|^2 = \sum_{i=1}^{n} |\langle e_i, x \rangle|^2$$

and the desired inequality follows by allowing n to tend to infinity.

The gist of the discussion can be summarized as follows.

Theorem 8.3 *A necessary and sufficient condition for a Hilbert space to have a Hilbert basis is its separability.*

The situation for Banach spaces is more delicate. The basis problem, posed by Stefan Banach, asks whether the necessary condition of separability is also sufficient for the existence of a Schauder basis in a Banach space. In 1973 Per Enflo provided a negative answer.

8.4 Fourier Series

The separable Hilbert space ℓ_2 is of fundamental importance. As the previous section shows, if $\mathcal{H}$ is any separable Hilbert space, then ℓ_2 can be used to provide coordinates $\psi: \ell_2 \to \mathcal{H}$ for all vectors in $\mathcal{H}$. Consequently, the space of square summable sequences, as humble as it may sound, implicitly contains the intricacies of any other separable Hilbert space, including, by Theorem 5.20, L_2 – the seemingly much more complicated space of square integrable functions.

In this section all functions are of the form $x: [-\pi, \pi] \to \mathbb{C}$. In particular, we write $L_2 = L_2([-\pi, \pi], \mathbb{C})$. Let us witness a first stark difference between ℓ_2 and L_2. Among the simplest members one finds in ℓ_2 is the harmonic sequence s with $s_n = 1/n$. Establishing that $s \in \ell_2$ is easily done by the integral test for the convergence of $\sum 1/n^2$. In L_2 we find the function $x(t) = t$ whose membership in L_2 is trivial: $t \mapsto t^2$ is continuous and thus integrable on $[-\pi, \pi]$. Now,

$$\|x\|_2^2 = \int_{-\pi}^{\pi} dt \ t^2 = \frac{2\pi^3}{3}$$

while trying to compute

$$\|s\|_2^2 = \sum_{n=1}^{\infty} \frac{1}{n^2}$$

we find ourselves with a lack of exact tools. These two examples are typical. For a large class of functions $x \in L_2$, e.g., all polynomials x, computing the norm $\|x\|_2$ is a simple matter of integration. By contrast, computing the norm of sequences in ℓ_2 directly is possible primarily for carefully designed sequences (that often feature as exercises when first encountering the theory of series).

The space ℓ_2, in a sense, is simpler than L_2. At the very least it is considerably easier to define. Nonetheless, we have just witnessed an aspect in which L_2 is simpler than ℓ_2 when it comes to computation. Let us thus make precise the intuitive sense in which ℓ_2 is simpler than L_2: it is trivial to exhibit a Hilbert basis (i.e., a complete orthonormal sequence) in ℓ_2. A moment's thought is all that is required to verify that the collection $\{e_n\}_{n\geq 1}$ of standard vectors in ℓ_2 is a Hilbert basis. A second pensive moment should convince the reader that the ease of defining ℓ_2 is inseparably and intimately linked to the ease of presentation of a Hilbert basis.

The main theorem of the previous section tells us that ℓ_2 and L_2 are abstractly the same Hilbert space. This should make us hopeful of the possibility of exploiting the ease of presentation of ℓ_2 with the computational tools present in L_2. In particular, if $\psi : \ell_2 \to L_2$ is an isomorphism, preserving the inner product, and thus norms, then ψ would serve as a bridge between the two formalisms. For this to be of any computational use it ought to be that ψ is explicit. And for that we must obtain an explicit description of a Hilbert basis in L_2.

For all $n \in \mathbb{Z}$ consider the function $e^{int} = \cos(nt) + i\sin(nt)$. For $n \neq m$

$$\langle e^{int}, e^{imt}\rangle = \int_{-\pi}^{\pi} dt \ e^{-int}e^{imt} = \int_{-\pi}^{\pi} dt \ e^{i(m-n)t} = 0$$

and so e^{int} and e^{imt} are orthogonal in L_2. Further, for all $n \in \mathbb{Z}$

$$\|e^{int}\|_2^2 = \int_{-\pi}^{\pi} dt \ e^{-int}e^{int} = 2\pi$$

and so the vectors

$$e_n = \frac{1}{\sqrt{2\pi}} e^{inx}$$

are an orthonormal sequence in L_2. Note that it is indexed by $\mathbb{Z}$ rather than $\mathbb{N}$, but that is immaterial since the two indexing sets are of the same cardinality so we proceed to work with ℓ_2 as a space of sequences indexed by $\mathbb{Z}$. It can be shown (by the Stone-Weierstrass theorem, a generalization of the Weierstrass approximation

theorem that was invoked in Theorem 5.16) that these functions span a dense subset
of L_2, implying they form a Hilbert basis.

We thus have a specific Hilbert basis in L_2 and with it the coordinates function
$\psi : \ell_2 \to L_2$ given by

$$\psi(a) = \sum_{n=-\infty}^{\infty} a_n \cdot e_n$$

For all vectors $a = (\ldots, a_{-n}, \ldots, a_{-1}, a_0, a_1, \ldots, a_n \ldots) \in \ell_2$. Finding the coordinates of a given $x \in L_2$, namely solving the inverse problem $\psi(a) = x$, is as simple
as computing the inner product in L_2: the coefficient of e_n is $\langle e_n, x \rangle$. In other words,

$$x = \sum_{n=-\infty}^{\infty} \langle e_n, x \rangle \cdot e_n$$

for all $x \in L_2$. The precise meaning of this equality is that the series on the right
converges to x with respect to the L_2 norm, i.e.,

$$\lim_{n \to \infty} \left\| x - \sum_{k=-n}^{n} \langle e_k, x \rangle \cdot e_k \right\|_2 = 0.$$

It is convenient to express x in terms of the un-normalized orthogonal vectors
$\{e^{inx}\}_{n \in \mathbb{Z}}$ as in

$$\langle e_n, x \rangle \cdot e_n = \frac{1}{\sqrt{2\pi}} \langle e^{int}, x \rangle \cdot \frac{1}{\sqrt{2\pi}} e^{int} = c_n \cdot e^{int}$$

with

$$c_n = \frac{1}{2\pi} \langle e^{int}, x \rangle$$

and write

$$x = \sum_{n=-\infty}^{\infty} c_n e^{int}.$$

The coefficients c_n are then known as the complex *Fourier coefficients* of x and
the series is called the *Fourier series* of x. Written directly in terms of the integral
formula for the inner product in L_2 the Fourier coefficients are

$$c_n = \frac{1}{2\pi} \int_{-\pi}^{\pi} dt \, x(t) e^{-int}$$

and the convergence of the series to x is in the sense that

$$\lim_{n \to \infty} \int_{-\pi}^{\pi} dt \left| x(t) - \sum_{k=-n}^{n} c_k \cdot e^{ikt} \right|^2 = 0.$$

In particular, no claim of point-wise convergence is made, i.e., at any particular $s \in [-\pi, \pi]$ there is no guarantee that $x(s) = \sum_{n=-\infty}^{\infty} c_n e^{ins}$.

The coordinates function ψ preserves norms, namely

$$\|x\|_2 = \|(\langle e_n, x \rangle)\|_2$$

where on the left-hand side we have the L_2 norm and on the right-hand side the ℓ_2 norm of the coordinates sequence. When this equality is written directly in terms of integrals and using the Fourier coefficients it reads as

$$\frac{1}{2\pi} \int_{-\pi}^{\pi} dt \, |x(t)|^2 = \sum_{n=-\infty}^{\infty} |c_n|^2$$

and is known as *Parseval's Identity*.

Let us now specialize to the function $x(t) = t$. A simple computation reveals its Fourier coefficients to be $c_0 = 0$ and, for all $n \in \mathbb{Z} \setminus \{0\}$,

$$c_n = \frac{1}{2\pi} \int_{-\pi}^{\pi} dt \ te^{-int} = \frac{i}{n} \cdot (-1)^n$$

so

$$t = 2 \cdot \sum_{n=1}^{\infty} \frac{i}{n} \cdot (-1)^n \cdot e^{int}$$

with convergence in the L_2 norm. Parseval's Identity tells us that

$$\frac{1}{2\pi} \int_{-\pi}^{\pi} dt \, t^2 = \sum_{n=-\infty}^{\infty} |c_n|^2$$

where we easily compute the left-hand side to be $\pi^2/3$ while $|c_0|^2 = 0$ and $|c_n|^2 = 1/n^2$ for $n \in \mathbb{Z} \setminus \{0\}$. With these values we obtain

$$\frac{\pi^2}{3} = 2 \sum_{n=1}^{\infty} \frac{1}{n^2}$$

and so

$$\|s\|_2 = \left(\sum_{n=1}^{\infty} \frac{1}{n^2} \right)^{1/2} = \frac{\pi}{\sqrt{6}}.$$

We have successfully computed the ℓ_2 norm of the harmonic sequence.

The quality of the convergence of the Fourier series of x to x is as subtle as it is important. The convergence in the L_2 norm is a rather weak form of convergence, yet, as we saw, it is strong enough to yield interesting non-trivial results. However, point-wise convergence is the holy grail and it is desirable to know, at least in theory, when it can be reached. The extent to which the Fourier series of an L_2 function x converges to x point-wisely intrigued mathematicians from Fourier himself to the giants of 20th century analysis. For well-behaved functions, for instance if x is continuously differentiable, point-wise convergence is guaranteed at all points. Hopes, at the dawn of the 20th century, that such was the case for all continuous functions were proven to be too optimistic and in 1915 Luzin's Conjecture stated that point-wise convergence for a continuous function was guaranteed with the exception of a negligible set of points (in the measure-theoretic sense, i.e., of measure 0). In 1966 the conjecture was proved true by Lennart Carleson. The proof, despite several simplifications, is quite involved. The situation gets even more involved when considering the Fourier series for $x \in L_p$ with $p \neq 2$. The extension of Carleson's Theorem to the case $p > 1$ was stated in his original 1966 article and proved rigorously by Richard Hunt in 1968. The situation for L_1 functions was known already several decades earlier to be beyond hope: there exist continuous such functions whose Fourier series diverge everywhere.

8.5   The Riesz Representation Theorem

The dual space $\mathcal{B}^*$ of a Banach space $\mathcal{B}$, consisting of all bounded linear functionals $\mathcal{B} \to K$ to the ground field, is an object of great interest. Important constructs are members of the dual space, e.g., taking the limit of a sequence is a function $\lim : c \to K$. This function is linear and bounded, so $\lim \in c^*$. The summation operation is a function $\sigma_\infty : \ell_1 \to K$ which is linear and bounded, so $\sigma_\infty \in (\ell_1)^*$. Similarly, integration is linear and bounded so $\int \in (L_1([a, b], K))^*$. For a general Banach space $\mathcal{B}$ it may be difficult to exhibit even a single non-trivial element in the dual space, attesting, in a sense, to the complexity of $\mathcal{B}$. The Hahn-Banach Theorem guarantees the existence of sufficiently many such elements, at least to separate the points of $\mathcal{B}$. In some cases the dual space can be identified up to a linear isometry, e.g., for all $1 \leq p < \infty$

$$\ell_p^* \cong \ell_q \text{ and } L_p^* \cong L_q$$

where q satisfies $1/p + 1/q = 1$. Among these we see that ℓ_2 and L_2 are self-dual. These are also the only Hilbert spaces in their family. This is no coincidence.

In a Hilbert space $\mathcal{H}$ there is an abundance of explicit bounded linear functionals. For arbitrary $x \in \mathcal{H}$ the function $F_x : \mathcal{H} \to K$ given by

$$F_x(y) = \langle x, y \rangle$$

is a linear operator (since by definition the inner product is linear in its second argument). Let us compute its norm $\|F_x\|$. Using the Cauchy-Schwarz Inequality

$$|F_x(y))| = |\langle x, y \rangle| \leq \|x\|\|y\|$$

and thus $\|F_x\| \leq \|x\|$. Taking $x = y$ shows that $\|F_x\| = \|x\|$ and in particular the norm is attained.

What we now have is a function $\Psi \colon X \to X^*$ given by $\Psi(x) = F_x$, or, directly in terms of the inner product, $\Psi(x)(y) = \langle x, y \rangle$. Moreover, $\|\Psi(x)\| = \|x\|$, i.e., Ψ preserves the norms and thus is an isometry. Note that Ψ also preserves much of the linear structure (but is not quite homogenous if $K = \mathbb{C}$) but we shall concentrate just on its metric aspects. What we have so far is an isometric embedding (injectivity follows from norm preservation) of $\mathcal{H}$ in its dual $\mathcal{H}^*$. We are about to establish that Ψ is surjective. In other words, we shall show that any bounded linear functional $F \in \mathcal{H}^*$ is of the form F_x for (a necessarily unique) $x \in \mathcal{H}$.

Let us fix a bounded linear functional $F \colon \mathcal{H} \to K$. Let us assume $F = F_x$ and see what we can extract about x. Let $M = \mathrm{Ker}(F)$ be the kernel of F, i.e., $y \in M \iff F(y) = 0$. But if $F = F_x$, then $F(y) = 0 \iff \langle x, y \rangle = 0$, namely $M = x^\perp$. Therefore, the x we are after resides in $M^\perp$, the orthogonal complement of $\mathrm{Ker}(F)$. That narrows down the search space considerably and allows us to divide and conquer according to Sect. 8.2. Indeed, we may write

$$\mathcal{H} = M \oplus M^\perp$$

since F is continuous and thus $M = F^{-1}(\{1\})$, as a continuous inverse image of a closed set, is a closed subspace of $\mathcal{H}$.

We still need to find an appropriate $x \in M^\perp$ so we should hope that space is not too large. At this point note that if $F = 0$, then we may take $x = 0$ so we proceed under the assumption that $F \neq 0$. We then observe that $\dim(M^\perp) = 1$. To see that, first note that Theorem 2.11 and $\mathcal{H} = M \oplus M^\perp$ imply $\mathcal{H}/M \cong M^\perp$. Since we are assuming $F \neq 0$ it follows that $F \colon \mathcal{H} \to K$ must be surjective and so Exercise 2.65 implies that $\mathcal{H}/M \cong K$. Therefore, $M^\perp \cong \mathcal{H}/M \cong K$, and thus has dimension equal to 1. Therefore, we may choose a vector $x \neq 0$ that spans $M^\perp$.

We now know that every vector $y \in \mathcal{H}$ can be written uniquely as $y = m + tx$ with $m \in M = \mathrm{Ker}(F)$ and $t \in K$. By computing we see that

$$F(y) = F(m) + t F(x) = t F(x)$$

while

$$F_x(y) = \langle x, y \rangle = \langle x, m \rangle + t \langle x, x \rangle = t \|x\|^2$$

so the condition for equality is $F(x) = \|x\|^2$. Since the choice of x can safely be altered up to a scalar multiple we can easily arrange to choose x so that $F(x) = \|x\|^2$, and thus conclude that $F = F_x$.

The fact that in a Hilbert space every bounded linear functional F is uniquely of the form F_x is known as the Riesz Representation Theorem proved by Frigyes Riesz in 1909. The precise properties of Ψ imply that a Hilbert space is congruent to its dual. This theorem is behind the following slogan.

Theorem 8.4   *Every Hilbert space $\mathcal{H}$ is self-dual.*

In the course of the discussion we noted that the linear functionals of the form F_x attain their norm. Since in a Hilbert space these functionals exhaust all of the bounded ones it follows that in a Hilbert space every bounded linear functional attains its norm. Another result of Robert Clarke James states that a Banach space has this same property if, and only if, it is reflexive. He first gave a proof in the separable case in 1957 and in the general case in 1972.

Further Reading

Hopefully the material of this chapter had whet the appetite of the reader. There are many introductory textbooks that go much further into Hilbert space theory. A mathematically oriented one is [4] and it is also masterfully short without being dense or imprecise. A very different book in style is the equally masterfully written [1] spanning more than a handful of modern applications of Hilbert space theory in great detail, sprawling over many hundreds of pages. Finally, we mention the classical [2] written by one of the greatest expositors of mathematics.

References

1. Debnath, L., Mikusinski, P.: Introduction to Hilbert Spaces With Applications, 3rd edn. Academic Press (2005)
2. Halmos, P.R.: A Hilbert Space Problem Book. Graduate Texts in Mathematics, vol. 19. Springer (1982)
3. Whitley, R.: Projecting m onto c₀. Am. Math. Mon. **73**(3), 285–286 (1966)
4. Young, N.: An Introduction to Hilbert Space. Cambridge University Press, Cambridge (1988)

Chapter 9
Solved Problems

9.1 Linear Spaces

9.1 Consider the linear space $C(\mathbb{R}, \mathbb{R})$ and the vectors $x(t) = \sin(t)$, $y(t) = \cos(t)$, and $z(t) = 1$. Show that x, y, z are linearly independent, while x^2, y^2, z^2 are not.

9.2 Let V be a linear space and $S \subseteq V$ a set of vectors. Prove that S is linearly independent if, and only if, every finite subset of S is linearly independent.

9.3 Consider $\mathbb{R}$ as a linear space over itself. Prove that $\mathbb{R}$ has precisely two linear subspaces. Now consider $\mathbb{R}$ as a linear space over $\mathbb{Q}$ and prove that it has infinitely many linear subspaces.

9.4 Let V and W be linear spaces over the field K, and $X \subseteq V$ an arbitrary subset of V. A *linear relation* in X is any expression

$$0 = \sum_{k=1}^{m} \alpha_k x_k,$$

where $m > 0$, $\alpha_1, \ldots, \alpha_m \in K$, and $x_1, \ldots, x_m \in X$. Let S be the span of X and consider an arbitrary function $f : X \to W$. Prove that f extends to a linear operator $F : S \to W$ if, and only if, for any linear relation in X as above, one has

$$0 = \sum_{k=1}^{m} \alpha_k f(x_k).$$

Conclude that if $\mathcal{B}$ is a basis for V, then any function $f : \mathcal{B} \to W$ extends to a linear operator $F : V \to W$.

9.5 Let $T : V \to W$ be a linear isomorphism between linear spaces. Prove that T maps a basis for V to a basis for W.

© Springer Nature Switzerland AG 2021
C. Alabiso and I. Weiss, *A Primer on Hilbert Space Theory*, UNITEXT for Physics,
https://doi.org/10.1007/978-3-030-67417-5_9

9.6 Prove that every infinite dimensional linear space over $\mathbb{R}$ contains an isomorphic copy of $\mathbb{R}^n$, for all $n \geq 0$. Is there a linear space that contains an isomorphic copy of every linear space over $\mathbb{R}$?

9.7 Prove that the space M of 3×3 matrices with entries in $\mathbb{R}$ is spanned by the subspace of matrices of rank 2.

9.8 Compute the dimension of the linear space P_n, $n \geq 0$, of polynomials with real coefficients and degree at most n, and of the linear space P of all polynomials with real coefficients.

9.9 Let $A, B : V \to V$ be two invertible linear operators from a linear space to itself. Prove that if A and B commute, i.e., $AB = BA$, then so do A^{-1} and B^{-1}.

9.10 Prove that a linear space V over an arbitrary field K (if you like you can take $K = \mathbb{R}$ or $K = \mathbb{C}$) is infinite dimensional if, and only if, V is isomorphic to a proper linear subspace of it.

9.2 Topological Spaces

9.11 (Separation properties) A topological space X is said to satisfy:

- the T_0 *separation axiom* if for all distinct points $x, y \in X$ there exists an open set U such that either $x \in U$ and $y \notin U$, or $y \in U$ and $x \notin U$;
- the T_1 *separation axiom* if for all distinct points $x, y \in X$ there exist two open sets U and V such that $x \in U$ and $y \notin U$, and $y \in V$ and $x \notin V$;
- the T_2 *separation axiom* if X satisfies the Hausdorff separations property, that is for all distinct points $x, y \in X$ there exist two open sets U and V such that $x \in U$, $y \in V$, and $U \cap V = \emptyset$.

Clearly, the separation properties are increasing in strength. For every separation property, give an example of a topological space satisfying it, but not the next one.

9.12 (Locally connected spaces) A topological space X is called *locally (path) connected* at x if, for every open set V with $x \in V$, there exists a (path) connected open set U with $x \in U \subseteq V$. The space X is said to be *locally (path) connected* if it is locally (path) connected at every $x \in X$.

Show that a locally (path) connected space need not be (path) connected, nor does a (path) connected space need be locally (path) connected. For the latter, consider the *comb space*, the subset C of $\mathbb{R}^2$ given by

$$C = \{(x, 0) \mid 0 \leq x \leq 1\} \cup \{(0, y) \mid 0 \leq y \leq 1\} \cup A_1 \cup A_2 \cup \cdots \cup A_n \cup \cdots$$

where $A_n = \{(1/n, t) \mid 0 \leq t \leq 1\}$. The set C is given the subspace topology of the Euclidean topology on $\mathbb{R}^2$. (Hint: Draw a picture of this space!)

9.13 (**Locally compact spaces**) A topological space X is called *locally compact* at $x \in X$ if there exists a compact set C and an open set U such that $x \in U \subseteq C$. The space X is *locally compact* when it is locally compact at every $x \in X$.

Show that every compact space is locally compact, but not every locally compact space is compact. Provide an example of a space that is not locally compact.

9.14 Consider

$$X = \left\{ A \in M_2(\mathbb{R}) \mid A^2 = 0 \right\}.$$

Endow X with the subspace topology of $M_2(\mathbb{R})$, where $M_2(\mathbb{R})$ is endowed with the topology induced by the identification $\Phi \colon M_2(\mathbb{R}) \stackrel{\simeq}{\to} \mathbb{R}^4$ given by

$$\begin{pmatrix} a & b \\ c & d \end{pmatrix} \stackrel{\Phi}{\mapsto} (a, b, c, d),$$

and $\mathbb{R}^4$ is endowed with the Euclidean topology. Decide whether X is closed in $M_2(\mathbb{R})$, whether X is compact, and whether X is connected.

9.15 Consider the topological spaces

$$X_1 = \left\{ (x, 1) \in \mathbb{R}^2 \mid x \in \mathbb{R} \right\} \quad \text{and} \quad X_2 = \left\{ (x, 2) \in \mathbb{R}^2 \mid x \in \mathbb{R} \right\}$$

endowed with the subspace topology from the Euclidean topology on $\mathbb{R}^2$. On the space

$$X = X_1 \cup X_2,$$

the coproduct of X_1 and X_2, consider the equivalence relation

$$(x, 1) \sim (x', 2) \quad \stackrel{\text{def}}{\Leftrightarrow} \quad x = x' \neq 0.$$

Let $\tilde{X} = X \mid / \sim$ be the quotient space. Decide whether:

1. $\tilde{X}$ is T_0, T_1, T_2;
2. $\tilde{X}$ is compact;
3. $\tilde{X}$ is path-connected.

9.16 Consider the spaces

$$X_1 = \mathbb{R}^2,$$

$$X_2 = \left\{ (x, y, z) \in \mathbb{R}^3 \mid x^2 + y^2 + z^2 = 1 \right\} - \left\{ \begin{pmatrix} 0 \\ 0 \\ 1 \end{pmatrix} \right\},$$

$$X_3 = \left\{ (y_1, y_2) \in \mathbb{R}^2 \mid y_1^2 + y_2^2 = 1 \right\},$$

$$X_4 = \left\{ (y_1, y_2) \in \mathbb{R}^2 \mid y_1^2 + y_2^2 + 2y_1 = 0 \right\} \cup \left\{ (y_1, y_2) \in \mathbb{R}^2 \mid y_1^2 + y_2^2 - 2y_1 = 0 \right\}.$$

For every pair of spaces, decide whether the spaces are homeomorphic or not.

9.17 Construct a topological space where the closed sets are stable under countable unions, but not under all arbitrary unions.

9.18 Let $A \in M_n(\mathbb{R})$ be a symmetric matrix whose eigenvalues are positive. Decide whether

$$W = \{x \in \mathbb{R}^n \mid x^t A x = 1\}$$

is a compact subset of $\mathbb{R}^n$ with the Euclidean topology.

9.19 Consider a topological Hausdorff space X and a subset $D \subseteq X$ which is dense and locally compact. Is D necessarily open in X?

9.20 Consider the ring $\mathbb{Z}$ of integer numbers. The *spectrum* of $\mathbb{Z}$ is the set

$$X = \{p \cdot \mathbb{Z} \mid p \text{ prime or } p = 0\},$$

where $p \cdot \mathbb{Z} = \{p \cdot k \mid k \in \mathbb{Z}\}$. For all $a \in \mathbb{Z}$ define the set

$$V(a \cdot \mathbb{Z}) = \{p \cdot \mathbb{Z} \in X \mid a \cdot \mathbb{Z} \subseteq p \cdot \mathbb{Z}\}.$$

- Prove that the collection $\mathcal{V} = \{V(a \cdot \mathbb{Z}) \mid a \in \mathbb{Z}\}$ is the set of closed subsets of a topology on X. That topology is known as the *Zarisky topology* on $\mathbb{Z}$.
- Does there exists a single point in X whose closure is the whole space?
- Is X Hausdorff?

9.3 Metric Spaces

9.21 Let (X, d) be a metric space. Suppose that there exists an infinite countable set $\{x_n\}_{n \in \mathbb{N}}$ of points such that $d(x_k, x_m) = 1$ for all $k, m \in \mathbb{N}$ with $k \neq m$. Prove that X is not compact.

9.22 Consider the set ℓ_∞ of all bounded sequences of real numbers with the metric induced by the ℓ_∞ norm, i.e.,

$$d(x, y) = \sum_n |x_n - y_n|.$$

Prove that ℓ_∞ is connected and not compact.

9.23 Let $f : \mathbb{R} \to \mathbb{R}$ be an infinitely differentiable function with the property that for all $x_0 \in \mathbb{R}$ there exists a natural number $n \geq 0$ with $f^{(n)}(x) = 0$. Prove that there exists an open interval $I = (a, b)$, with $a < b$, such that f agrees with a polynomial function on I.

9.24 Consider the space $C([0, 1], \mathbb{R})$ of all continuous real-valued functions on the interval $[0, 1]$, endowed with the L_∞ norm

$$\|f\|_\infty = \max_{t \in [0,1]} |f(t)| .$$

Is this norm induced by any inner product on $C([0, 1], \mathbb{R})$?

9.25 Prove that the empty set and every singleton set admit a unique metric, while every other set admits infinitely many non-isometric metric structures.

9.26 Let (X, d) be a metric space. Consider a subset S of X. Prove that the closure $\overline{S}$ of S in X is the set

$$\{x \in X \mid d(x, S) = 0\},$$

where $d(x, S) = \inf_{s \in S} d(x, s)$.

9.27 Let (X, d) be a metric space. Prove that the function $\tilde{d} \colon X \times X \to \mathbb{R}$ defined by $\tilde{d}(x, y) = \min\{1, d(x, y)\}$ is a metric on X which induces the same topology as d does.

9.28 Consider the spaces $\mathbb{R}$ and $\mathbb{R}^2$ endowed with the Euclidean metrics. Decide whether there exists an isometry $f \colon \mathbb{R}^2 \to \mathbb{R}$.

9.29 Let $\{d_i\}_{i \in I}$ be a family of metric functions on a given set X. Prove that the supremum $d_s \colon X \times X \to \mathbb{R}_+$ given by

$$d_s(x, y) = \sup_{i \in I}\{d_i(x, y)\}$$

is a metric function on X. In contrast, show that the similarly defined infimum of metric functions need not be a metric function.

9.30 Let $f \colon X \to Y$ be a function between metric spaces.

1. Is continuity, or uniform continuity, a sufficient condition to ensure that if X is complete then $F(X)$ is complete?
2. Is continuity, or uniform continuity, a sufficient condition to ensure that if X is totally bounded then $F(X)$ is totally bounded?
3. Is continuity, or uniform continuity, a sufficient condition to ensure that if X is complete and totally bounded then $F(X)$ is complete and totally bounded?

9.4 Normed Spaces and Banach Spaces

9.31 A metric function d on a linear space V, with ground field $K = \mathbb{R}$ or $K = \mathbb{C}$, is said to be *translation invariant* when

$$d(x, y) = d(x + z, y + z)$$

for all $x, y, z \in V$ and is said to be *scale homogenous* when

$$d(\alpha x, \alpha y) = |\alpha| d(x, y)$$

for all $x, y \in V$ and $\alpha \in K$. Prove that V is a normed space if, and only if, it is endowed with a translation invariant and scale homogenous metric.

9.32 Let $X = C([a, b], \mathbb{R})$ be the space of continuous functions $f : [a, b] \to \mathbb{R}$ and consider the assignment

$$f \mapsto \|f\| = \int_a^b |f(x)| \, dx$$

for all $f \in X$.

1. Give a direct proof that $f \mapsto \|f\|$ is a norm on X.
2. Decide whether X with the induced metric $d(f, g) = \|f - g\|$ is a complete metric space.

9.33 Consider $\mathbb{R}$ with the function $d : \mathbb{R} \times \mathbb{R} \to \mathbb{R}_+$ given by

$$d(x, y) = \frac{|x - y|}{1 + |x - y|}$$

for all $x, y \in \mathbb{R}$. Decide whether d is a distance function on $\mathbb{R}$ and if so whether the metric is induced by a norm.

9.34 Let X be a linear space endowed with two norms, $\|\cdot\|_1$ and $\|\cdot\|_2$. Suppose that there exists $K > 0$ such that

$$\|v\|_1 \le K \cdot \|v\|_2$$

holds for all $v \in X$. Prove that the topology induced by $\|\cdot\|_2$ is finer than the topology induced by $\|\cdot\|_1$.

9.35 For vectors x, y in a linear space V, let $L(x, y)$ be the *affine line* connecting x and y, that is

$$L(x, y) = \{\alpha x + (1 - \alpha)y \mid 0 \le \alpha \le 1\}.$$

A normed space V is said to be *strictly convex* if for all distinct vectors x, y on the unit sphere, i.e., $\|x\| = \|y\| = 1$, the line $L(x, y)$ intersects the unit sphere only at x and y.

1. Prove that for x, y with $\|x\| = \|y\| = 1$ the line $L(x, y)$ is fully contained in the unit ball $\{z \in V \mid \|z\| \le 1\}$.
2. Prove that a normed space induced by an inner product is strictly convex.

3. Prove that in a strictly convex normed space, if $x \neq y$ are vectors which satisfy $\|x\| = \|y\| = 1$, then $\|x + y\| < 2$.
4. Of the ℓ_1, ℓ_2, and ℓ_∞ norms on $\mathbb{R}^2$, decide which, if any, are strictly convex.

9.36 Prove that the kernel of a bounded linear operator $A : V \to W$ is a closed linear subspace of V.

9.37 Consider the normed space ℓ_2 and the linear operator $A : \ell_2 \to \ell_2$ given by

$$A(\{x_n\}_{n \geq 1}) = \left\{ \left(1 - \frac{1}{n}\right) x_n \right\}.$$

Prove that A is a bounded linear operator that never attains its norm, i.e., there exists no x_0 such that

$$\|Ax_0\| = \|A\|\|x_0\|.$$

9.38 Let $A : U \to V$ and $B : V \to W$ be bounded linear operators between normed spaces. Prove that the composition BA is a bounded linear operator and that

$$\|BA\| \leq \|B\|\|A\|.$$

9.39 Let $\mathcal{B}$ be a Banach space and recall that $\mathbf{B}(\mathcal{B})$, the set of all bounded linear operators $A : \mathcal{B} \to \mathcal{B}$, is a normed space when endowed with the operator norm, and thus is a topological space. Prove that the set $G \subseteq \mathbf{B}(\mathcal{B})$ consisting of the invertible operators is an open subset of $\mathbf{B}(\mathcal{B})$.

9.40 Let $\mathcal{B}_1$, $\mathcal{B}_2$, $\mathcal{B}_3$ be Banach spaces with $T : \mathcal{B}_1 \to \mathcal{B}_2$ a bounded linear operator (whose domain is all of $\mathcal{B}_1$) and let $S : \mathcal{B}_2 \to \mathcal{B}_3$ be a closed linear operator whose domain is $\mathcal{D}(S) \subseteq \mathcal{B}_2$. Prove that ST is a closed linear operator.

9.5 Topological Groups

9.41 Prove that the group $(\mathbb{R}^n, +)$ with the Euclidean topology is a Hausdorff topological group. In fact, it is a topological vector space, i.e., all of the linear space structure mappings are continuous.

9.42 Prove that the *special linear group*

$$SL_n(\mathbb{R}) = \{A \in GL_n(\mathbb{R}) \mid \det(A) = 1\}$$

is a topological group.

9.43 Prove that the *orthogonal group*

$$O(n) = \left\{A \in GL_n(\mathbb{R}) \mid A^{-1} = A^t\right\}$$

is a topological group.

9.44 Prove that the *special orthogonal group*

$$SO(n) = \{A \in O(n) \mid \det(A) = 1\}$$

is a topological group.

9.45 Prove that the *symplectic group*

$$Sp_{2n}(\mathbb{R}) = \left\{A \in GL_{2n}(\mathbb{R}) \mid A^t \cdot J_0 \cdot A = J_0\right\},$$

where

$$J_0 = \left(\begin{array}{c|c} 0 & \mathrm{id}_n \\ \hline -\mathrm{id}_n & 0 \end{array}\right),$$

is a topological group.

9.46 Prove that the topological group $SO(n)$ is path-connected.

9.47 Prove that the topological group $O(n)$ is the disjoint union of two path-connected components, $O^+(n)$ and $O^-(n)$.

9.48 Prove that

$$GL_n^+(\mathbb{R}) = \{A \in GL_n(\mathbb{R}) \mid \det(A) > 0\}$$

is a path-connected component of the topological group $GL(n; \mathbb{R})$.

9.49 Prove that the topological group $GL_n(\mathbb{R})$ has two connected components, more precisely,

$$GL_n^+(\mathbb{R}) = \{A \in GL_n(\mathbb{R}) \mid \det(A) > 0\}$$

and

$$GL_n^-(\mathbb{R}) = \{A \in GL_n(\mathbb{R}) \mid \det(A) < 0\}.$$

9.50 Prove that the special linear group

$$SL_n(\mathbb{R}) = \{A \in GL_n(\mathbb{R}) \mid \det(A) = 1\}$$

is path-connected.

Solutions

Linear Spaces

9.1 Suppose that

$$\alpha x + \beta y + \gamma z = 0$$

for some scalars $\alpha, \beta, \gamma \in \mathbb{R}$. The equation is a functional equality, and thus for any choice of $t \in \mathbb{R}$ we have that

$$\alpha x(t) + \beta y(t) + \gamma z(t) = 0.$$

However, by considering the values $t = 0, t = \pi/2, t = \pi$, we obtain the equations: $\beta + \gamma = 0, \alpha + \gamma = 0$, and $-\beta + \gamma = 0$, from which $\alpha = \beta = \gamma = 0$ follows easily. Thus the only linear combination resulting in the constantly zero function is the trivial combination, showing the desired linear independence. Now, as for x^2, y^2, and z^2, noting that $x^2(t) + y^2(t) = 1 = z^2(t)$, we see that $x^2 + y^2 - z^2 = 0$, establishing the desired linear dependence.

9.2 Given a set S of vectors in V, suppose first that S is linearly independent. We must show that every finite subset of it is linearly independent, thus let $S_f \subseteq S$ be a finite subset of S. Suppose that

$$0 = \sum_{k=1}^{m} \alpha_k x_k,$$

where $\alpha_1, \ldots, \alpha_m$ are scalars and $x_1, \ldots, x_m \in S_f$. But then the exact same linear combination is also a linear combination of elements from the linearly independent set S, and thus all scalars must be 0. This shows S_f is linearly independent. In the other direction, suppose every finite subset of S is linearly independent. To show that S is linearly independent suppose that 0 is obtained as a linear combination (necessarily a finite one!) as above, with $x_1 \ldots, x_m \in S$. Consider then the finite set

© Springer Nature Switzerland AG 2021

C. Alabiso and I. Weiss, *A Primer on Hilbert Space Theory*, UNITEXT for Physics,

https://doi.org/10.1007/978-3-030-67417-5

$$\{x_1, \ldots, x_m\} \subseteq S$$

and we thus see that 0 is written as linear combination of elements from that subset, which is linearly independent by assumption. Thus all scalars must be 0, and thus S is linearly independent.

9.3 Consider $\mathbb{R}$ as a linear space over itself and let W be a non-trivial linear subspace of $\mathbb{R}$. Thus W contains some non-zero vector, namely a real number $x \neq 0$. The dimension of $\mathbb{R}$ over itself is clearly 1, and thus any non-zero vector spans all of $\mathbb{R}$. Since W, as a linear subspace, is its own span, we conclude that $W = \mathbb{R}$. Thus, we showed that other than the trivial subspace $\{0\}$, the only other subspace of $\mathbb{R}$ is $\mathbb{R}$ itself, as required.

Now we consider $\mathbb{R}$ as a linear space over $\mathbb{Q}$, and note that this is an infinite dimensional linear space. Let $\mathcal{B}$ be a Hamel basis for $\mathbb{R}$ over $\mathbb{Q}$. For every subset X of $\mathcal{B}$ consider the span of X, and denote it by W_X. We claim that $W_X \neq W_Y$ for all subsets $X, Y \subseteq \mathcal{B}$ with $X \neq Y$. Indeed, suppose $W_X = W_Y$ and we may assume that X is not a subspace of Y. Choose a vector $x \in X - Y$. Since $x \in W_X$ it follows that $x \in W_Y$. Thus

$$x = \sum_{k=1}^{m} \alpha_k y_k,$$

where $\alpha_1, \ldots, \alpha_m \in \mathbb{Q}$ and $y_1, \ldots, y_m \in Y$. But then

$$0 = x - \sum_{k=1}^{m} \alpha_k y_k$$

is a non-trivial linear combination of vectors from the linearly independent set $\mathcal{B}$, an impossibility. To conclude then, for each subset of $\mathcal{B}$ we obtained a unique linear subspace of $\mathbb{R}$. There are thus infinitely many subspaces, as required. In fact, the cardinality of a Hamel basis in this case was shown to be equal to $|\mathbb{R}|$, the cardinality of the real numbers. We thus showed that $\mathbb{R}$ as a linear space over $\mathbb{Q}$ has at least $|\mathcal{P}(\mathcal{B})|$ many linear subspaces, a cardinality known to be strictly larger than the cardinality of $\mathcal{B}$, and thus of $\mathbb{R}$.

9.4 Suppose first that the given function $f : X \to W$ extends to a linear operator $F : S \to W$. If

$$0 = \sum_{k=1}^{m} \alpha_k x_k,$$

is a linear relation in X, then

$$0 = F(0) = F\left(\sum_{k=1}^{m} \alpha_k x_k\right) = \sum_{k=1}^{m} \alpha_k F(x_k) = \sum_{k=1}^{m} \alpha_k f(x_k),$$

as claimed. In the other direction, suppose the condition on linear relations in X is met. Given $x \in S$, write

$$y = \sum \alpha_x \cdot x$$

as a (finite!) linear combination of vectors from X. If

$$F(y) = \sum \alpha_x \cdot f(x)$$

is well-defined, then it clearly gives rise to the desired linear extension of f. So, suppose that

$$y = \sum \beta_x \cdot x$$

is another expression of y as a linear combination of vectors from X. But then, by subtracting the two expressions, we obtain

$$0 = \sum (\alpha_x - \beta_x) \cdot x$$

which is a linear relation in X. It thus follows that

$$0 = \sum (\alpha_x - \beta_x) \cdot f(x)$$

from which it follows that the formula for computing $F(y)$ is independent of the presentation of y as a linear combination of vectors from X, as desired.

As for the special case where $X = \mathcal{B}$ is a basis, note that the span of X is then the entire ambient space V, and that the linear independence of $\mathcal{B}$ implies that there are no linear relations in X, so the needed condition for the extension is vacuously satisfied.

9.5 Suppose that $T : V \to W$ is an isomorphism and that $\mathcal{B}$ is a basis for V. We will show that $T(\mathcal{B}) = \{T(b) \mid b \in \mathcal{B}\}$ is a basis for W. Given an arbitrary $w \in W$, consider $T^{-1}(w) \in V$. We may write $T^{-1}(w)$ as a linear combination

$$T^{-1}(w) = \sum \alpha_b \cdot b$$

and thus

$$w = T(T^{-1}(w)) = \sum \alpha_b \cdot T(b)$$

showing that $T(\mathcal{B})$ spans W. Next, suppose a linear combination

$$0 = \sum \alpha_b \cdot T(b)$$

is given. But

$$\sum \alpha_b \cdot T(b) = T\left(\sum \alpha_b \cdot b\right)$$

and thus $\sum \alpha_b \cdot b \in \mathrm{Ker}(T) = \{0\}$ (since T is injective). It follows that

$$\sum \alpha_b \cdot b = 0$$

and therefore that all of the coefficients are 0, as required.

9.6 Suppose V is an infinite dimensional linear space and let $\mathcal{B}$ be a basis for V. Choose a countable subset $\{y_1, y_2, \ldots\} \subseteq \mathcal{B}$ and let W_n be the span of $\{y_1, \ldots, y_n\}$. It is immediate to verify that $T : \mathbb{R}^n \to V$ given by

$$T(x) = T(x_1, \ldots, x_n) = \sum_{k=1}^{n} x_k y_k$$

is a linear isomorphism identifying a copy of $\mathbb{R}^n$ inside V, as claimed.

As for the existence of a linear space $\mathbf{U}$ which contains an isomorphic copy of every linear space over $\mathbb{R}$, the fact that no such linear space exists is a consequence of a set-theoretic result known as Cantor's Theorem. Suppose that such a $\mathbf{U}$ does exist. By constructing free linear spaces one sees that linear spaces of arbitrarily large cardinality exist. That is, given any set X, no matter how large, there always exists a linear space having X as a basis. Since $\mathbf{U}$ is assumed to contain a copy of every linear space, it follows that X is in bijection with a subset of $\mathbf{U}$. In other words, there exists an injection $X \to \mathbf{U}$, and thus $|X| \le |\mathbf{U}|$, for all sets X.

In particular, for the set $X = \mathcal{P}(\mathbf{U})$ of all subsets of $\mathbf{U}$ one has that

$$|\mathcal{P}(\mathbf{U})| \le |\mathbf{U}|.$$

However, the function $u \mapsto \{u\}$ is clearly an injection $\mathbf{U} \to \mathcal{P}(\mathbf{U})$, and thus

$$|\mathbf{U}| \le |\mathcal{P}(\mathbf{U})|.$$

We thus conclude, by the Cantor-Shröder-Bernstein Theorem (see Preliminaries if needed), that

$$|\mathbf{U}| = |\mathcal{P}(\mathbf{U})|.$$

However, this is known to be impossible by an argument we now present.

Theorem 9.1 (*Cantor's Theorem*) For all sets S there exists no surjective function $S \to \mathcal{P}(S)$. In particular $|S| < |\mathcal{P}(S)|$.

Proof Suppose a surjective function $f : S \to \mathcal{P}(S)$ exists. Consider then the set

$$S_! = \{s \in S \mid s \notin f(s)\}$$

which is clearly a subset of S, and thus $S_! \in \mathcal{P}(S)$. Since f is surjective, it follows that there exists $s_! \in S$ with $f(s_!) = S_!$. To obtain a contradiction let us consider whether $s_! \in S_!$ or not. If $s_! \in S_!$, then $s_! \notin f(s_!) = S_!$, which is absurd. However, if $s_! \notin S_! = f(s_!)$, then $s_!$ fulfills the condition for being an element in $S_!$, and thus $s_! \in S_!$, again an absurdity. We must conclude that no such f exists.

Since no surjection from S to $\mathcal{P}(S)$ exists, no bijection exists either, and thus the sets have different cardinalities. Since $s \mapsto \{s\}$ is clearly an injection $S \to \mathcal{P}(S)$ it follows that $|S| < |\mathcal{P}(S)|$. □

9.7 Let E_{ab} be the 3×3 matrix whose entries are all 0 except for the (a, b) entry being 1. Clearly, the set $\{E_{ab}\}_{a,b \in \{1,2,3\}}$ spans M. We have to find a basis composed of rank 2 matrices. For $a, b, c, d \in \{1, 2, 3\}$, define

$$F_{a,b;c,d} = E_{ab} + E_{cd} \quad \text{and} \quad G_{a,b;c,d} = E_{ab} - E_{cd}.$$

Note that, if $a \neq c$ and $b \neq d$, then $F_{a,b;c,d}$ and $G_{a,b;c,d}$ have rank 2. Note also that

$$E_{ab} = \frac{1}{2} F_{a,b;c,d} + \frac{1}{2} G_{a,b;c,d},$$

where we can choose $a \neq c$ and $b \neq d$. The reader can now easily identify a set of rank 2 matrices that span M.

9.8 We prove that $\dim(P_n) = n + 1$ by showing that

$$\{1, x, x^2, \ldots, x^n\}$$

is a basis. Clearly these vectors span P_n and the fact that they are linearly independent follows easily from the more general argument given next, where the dimension of P is computed.

We show that the dimension of P is countably infinite by showing that

$$\{1, x, x^2, \ldots, x^n, \ldots\}$$

is a basis. It is obvious that these vectors span P. To see that they are linearly independent suppose that a linear combination of the vectors results in the zero vector. In other words, $0 = \sum_{k=0}^{n} \alpha_k x^k$ for some real numbers $\alpha_0, \ldots, \alpha_n$. But this linear combination is a polynomial and it is well-known that the only polynomial that evaluates to 0 on all real numbers is the 0 polynomial. Thus all coefficients must be equal to 0, and therefore the only linear combination resulting in the zero vector is the trivial linear combination.

9.9 We note first the general fact that if $A : U \to V$ and $B : V \to W$ are invertible functions, then $(B \circ A)^{-1} = A^{-1} \circ B^{-1}$. This is nothing but the computation

$$B(A(A^{-1}(B^{-1}(w)))) = B(B^{-1}(w)) = w$$

and similarly

$$A^{-1}(B^{-1}(B(A(u)))) = A^{-1}(A(u)) = u$$

for all $w \in W$ and all $u \in U$. Now, going back to the linear operators A and B, assumed to commute, we have

$$A^{-1}B^{-1} = (BA)^{-1} = (AB)^{-1} = B^{-1}A^{-1},$$

as required.

9.10 Assume first that V is infinite dimensional, and let $\mathcal{B}$ be an infinite basis of V. Fix a vector $x_0 \in \mathcal{B}$ and consider the set $S = \mathcal{B} - \{x_0\}$. As a subset of a basis these vectors are linearly independent, and thus S is a basis of the subspace spanned by S. Clearly, x_0 is missing from that span (otherwise the original $\mathcal{B}$ would be linearly dependent, which is impossible since it is a basis), and thus S is a proper subspace of V. To show that $S \cong V$ it suffices to show they have equal dimensions, namely that $|\mathcal{B}| = |S|$. Clearly, $|S| \leq |\mathcal{B}|$, since the inclusion $S \to \mathcal{B}$ is an injection. If we can construct an injection in the other direction as well, then, using the Cantor-Shröder-Bernstein Theorem, the cardinalities will indeed be shown to be equal. To construct an injection $f : \mathcal{B} \to S$, let $\{x_m\}_{m \geq 1}$ be a countably infinite list of distinct vectors in S. Define now $f(x_m) = x_{m+1}$ for all $m \geq 1$, with $f(x_0) = x_1$, and $f(x) = x$ in all other cases. It is immediate to verify that f is indeed an injection. This completes the proof that if V is infinite dimensional, then it is isomorphic to a proper subspace of it.

In the other direction, suppose that $V \cong U$ for some proper subspace U. Since linear spaces over the same field are isomorphic if, and only if, their dimensions are equal, it follows that $\dim(V) = \dim(U)$. But in a finite dimensional space, if a subspace has the same dimension as the ambient space, then the subspace coincides with the ambient space, and is thus not proper. We conclude that V must be infinite dimensional.

Topological Spaces

Remark 9.1 In topology the term *neighborhood* of a point x in a topological space X may mean one of two things. Either an open set U in X with $x \in U$, or an arbitrary subset $N \subseteq X$ which contains an open set U with $x \in U$. The difference is only cosmetic, and one can easily translate between the two situations. In the solutions below we adhere to the first meaning of neighborhood.

9.11 Any indiscrete space with more than one element very clearly does not satisfy the T_0 property. For an example of a space that is T_0 but not T_1 consider the Sierpinski space $\mathbb{S} = \{0, 1\}$. For a T_1 space that is not T_2, let S be an infinite set endowed with the cofinite topology. Clearly the T_1 separation property is satisfied since given distinct points $x, y \in S$, the sets $S - \{x\}$ and $S - \{y\}$ are open and clearly satisfy the requirements. We show that S is not T_2. Indeed, let x, y be two distinct points in S and suppose $x \in U$ and $y \in V$ for some open sets U, V. Then the complements of U and of V are finite subsets of S, and

$$S - (U \cap V) = (S - U) \cup (S - V)$$

is also finite. Since S is infinite it follows that $U \cap V \neq \emptyset$.

9.12 A locally (path) connected space that is not (path) connected is, for instance, the space $(0, 1) \cup (3, 4)$ as a subspace of $\mathbb{R}$ with the Euclidean topology (as is easy to see). As for the comb space, we first show that it is path connected, and thus also connected. Indeed, any point admits a path to the point $(0, 0)$ by first traveling south to meet the X-axis and then traveling west. Thus any two points can be joined by a path in the space, so it is indeed path-connected. It is however not locally connected, and thus not locally path-connected. Indeed, given any point of the form $p = (0, y)$ with $0 < y < 1$, consider a small enough circle centered at p which does not intersect the X-axis. Any open set containing p must contain a subset of this form, but such a set is not connected.

9.13 Let X be a compact space. Given any $x \in X$ we may take $U = C = X$, and then $x \in U \subseteq X$ with U open and C compact, so X is locally compact. An example of a locally compact space that is not compact is, e.g., $\mathbb{R}$ with the Euclidean topology. Its non-compactness is obvious. As for it being locally compact, suppose $x \in \mathbb{R}$ is given. Then $x \in (x - 1, x + 1) \subseteq [x - 1, x + 1]$, as required.

As an example of a space that is not locally compact, consider $\mathbb{Q}$ as a subspace of $\mathbb{R}$ with the Euclidean topology. We will show that $\mathbb{Q}$ is not locally compact at any point. This will be done by showing that in fact no non-empty compact subset of $\mathbb{Q}$ contains a non-empty open set. To see that, suppose that $U \subseteq C$ where U is open and C is compact, and both are non-empty. It follows that $\mathbb{Q} \cap (a, b) \subseteq U$ for some non-empty open interval (a, b). Since $\mathbb{Q}$ is Hausdorff, the compact set C is closed, and thus, taking closures, we obtain that $[a, b] \cap \mathbb{Q} \subseteq C$. This is thus a closed subset of a compact set in a Hausdorff space, and thus $\mathbb{Q} \cap [a, b]$ is compact. But that is not the case. Indeed, let $y \in [a, b]$ be an irrational number and let $\{q_n\}_n \in \mathbb{N}$ be a strictly increasing sequence of rationals tending to y. Let $U_n = \mathbb{Q} \cap (-\infty, q_n)$ and consider also the set $\mathbb{Q} \cap (y, \infty)$. These are open sets in the subspace topology that cover $[a, b]$ but (as the reader may easily verify) they admit no finite subcovering.

9.14 We prove that X is closed. The condition

$$\begin{pmatrix} a & b \\ c & d \end{pmatrix} \in X$$

reads as

$$\begin{cases} a^2 + bc = 0 \\ ab + bd = 0 \\ bc + d^2 = 0 \\ ac + cd = 0. \end{cases}$$

In particular, by means of the homeomorphism $\Phi \colon M_2(\mathbb{R}) \to \mathbb{R}^4$, the set X can be identified with the set

$$S = \{u = (a, b, c, d) \in \mathbb{R}^4 \mid \psi_1(u) = \psi_2(u) = \psi_3(u) = \psi_4(u) = 0\},$$

where

$$\psi_1, \psi_2, \psi_3, \psi_4 \colon \mathbb{R}^4 \to \mathbb{R}$$

are the functions

$$\psi_1(x, y, z, w) = x^2 + yz, \qquad \psi_2(x, y, z, w) = xy + yw,$$

$$\psi_3(x, y, z, w) = yz + z^2, \qquad \psi_4(x, y, z, w) = xz + zw.$$

In other words,

$$S = \psi_1^{-1}(\{0\}) \cap \psi_2^{-1}(\{0\}) \cap \psi_3^{-1}(\{0\}) \cap \psi_4^{-1}(\{0\})$$

and since each ψ_i is a polynomial function, and thus continuous, the set $\phi_i^{-1}(\{0\})$ is closed (since $\{0\}$ is closed in $\mathbb{R}$). We see that S is the intersection of four closed sets, and hence it is itself closed.

We show now that X is not compact. Indeed, for all $n \in \mathbb{Z}$, consider the subset

$$U_n = \left\{ \begin{pmatrix} a & b \\ c & d \end{pmatrix} \in X \mid a \in (n-1, n+1), b \in \mathbb{R}, c \in \mathbb{R}, d \in \mathbb{R} \right\} \cap X.$$

Each U_n is an open set in X. Furthermore, one has that

$$\bigcup_{n \in \mathbb{Z}} U_n = X.$$

But there is no finite sub-family of $\{U_n\}_{n \in \mathbb{Z}}$ that still covers X. Indeed, for any $m \in \mathbb{Z}$, the matrix $\begin{pmatrix} m & 0 \\ 0 & 0 \end{pmatrix}$ belongs only to U_m.

Finally, we show that X is connected. In fact, we prove that X is path-connected. To do that, take any $A \in X$. We first notice that there exists a non-singular matrix $M \in M_2(\mathbb{R})$ such that

$$A = M^{-1} \cdot \begin{pmatrix} 0 & b \\ 0 & 0 \end{pmatrix} \cdot M,$$

for $b \in \mathbb{R}$. In other words, any matrix in X is similar to an upper-triangular matrix. Indeed, consider A as a matrix in $M_2(\mathbb{C})$. Since $\mathbb{C}$ is algebraically closed, A can be triangulated in $M_2(\mathbb{C})$, namely, it is similar to a matrix

$$\begin{pmatrix} a & b \\ 0 & c \end{pmatrix},$$

with a, b, c possibly complex numbers. The condition $A^2 = 0$ reads as

$$\begin{cases} a^2 = 0 \\ ab + bc = 0 \\ c^2 = 0, \end{cases}$$

hence the eigenvalues are $a = 0$ and $c = 0$. In particular, the eigenvalues belong to $\mathbb{R}$. Therefore A is triangulable also in $M_2(\mathbb{R})$, proving the claim.

Now, consider the path

$$\gamma: [0, 1] \to X \subset M_2(\mathbb{R}), \qquad \gamma(t) = M^{-1} \cdot \begin{pmatrix} 0 & tb \\ 0 & 0 \end{pmatrix} \cdot M.$$

Note that, for any $t \in [0, 1]$, $(\gamma(t))^2 = 0$, and hence $\gamma(t) \in X$. Note also that $\gamma: [0, 1] \to X$ is continuous and that

$$\gamma(0) = \begin{pmatrix} 0 & 0 \\ 0 & 0 \end{pmatrix} \qquad \text{and} \qquad \gamma(1) = A,$$

proving that, for any $A \in X$, there exists a path connecting it to the zero matrix. This clearly suffices to show that any two elements in X can be connected by a path in X, and thus X is path-connected, as claimed.

9.15

1. We prove that $\tilde{X}$ is T_1 and hence also T_0. We have to prove that, for any pair of distinct points $p, q \in \tilde{X}$, there exists a neighborhood U_p of p not containing q and a neighborhood U_q of q not containing p. Consider first the following case. Suppose that the points are $p = [(x, 1)] = [(x, 2)]$ with $x \in \mathbb{R} - \{0\}$ and $q = [(x', y')]$ with $x' \in \mathbb{R}$ and $y' \in \{1, 2\}$. Take $r \in \mathbb{R}$ such that $0 < r < |x - x'|$. Then the sets $U_p = \left\{ [(u, 1)] \in \tilde{X} \mid |u - x| < r \right\}$ and $U_q = \left\{ [(u', y')] \in \tilde{X} \mid |u' - x'| < r \right\}$ are open neighborhoods of p, respectively q, and it holds that $p \notin U_q$ and $q \notin U_p$. It remains to consider just the case $p = [(0, 1)]$ and $q = [(0, 2)]$. In this case, we consider the sets $U_p = \left\{ [(u, 1)] \in \tilde{X} \mid u \in \mathbb{R} \right\}$ and $U_q = \left\{ [(v, 2)] \in \tilde{X} \mid v \in \mathbb{R} \right\}$. These are clearly neighborhoods of p, respectively q, and $p \notin U_q$ and $q \notin U_p$. The space $\tilde{X}$ is thus T_1.

   We prove next that $\tilde{X}$ is not T_2. Consider the points $p = [(0, 1)], q = [(0, 2)] \in \tilde{X}$ for which we prove that no disjoint neighborhoods exist. Indeed, any neighborhood of p contains an open neighborhood of the type

$$\hat{U}_p = \left\{ [(u, 1)] \in \tilde{X} \mid |u| < r \right\}$$

for some $r > 0$, and any neighborhood of q contains an open neighborhood of the type $\hat{U}_q = \left\{ [(u', 2)] \in \tilde{X} \mid |u'| < r' \right\}$ for some $r' > 0$. We thus see that the intersection of any two neighborhoods of p and q is non-empty.

2. We prove next that $\tilde{X}$ is not compact. Indeed, for any $n \in \mathbb{Z} - \{0\}$, consider the open set

$$U_n = \left\{ [(x, 1)] = [(x, 2)] \in \tilde{X} \mid n - 1 < x < n + 1 \right\}$$

and consider also the open sets

$$U_0' = \left\{ [(x, 1)] \in \tilde{X} \mid -1 < x < 1 \right\}$$

and

$$U_0'' = \left\{ [(x, 2)] \in \tilde{X} \mid -1 < x < 1 \right\}.$$

The family $\mathcal{U} = \{U_n\}_{n \in \mathbb{Z} - \{0\}} \cup \{U_0', U_0''\}$ provides an open covering of $\tilde{X}$. But, for any $m \in \mathbb{Z}$, the point $[(m, y)] \in \tilde{X}$, for $y \in \{1, 2\}$, belongs to just one of the elements of the family $\mathcal{U}$. Hence there exists no finite sub-family of $\mathcal{U}$ covering $\tilde{X}$. This proves that $\tilde{X}$ is not compact.

3. We prove that $\tilde{X}$ is path-connected. Indeed, consider a point $[(x, y)] \in \tilde{X}$ with $x \in \mathbb{R}$ and $y \in \{1, 2\}$. Consider the path

$$\gamma : [0, 1] \to \tilde{X}, \quad \gamma(t) = [((1 - t)x + t, y)].$$

Note that $\gamma : [0, 1] \to \tilde{X}$ is continuous and that

$$\gamma(0) = [(x, y)]$$

while

$$\gamma(1) = [(1, 1)] = [(1, 2)].$$

Hence, any point $[(x, y)] \in \tilde{X}$ can be connected to the point $[(1, 1)] = [(1, 2)]$ by means of a continuous path in $\tilde{X}$.

9.16 Note that X_1 and X_2 are not compact, while X_3 and X_4 are. Hence $X_i \not\cong X_j$ for $i = 1, 2$ and $j = 3, 4$. Further, note that $X_1 \cong X_2$ by means of the stereographic projection

$$\varphi : X_1 \to X_2$$
$$(x, y) \mapsto \left(\frac{2x}{1 + x^2 + y^2}, \frac{2y}{1 + x^2 + y^2}, \frac{-1 + x^2 + y^2}{1 + x^2 + y^2} \right)$$
$$\varphi^{-1} : X_2 \to X_1$$
$$(x, y, z) \mapsto \left(\frac{x}{1 - z}, \frac{y}{1 - z} \right)$$

Finally, note that $X_3 \not\cong X_4$. Indeed, the property that a topological space remains connected after removing two points is a topological invariant. But if one removes two points from X_4, then the space obtained is connected while any two points one removes from X_3 results in a disconnected space.

9.17 Let S be an uncountable set and endow it with the cocountable topology. In this topology every countable subset is closed since its complement obviously has countable complement. In the other direction, if F is a closed subset of S, then its complement $S - F$ is open, and thus either $F = S$, or the complement of $S - F$, which is F, is countable. We thus identifies the closed sets as the collection

$$\mathcal{F} = \{F \subseteq S \mid F \text{ is countable}\} \cup \{S\}.$$

Since a countable union of countable sets is countable, it follows that the collection $\mathcal{F}$ is stable under countable unions. To see that taking arbitrary unions may take one outside of the collection $\mathcal{F}$ it suffices to observe that a proper uncountable subset of S exists. Indeed, let $X \subset S$ be such a set. Since

$$X = \bigcup_{x \in X} \{x\}$$

the set X is certainly a union of closed sets. However, since X is uncountable and $X \neq S$, it follows that S is not closed. We leave it to the reader to exhibit proper uncountable subsets of S.

9.18 Since A is a real symmetric matrix, it is diagonalizable. It follows that, up to a homeomorphism induced by a change of coordinates of $\mathbb{R}^n$, we can suppose that A is a diagonal matrix with positive entries $\lambda_1 > 0, \ldots, \lambda_n > 0$,

$$A = \begin{pmatrix} \lambda_1 & & \\ & \ddots & \\ & & \lambda_n \end{pmatrix},$$

and so

$$W = \left\{ \begin{pmatrix} x_1 \\ \vdots \\ x_n \end{pmatrix} \in \mathbb{R}^n \mid \lambda_1 x_1^2 + \cdots + \lambda_n x_n^2 = 1 \right\}.$$

The set W is closed and bounded, and hence compact, in $\mathbb{R}^n$ with the Euclidean topology.

9.19 We prove that D is open in X. Take any point $p \in D$. Since D is locally compact, there exists a compact set $K \subseteq D$ in D and an open set U in X such that

$$p \in U \cap D \subseteq K.$$

Consider the set $U - K \subseteq X$, which is open in X since K is compact in the Hausdorff space X, and is thus closed. Since D is dense, if the open set $U - K$ were non-empty, it would intersect D, but this contradicts with

$$U \cap D \subseteq K.$$

Hence $U - K = \emptyset$, that is
$$p \in U \subseteq K \subseteq D.$$

Hence the point p is an interior point of D. As p was arbitrary, we conclude that all of the points of D are interior, and thus that D is open.

9.20

1. We have to show that $\mathcal{V}$ contains the empty set and the whole space X, and that it is closed under finite unions and under arbitrary intersections. Indeed, $\emptyset = V(1 \cdot \mathbb{Z}) \in \mathcal{V}$ and $X = V(0 \cdot \mathbb{Z}) \in \mathcal{V}$. Take $V(a_1 \cdot \mathbb{Z}) \in \mathcal{V}$ and $V(a_2 \cdot \mathbb{Z}) \in \mathcal{V}$. Note that

   $$V(a_1 \cdot \mathbb{Z}) \cup V(a_2 \cdot \mathbb{Z}) = V(\text{lcm}\{a, b\} \cdot \mathbb{Z}) \in \mathcal{V}.$$

   Consider the family $\{V(a_n \cdot \mathbb{Z})\}_{n \in \mathbb{Z}} \subseteq \mathcal{V}$. Note that

   $$\bigcap_{n \in \mathbb{Z}} V(a_n \cdot \mathbb{Z}) = V(\gcd\{a_n : n \in \mathbb{Z}\}) \in \mathcal{V}.$$

2. We prove that $\{0 \cdot \mathbb{Z}\} \subset X$ is dense in X, that is, $\overline{\{0 \cdot \mathbb{Z}\}} = X$. Indeed, the only closed set containing $\{0 \cdot \mathbb{Z}\}$ is $V(0 \cdot \mathbb{Z}) = X$. In particular, X is the smallest closed set containing $\{0 \cdot \mathbb{Z}\}$, which is precisely the closure.

3. We prove now that X is not Hausdorff. Indeed, note that for any $a \in \mathbb{Z}$ the closed set $V(a \cdot \mathbb{Z}) = \{p \cdot \mathbb{Z} \mid p \text{ is a prime dividing } a \text{ or } p = a = 0\}$ is a finite set. Hence, any two open sets in X intersect and hence X is not T_2.

Metric Spaces

9.21 Recall that in a metric space every compact set is also sequentially compact. Since $\{x_n\}_{n \in \mathbb{N}}$ does not admit any convergent subsequence, the space (X, d) cannot be sequentially compact, and consequently is not compact. Another approach to this problem is to use the fact that a metric space is compact if, and only if, it is complete and totally bounded. But clearly X is not totally bounded (though it may be bounded and it may be complete).

9.22 To show that ℓ_∞ is connected we actually show that it is path-connected and thus consider an arbitrary bounded real sequence $x = (x_n)_n \in \ell_\infty$, for which it suffices to construct a path to the 0 vector, namely the constantly 0 sequence. To that end, consider the function

$$\gamma : [0, 1] \to \ell_\infty, \qquad \gamma(t) = (tx_n)_n .$$

Notice that the codomain is indeed ℓ_∞, as is immediately seen (in fact this is nothing but the closure of ℓ_∞ under scalar products). Since $\gamma(1) = x$ and $\gamma(0) = 0$, it remains to show that γ is continuous with respect to the relevant metrics. And indeed, for all $t, s \in [0, 1]$, it holds that

$$d(\gamma(t), \gamma(s)) = d((tx_n)_n, (sx_n)_n) = \sup_n |(t - s) \cdot x_n| = |t - s| \cdot \sup_n |x_n| = \|x\|_\infty \cdot d(s, t)$$

and thus γ is uniformly continuous (notice that we actually proved γ is Lipschitz), and thus continuous, as was needed.

To see that ℓ_∞ as not compact it suffices to construct a sequence of infinitely many vectors with all pair-wise distances equal to 1. This can easily be done in many ways. We mention here another approach. Notice that the set of all vectors of the form $x = (x_1, 0, 0, 0, \ldots)$, with $x_1 \in \mathbb{R}$ arbitrary, in other words the kernel of the shift mapping $f : \ell_\infty \to \ell_\infty$, is a closed subspace of ℓ_∞ which with the induced metric is isometric to $\mathbb{R}$ with the Euclidean metric. Since ℓ_∞ is Hausdorff (any metric space is Hausdorff), if ℓ_∞ were compact, then $\mathbb{R}$ would be compact as well (since any compact subspace of a compact Hausdorff space is compact). But $\mathbb{R}$ is of course not compact.

9.23 For all $n \geq 0$ let $X_n = \{x \in \mathbb{R} \mid f^{(n)} = 0\}$. Since f is infinitely differentiable, $f^{(n)}$ is continuous, and thus the set X_n, being the inverse image of the closed set $\{0\}$, is closed. Moreover, the condition on the function f states precisely that

$$\mathbb{R} = \bigcup_{n \geq 0} X_n$$

and thus we have expressed the non-empty and complete metric space $\mathbb{R}$ (with the Euclidean metric) as a countable union of closed sets. By Baire's Theorem at least one of the sets X_n contains a non-empty open set. Thus, there exists an $n \geq 0$ and an open set $U \neq \emptyset$ with $U \subseteq X_n$. But any open set in $\mathbb{R}$, by definition, contains an open interval. Thus, there exist $a < b$ with $(a, b) \subseteq X_n$. In other words, $f^{(n)}(x) = 0$ for all $x \in (a, b)$. Integrating n times in the range a to b reveals that f is a polynomial of degree at most n.

9.24 Recall that in any inner product space the parallelogram law

$$\|x + y\|^2 + \|x - y\|^2 = 2\|x\|^2 + 2\|y\|^2$$

holds for all vectors x and y. We prove that there exists no inner product on $C([0, 1], \mathbb{R})$ whose associated norm is $\|\cdot\|_\infty$ by showing the parallelogram law fails. Indeed, consider, for example, the continuous functions $x, y : [0, 1] \to \mathbb{R}$ given by

$$x(t) = 1 \quad \text{and} \quad x(t) = t.$$

One computes that

$$\|x + y\|_\infty^2 + \|x - y\|_\infty^2 = \max_{t \in [0,1]} |x(t) + y(t)|^2 + \max_{t \in [0,1]} |x(t) - y(t)|^2$$

$$= \max_{t \in [0,1]} |1 + t|^2 + \max_{t \in [0,1]} |1 - t|^2 = 4 + 1 = 5$$

while

$$2\|x\|_\infty^2 + 2\|y\|_\infty^2 = 2 \max_{t \in [0,1]} |x(t)|^2 + 2 \max_{t \in [0,1]} |y(t)|^2$$

$$= 2 \max_{t \in [0,1]} |1|^2 + 2 \max_{t \in [0,1]} |t|^2 = 2 \cdot 1 + 2 \cdot 1 = 4.$$

Therefore, the norm $\|\cdot\|$ does not satisfy the parallelogram law and is thus not induced by any inner product.

9.25 A metric on the empty set is a function $d : \emptyset \times \emptyset \to \mathbb{R}_+$ satisfying the metric axioms. The domain is the empty set, and thus there is a unique such function, which is vacuously satisfying the metric axioms. As for a metric on a singleton set $X = \{p\}$, that is a function $X \times X \to \mathbb{R}_+$, notice that the axioms require that $d(p, p) = 0$, and thus d is already determined. It is trivially seen that this d indeed determines a metric on X. Finally, if X contains at least two distinct points p an q, then noting first that there always exists a metric structure on d, for instance $d(x, y) = 1$ for all $x \neq y$ and $d(x, x) = 0$ for all x, the existence of infinitely many non-isometric metric structures follows by noting that if d is any metric function and $\alpha > 0$ an arbitrary positive real number, then defining

$$\alpha d : X \times X \to \mathbb{R}_+$$

by $(\alpha d)(x, y) = \alpha d(x, y)$ is also a metric on X. Clearly the spaces (X, d) and $(X, \alpha d)$ are isometric if, and only if, $\alpha = 1$ or if $d(x, y) = 0$ for all points. However, since $p \neq q$ we have that $d(p, q) \neq 0$. We thus obtain infinitely many non-isometric metrics on X.

9.26 Suppose that $x \in X$ satisfies $d(x, S) = 0$ and let U be an arbitrary open set with $x \in U$. It then holds that $B_\varepsilon(x) \subseteq U$ for some $\varepsilon > 0$. Since $d(x, S) = 0$ it follows that there is an element $s \in S$ with $d(x, s) < \varepsilon$. In particular then $s \in B_\varepsilon(x) \subseteq U$. We thus showed that every open set containing x intersects S, and thus $x \in \overline{S}$.

Conversely, if $x \notin \overline{S}$, then (noting that $\overline{S}$ is closed, and thus its complement is open) there exists an $\varepsilon > 0$ with $B_\varepsilon(x) \subseteq X - S$. But then $d(x, s) \geq \varepsilon$ for all $s \in S$, and thus $d(x, S) \geq \varepsilon$.

9.27 We establish the metric space axioms for $\tilde{d}$. Clearly, $\tilde{d}(x, y) \geq 0$ for all $x, y \in X$. If $\tilde{d}(x, y) = 0$, then $d(x, y) = 0$ and thus $x = y$. For all $x, y \in X$ if $d(x, y) < 1$, then

$$\tilde{d}(x, y) = d(x, y) = d(y, x) = \tilde{d}(y, x)$$

while if $d(x, y) \geq 1$, then $d(y, x) \geq 1$ too, and thus $\tilde{d}(x, y) = 1 = \tilde{d}(y, x)$. Finally, the verification of the triangle inequality follows similarly by case splitting.

To show that d and $\tilde{d}$ induce the same topology, it suffices to notice that an open ball $B_\varepsilon(x)$ of radius $\varepsilon < 1$ is the same set when computed in (X, d) as it is when computing in $(X, \tilde{d})$.

9.28 We prove that there exists no isometry from $\mathbb{R}^2$ to $\mathbb{R}$ when endowed with the Euclidean metrics $d_{\mathbb{R}^2}$ and $d_{\mathbb{R}}$ respectively. Indeed, suppose that such an isometry $f : \mathbb{R}^2 \to \mathbb{R}$ exists. Take $P, Q, R \in \mathbb{R}^2$ such that

$$d_{\mathbb{R}^2}(P, Q) = d_{\mathbb{R}^2}(P, R) = d_{\mathbb{R}^2}(Q, R) = 1.$$

Denote by $p = f(P), q = f(Q)$, and $r = f(R)$ the images in $\mathbb{R}$ of the three chosen points. Since f is an isometry, it follows that

$$|p - q| = d_{\mathbb{R}}(p, q) = d_{\mathbb{R}}(f(P), f(Q)) = d_{\mathbb{R}^2}(P, Q) = 1,$$

and analogously that $|p - r| = 1$ and $|q - r| = 1$. However, elementary algebra reveals this to an impossibility.

9.29 We show that d_s satisfies the metric axioms, remembering that each d_i does. Firstly, for all $x \in X$

$$d_s(x, x) = \sup_{i \in I} d_i(x, x) = \sup_{i \in I}\{0\} = 0.$$

Next, for all $x, y \in X$

$$d_s(x, y) = \sup_{i \in I}\{d_i(x, y)\} = \sup_{i \in I}\{d_i(y, x)\} = d_s(y, x).$$

Finally, for all $x, y, z \in X$, we need to show that

$$\sup_{i \in I}\{d_i(x, z)\} \leq \sup_{i \in I}\{d_i(x, y)\} + \sup_{i \in I}\{d_i(y, z)\}$$

for which it suffices to show, for each $i \in I$, that

$$d_i(x, z) \leq \sup_{i \in I}\{d_i(x, y)\} + \sup_{i \in I}\{d_i(y, z)\}$$

And indeed, given $i \in I$, using the triangle inequality for d_i

$$d_i(x, z) \leq d_i(x, y) + d_i(y, z) \leq \sup_{i \in I}\{d_i(x, y)\} + \sup_{i \in I}\{d_i(y, z)\},$$

completing the proof.

For an example showing that the infimum of metric functions need not be a metric function, consider a set $X = \{a, b, c\}$ with three distinct points. On it consider the following assignments of distances

$$d_1(a, b) = 1 \quad d_2(a, b) = 2$$
$$d_1(b, c) = 2 \quad d_2(b, c) = 1$$
$$d_1(a, c) = 3 \quad d_2(a, c) = 3$$

which are easily seen to endow X with two structures of metric space. However, when computing the minima of the distances one obtains the function e with $e(a, b) = e(b, c) = 1$, and $e(a, c) = 3$. Therefore $e(a, c) > e(a, b) + e(b, c)$ and so the triangle inequality fails.

9.30

1. No, not continuity nor uniform continuity suffices. For instance, let $X = \mathbb{R}$ with the Euclidean topology and consider $f : \mathbb{R} \to \mathbb{R}$ with $f(x) = \arctan(x)$, which is uniformly continuous, and thus also continuous. Then $\mathbb{R}$ is complete but the set $F(X) = (-\pi/2, \pi/2)$ is not complete.
2. Uniform continuity is a sufficient condition. Indeed, to show that $F(X)$ is totally bounded, let $\varepsilon > 0$ be given and we will find a covering of $F(X)$ by sets whose diameters do not exceed ε. Let $\delta > 0$ correspond to the given ε, that is $\mathrm{diam}(F(S)) < \varepsilon$ whenever $S \subseteq X$ satisfies $\mathrm{diam}(S) < \delta$. Since X is totally bounded, one can cover X by sets $S_1, \ldots, S_m$ whose diameters do not exceed δ. Each of the images $f(S_1), \ldots, f(S_m)$ has diameter not exceeding ε, and they cover $F(X)$, as needed. Finally, we note that continuity alone does not suffice. For instance, consider the function $\tan : (-\pi/2, \pi/2) \to \mathbb{R}$, with $\mathbb{R}$ endowed with the Euclidean metric and the interval with the induced metric. The interval is totally bounded but its image, $\mathbb{R}$, is not even bounded.
3. Continuity, and thus also uniform continuity, suffices. Indeed, in a metric space a set is compact if, and only if, it is complete and totally bounded. Moreover, for a function f, the image $f(C)$ is compact whenever C is.

Normed Spaces and Banach Spaces

9.31 First we show that the metric $d(x, y) = \|x - y\|$ induced by the norm of a normed space V is translation invariant and scale homogeneous. Indeed,

$$d(x + z, y + z) = \|(x + z) - (y + z)\| = \|x - z\| = d(x, z)$$

holds for all $x, y, z \in V$ and for any scalar α

$$d(\alpha x, \alpha y) = \|\alpha x - \alpha y\| = \|\alpha(x - y)\| = |\alpha|\|x - y\| = |\alpha| d(x, y).$$

In the other direction, suppose that V is endowed with a translation invariant and scale homogeneous metric d. We then define, for all $x \in V$

$$\|x\| = d(0, x)$$

and proceed to show that the axioms for a normed space are satisfied. It is clear that $\|x\| \geq 0$ for all $x \in V$ and that if $\|x\| = 0$, then $d(0, x) = 0$, and thus $x = 0$. For any scalar α and $x \in V$ one has

$$\|\alpha x\| = d(0, \alpha x) = d(\alpha \cdot 0, \alpha x) = |\alpha| d(0, x) = |\alpha| \|x\|.$$

Finally, for all $x, y \in V$ we have, by translation invariance, that $d(x, x + y) = d(0, y)$. It then follows that

$$\|x + y\| = d(0, x + y) \leq d(0, x) + d(x, x + y) = d(0, x) + d(0, y) = \|x\| + \|y\|,$$

as required.

9.32

1. We establish the norm axioms for the given proposed norm function. Clearly $\|f\| \geq 0$ since the integral of a non-negative function is non-negative. Moreover, suppose that $\|f\| = 0$ for a continuous function $f : [a, b] \to \mathbb{R}$. If $f \neq 0$, then $f(x_0) \neq 0$ for some $x_0 \in [a, b]$. Since $|f(x)|$ is continuous and $|f(x_0)| > 0$ it follows that there is an interval $(c, d) \subseteq [a, b]$ upon which $|f(x)|$ attains values greater than $|f(x_0)|/2 > 0$. It follows that

$$\int_a^b |f(x)| dx \geq \int_c^d |f(x)| dx \geq (d - c)|f(x_0)|/2 > 0$$

which is a contradiction. We conclude thus that $\|f\| = 0$ implies $f = 0$. Next, for all $\lambda \in \mathbb{R}$

$$\|\lambda \cdot f\| = \int_a^b |\lambda \cdot f(x)| dx = \lambda \cdot \int_a^b |f(x)| dx = \lambda \cdot \|f\|.$$

Finally, for $f : [a, b] \to \mathbb{R}$ and $g : [a, b] \to \mathbb{R}$, it holds that

$$\|f + g\| = \int_a^b |f(x) + g(x)| dx \leq \int_a^b |f(x)| dx + \int_a^b |g(x)| dx = \|f\| + \|g\|,$$

establishing the triangle inequality.

2. We prove that the metric space (X, d) with $d(f, g) = \|f - g\|$ is not complete. For simplicity, assume that $[a, b] = [-1, 1]$. We provide an example of a Cauchy sequence in X having no limit point in X. For all $n \in \mathbb{N}$, consider the function

$$f_n : [0, 1] \to \mathbb{R}, \qquad f_n(x) = \begin{cases} -1 & \text{for } -1 \le x < -\frac{1}{n} \\ nx & \text{for } -\frac{1}{n} \le x < \frac{1}{n} \\ 1 & \text{for } \frac{1}{n} \le x \le 1. \end{cases}$$

Note that $f_n(x) \in X$ for all $n \in \mathbb{N}$. Furthermore, for $n, m \in \mathbb{N}$ with $n < m$ one computes

$$\|f_n - f_m\| = \int_{-1}^{1} |f_n(x) - f_m(x)| \, dx = 2 \left(\int_0^{\frac{1}{m}} (m-n)x \, dx + \int_{\frac{1}{m}}^{\frac{1}{n}} (1-nx) \, dx \right)$$

$$\le 2 \int_0^{\frac{1}{n}} (1-nx) \, dx = \frac{1}{n}.$$

It thus follows that the sequence $(f_n)_{n \in \mathbb{N}}$ is a Cauchy sequence in (X, d).
We prove now that $(f_n)_{n \in \mathbb{N}}$ has no limit point in X. Indeed, we claim that, if such a limit point $f \in X$ were to exist, then it would be that

$$f(x) = -1 \text{ for all } x \in [-1, 0) \qquad \text{and} \qquad f(x) = 1 \text{ for all } x \in (0, 1],$$

which is absurd since such a function f is not continuous on $[-1, 1]$. To prove the claim, suppose that there were a point $x \in [-1, 0)$ such that $|f(x) - (-1)| = \delta$ for some $\delta > 0$. For simplicity, we assume $x \ne -1$: otherwise, by continuity, if $x = -1$, then there exists a point $x' > -1$ such that $|f(x') - (-1)| = \delta' > 0$. Then, by continuity, there would be a neighbourhood $(x - \varepsilon, x + \varepsilon) \subseteq [-1, 0)$, where $\varepsilon > 0$, of x in $[-1, 0)$ such that $|f(y) - (-1)| > \delta/2$ for all $y \in (x - \varepsilon, x + \varepsilon)$. Choose $N > -1/(x + \varepsilon)$, so $x + \varepsilon < -1/N$. Hence, for all $n \ge N$, it holds that

$$\|f_n - f\| = \int_{-1}^{1} |f_n(x) - f(x)| \, dx \ge \int_{x-\varepsilon}^{x+\varepsilon} |f_n(x) - f(x)| \, dx \ge 2\varepsilon\delta > 0,$$

which is not possible.

9.33 We first prove that d is a distance on $\mathbb{R}$. Firstly, it is clear that for all $x, y \in \mathbb{R}$

$$0 \le d(x, y) = d(y, x).$$

Furthermore, if $d(x, y) = 0$ then $|x - y| = 0$, and thus $x = y$. For the triangle inequality, consider the function $f : [0, \infty)] \to [0, \infty]$ given by

$$f(t) = \frac{t}{1+t}$$

and notice that the proposed distance function is

$$d(x, y) = f(d_E(x, y))$$

where $d_E(x, y) = |x - y|$ is the Euclidean metric on $\mathbb{R}$. Using elementary analysis it is seen that f is monotonically increasing and that it is concave and thus it is also subadditive, i.e.,

$$f(s + t) \le f(s) + f(t).$$

We may now argue as follows. For all $x, y, z \in \mathbb{R}$

$$d(x, z) = f(d_E(x, z)) \le f(d_E(x, y) + d_E(y, z))$$

where we used the monotonicity of f. The subadditivity of f now implies that

$$f(d_E(x, y) + d_E(y, z)) \le f(d_E(x, y)) + f(d_E(y, z)) = d(x, y) + d(y, z)$$

establishing the triangle inequality, and thus that $(\mathbb{R}, d)$ is a metric space.

We now show that d is not induced by any norm. Indeed, if there were such a norm $x \mapsto \|x\|$ for which $d(x, y) = \|x - y\|$, then, for all $x, y \in \mathbb{R}$ and all $\lambda \in \mathbb{R}$, it would hold that

$$d(\lambda x, \lambda y) = \|\lambda x - \lambda y\| = |\lambda| \cdot \|x - y\| = |\lambda| \cdot d(x, y).$$

However, By using the explicit expression of d, we get

$$\frac{|\lambda x - \lambda y|}{1 + |\lambda x - \lambda y|} = \frac{|x - y|}{1 + |x - y|}$$

and taking, e.g., $x = 1$, $y = 0$, and $\lambda = 2$, the left-hand-side is $2/3$ while the right-hand-side is $1/2$, yielding an absurdity.

9.34 To show that the topology induced by $\| \cdot \|_2$ is finer than the topology induced by $\| \cdot \|_1$ it suffices to show that the identity function

$$\mathrm{id} : (X, \| \cdot \|_2) \to (X, \| \cdot \|_1)$$

is continuous. And indeed, given $\varepsilon > 0$ let $\delta = \varepsilon / K$. Then, for all $x_1, x_2 \in X$ such that $\|x_1 - x_2\|_2 < \delta$, one has $\|\mathrm{id}(x_1) - \mathrm{id}(x_2)\|_1 = \|x_1 - x_2\|_1 \le K \cdot \|x_1 - x_2\|_2 < K\delta = \varepsilon$.

9.35

1. Given $x, y \in V$ with $\|x\| = \|y\| = 1$, for every $0 \le \alpha \le 1$ the triangle inequality gives

$$\|\alpha x + (1 - \alpha)y\| \le \|\alpha x\| + \|(1 - \alpha)y\| = \alpha\|x\| + (1 - \alpha)\|y\| = 1.$$

2. Suppose now that $\|x\|^2 = \langle x, x \rangle$ for an inner product $\langle \cdot, \cdot \rangle$ on V, and let's assume the ground field is $\mathbb{R}$ (the case $K = \mathbb{C}$ is left as an extra exercise). If $x \ne y$ satisfy $\|x\| = \|y\| = 1$, then for all $0 < \alpha < 1$

$$\|\alpha x + (1-\alpha)y\|^2 = \alpha^2 + (1-\alpha)^2 + 2\alpha(1-\alpha)\langle x, y\rangle.$$

By the Cauchy-Schwarz Inequality we have that

$$|\langle x, y\rangle| \le \|x\|\|y\| = 1.$$

Since $x \ne y$ and $\|x\| = \|y\|$, it follows that x is not a scalar multiple of y, and thus the Cauchy-Schwarz Inequality holds strictly. Thus $\langle x, y\rangle < 1$ and it follows that

$$\|\alpha x + (1-\alpha)y\|^2 < 1 + 2\alpha^2 - 2\alpha + 2\alpha - 2\alpha^2 = 1$$

showing that $\|\alpha x + (1-\alpha)y\| < 1$ and thus the only points of intersection of $L(x, y)$ with the unit sphere occur when $\alpha = 0$ or $\alpha = 1$ at the points y and x, respectively.

3. Assume that $x \ne y$ satisfy $\|x\| = \|y\| = 1$ in a strictly convex space. By the triangle inequality $\|x + y\| \le \|x\| + \|y\| \le 2$ so we just need to show that $\|x + y\| = 2$ is impossible. Indeed, if $\|x + y\| = 2$, then

$$\|(1/2)x + (1/2)y\| = (1/2)\|x + y\| = 1$$

contradicting strict convexity.

4. $\mathbb{R}^2$ with the ℓ_2 norm is strictly convex since the ℓ_2 norm is induced by an inner product (the standard inner product). That $\mathbb{R}^2$ with the ℓ_1 norm is not strictly convex is seen by choosing $x = (1, 0)$ and $y = (0, 1)$. Then $\|x\|_1 = \|y\|_1 = 1$ but $\|x + y\| = 2$. Similarly, for $x = (1, 1)$ and $y = (1, -1)$ it holds that $\|x\|_\infty = \|y\|_\infty$, but $\|x + y\|_\infty = 2$, showing that the ℓ_∞ norm is not strictly convex either.

9.36 To show that the kernel of A is a closed linear subspace of V recall that the kernel, by definition, is the set $A^{-1}(\{0\})$. The fact that the kernel is a linear subspace of the domain was established in the main text and in any case is easy to re-establish if needed. To show that the kernel is closed recall that the inverse image under a continuous function of a closed set is closed. Since a bounded linear operator is continuous, it suffices to show that $\{0\}$ is a closed set in W. Indeed, the metric induced on W by the norm is a metrizable topology and thus is Hausdorff, and thus every singleton set is closed.

9.37 The verification that A is a linear operator is trivial, and thus we omit it. Next we observe that for all $x \in \ell_2$ with $x \ne 0$

$$\|Ax\|^2 = \sum_{k=1}^{\infty}\left|\left(1 - \frac{1}{n}\right)x_n\right|^2 < \sum_{k=1}^{\infty}|a_k|^2 = \|x\|^2$$

and thus

$$\|Ax\| < \|x\|$$

holds for all $x \neq 0$. It follows that $\|A\| \leq 1$ and that if $\|A\| = 1$, then the norm is never attained. To see that the operator norm of A is indeed precisely 1, consider

$$e_n = (0, 0, \ldots, 0, 1, 0, 0, \ldots)$$

with 1 in the n-th position. It holds that

$$\|Ae_n\| = 1 - \frac{1}{n}$$

and since $\|e_n\| = 1$ it follows that $\|A\| \geq \|Ae_n\|$, for all $n \geq 1$, and so we conclude that indeed $\|A\| = 1$.

9.38 The composition of linear operators is always a linear operator, so all we need to do is show the composition BA is bounded, which will be done by showing the requested upper bound of the operator norm. Indeed, for all $x \in U$

$$\|BAx\| = \|B(Ax)\| \leq \|B\|\|Ax\| \leq \|B\|\|A\|\|x\|$$

and the claim follows.

9.39 In an exercise the reader was requested to prove that for bounded linear operators $A, B \in \mathbf{B}(\mathcal{B})$, if A is invertible and $\|A - B\| < \|A\|$, then B is invertible. With this result in mind one just needs to correctly interpret the meaning of the claim that G is open. Let $A \in G$ be given. To show that G is open it suffices to show that G contains an open ball with positive radius and centre A. In other words, we need to find $\varepsilon > 0$ such that every $B \in \mathbf{B}(\mathcal{B})$ with $\|A - B\| < \varepsilon$ is invertible. By the above result, one may take $\varepsilon = \|A\|$, which is positive (since A is invertible, so clearly $A \neq 0$).

9.40 Notice that the domain of ST is the set $\{x \in \mathcal{B}_1 \mid Tx \in \mathcal{D}(S)\}$. Given a sequence $\{x_n\} \in$ in the domain of ST, suppose that $x_n \to x_0$ and that $STx_n \to z$. We need to show that x_0 is in the domain of ST and that $STx_0 = z$. Indeed, since T is continuous, it follows that

$$Tx_n \to Tx_0.$$

Since Tx_n is in the domain of S and since $S(Tx_n) \to z$, it follows from the fact that S is closed that Tx_0 is in the domain of S and that $STx_0 = z$. Thus, as was required, we see that x_0 is in the domain of ST and that $STx_0 = z$, showing ST is closed.

Topological Groups

9.41 Firstly, note that $\mathbb{R}^n$ with the Euclidean topology is a Hausdorff topological space since the topology is metrizable. Both the group sum

$$+ : \mathbb{R}^n \times \mathbb{R}^n \to \mathbb{R}^n, \qquad ((a_1, \ldots, a_n), (b_1, \ldots, b_n)) \overset{+}{\mapsto} (a_1 + b_1, \ldots, a_n + b_n)$$

and the group inverse

$$- : \mathbb{R}^n \to \mathbb{R}^n, \qquad (a_1, \ldots, a_n) \mapsto (-a_1, \ldots, -a_n)$$

are continuous with respect to the Euclidean topology on $\mathbb{R}^n$ and to the product topology on $\mathbb{R}^n \times \mathbb{R}^n$. Hence, the additive group $\mathbb{R}^n$ with the Euclidean topology is a topological group. In fact, the map

$$\mathbb{R} \times \mathbb{R}^n \to \mathbb{R}^n, \qquad (\lambda, (a_1, \ldots, a_n)) \mapsto (\lambda \cdot a_1, \ldots, \lambda \cdot a_n)$$

is also continuous with respect to the Euclidean topology of $\mathbb{R}^n$ and the product topology of $\mathbb{R} \times \mathbb{R}^n$. That is to say, the $\mathbb{R}$-vector space $\mathbb{R}^n$ is actually a topological vector space.

9.42 Note that $\mathrm{SL}_n(\mathbb{R}) = \det^{-1}(\{1\})$, and $\det : \mathrm{GL}_n(\mathbb{R}) \to \mathbb{R}$ is continuous. Hence $\mathrm{SL}_n(\mathbb{R})$ is a closed subgroup of the topological group $\mathrm{GL}_n(\mathbb{R})$. In particular, $\mathrm{SL}_n(\mathbb{R})$ is a topological group.

9.43 Note that $\mathrm{O}(n) = \varphi^{-1}(\{\mathrm{id}_n\})$, where $\varphi : \mathrm{GL}_n(\mathbb{R}) \to \mathrm{GL}_n(\mathbb{R})$ is given by

$$\varphi(A) = A^t A$$

and is a continuous map. Hence $\mathrm{O}(n)$ is a closed subgroup of the topological group $\mathrm{GL}_n(\mathbb{R})$. In particular, $\mathrm{O}(n)$ is a topological group.

9.44 Note that $\mathrm{SO}(n) = \det^{-1}(\{1\})$, where $\det : \mathrm{O}(n) \to \{1, -1\}$ is continuous. Hence $\mathrm{SO}(n)$ is a closed subgroup of the topological group $\mathrm{O}(n)$ and is thus a topological group.

9.45 Note that $\mathrm{Sp}_{2n}(\mathbb{R}) = \varphi^{-1}(\{J_0\})$, where $\varphi : \mathrm{GL}_{2n}(\mathbb{R}) \to \mathrm{GL}_{2n}(\mathbb{R})$ is given by

$$\varphi(A) = A^t \cdot J_0 \cdot A$$

is a continuous map. Hence $\mathrm{Sp}_{2n}(\mathbb{R})$ is a closed subgroup of the topological group $\mathrm{GL}_{2n}(\mathbb{R})$, and thus is itself a topological group.

9.46 For a given $B \in \mathrm{SO}(n)$, we construct a path

$$\gamma : [0, 1] \to \mathrm{SO}(n) \qquad \text{such that} \qquad \gamma(0) = \mathrm{id}_n \text{ and } \gamma(1) = B.$$

The matrix B being orthogonal, there exists an orthogonal matrix M such that

$$B = M^t K M$$

where

$$K = \left(\begin{array}{c|c|c} \mathrm{id}_p & & \\ \hline & -\mathrm{id}_q & \\ \hline & & S \end{array} \right),$$

with $q = 2r$ and, for $k = n - (p + q)$,

$$S = \begin{pmatrix} \begin{array}{cc} \cos\alpha_1 & -\sin\alpha_1 \\ \sin\alpha_1 & \cos\alpha_1 \end{array} & & \\ & \ddots & \\ & & \begin{array}{cc} \cos\alpha_k & -\sin\alpha_k \\ \sin\alpha_k & \cos\alpha_k \end{array} \end{pmatrix}$$

for some $\alpha_1, \ldots, \alpha_k \in (0, \pi) \cup (\pi, 2\pi)$. For $t \in [0, 1]$, define

$$K_t = \begin{pmatrix} \mathrm{id}_p & & \\ & R_t & \\ & & S_t \end{pmatrix},$$

where

$$R_t = \begin{pmatrix} \begin{array}{cc} \cos(t\pi) & -\sin(t\pi) \\ \sin(t\pi) & \cos(t\pi) \end{array} & & \\ & \ddots & \\ & & \begin{array}{cc} \cos(t\pi) & -\sin(t\pi) \\ \sin(t\pi) & \cos(t\pi) \end{array} \end{pmatrix}.$$

and

$$S_t = \begin{pmatrix} \begin{array}{cc} \cos(t\alpha_1) & -\sin(t\alpha_1) \\ \sin(t\alpha_1) & \cos(t\alpha_1) \end{array} & & \\ & \ddots & \\ & & \begin{array}{cc} \cos(t\alpha_k) & -\sin(t\alpha_k) \\ \sin(t\alpha_k) & \cos(t\alpha_k) \end{array} \end{pmatrix}.$$

Hence define, for all $t \in [0, 1]$,

$$\gamma(t) = M^t K_t M \in SO(n).$$

Since $\gamma : [0, 1] \to SO(n)$ connects id_n to B, we get that $SO(n)$ is path-connected and hence also connected.

9.47 Note that $O(n)$ is the disjoint union of $O^+(n)$ and $O^-(n)$, where

$$O^+(n) = \{A \in O(n) \mid \det(A) > 0\}$$

and

$$O^-(n) = \{A \in O(n) \mid \det(A) < 0\}.$$

Furthermore, $O^+(n)$ and $O^-(n)$ are homeomorphic. In fact, the map

$$\text{Or}^+(n) \to \text{Or}^-(n), \qquad \begin{pmatrix} x_{11} & x_{12} & \cdots & x_{1n} \\ \vdots & \vdots & \ddots & \vdots \\ x_{n1} & x_{n2} & \cdots & x_{nn} \end{pmatrix} \mapsto \begin{pmatrix} -x_{11} & x_{12} & \cdots & x_{1n} \\ \vdots & \vdots & \ddots & \vdots \\ -x_{n1} & x_{n2} & \cdots & x_{nn} \end{pmatrix}$$

provides a homeomorphism. Therefore, it suffices to show that $\text{O}^+(n)$ is path-connected. But $\text{O}^+(n) = \text{SO}(n)$, and the claim follows since it was already established that $\text{SO}(n)$ is path-connected.

9.48 For a given $B \in \text{GL}_n^+(\mathbb{R})$, we construct a path

$$\gamma \colon [0, 1] \to \text{GL}_n^+(\mathbb{R}) \qquad \text{such that} \qquad \gamma(0) = \text{id}_n \text{ and } \gamma(1) = B.$$

By the polar decomposition theorem there are a symmetric positive-definite matrix L and an orthogonal matrix P such that $B = LP$. Since $\det(L) > 0$ and P is orthogonal, it follows that $\det(P) = 1$, and hence $P \in \text{O}^+(n) = \text{SO}(n)$. The matrix L being symmetric, there exists an orthogonal matrix C such that

$$L = C^t \Lambda C \qquad \text{where} \qquad \Lambda = \begin{pmatrix} \lambda_1 & & \\ & \ddots & \\ & & \lambda_n \end{pmatrix}.$$

For all $t \in [0, 1]$ define

$$\Lambda_t = \begin{pmatrix} t\lambda_1 + (1-t) & & \\ & \ddots & \\ & & t\lambda_n + (1-t) \end{pmatrix}.$$

Since the topological group $\text{SO}(n)$ is path-connected, there exists a path

$$\mu \colon [0, 1] \to \text{SO}(n) \qquad \text{such that} \qquad \mu(0) = \text{id}_n \text{ and } \mu(1) = P.$$

Hence define, for all $t \in [0, 1]$,

$$\gamma(t) = C^t \Lambda_t C \mu(t).$$

Since $\gamma \colon [0, 1] \to \text{GL}_n^+(\mathbb{R})$ connects id_n to B, we get that $\text{GL}_n^+(\mathbb{R})$ is path-connected and hence also connected.

9.49 Note that the function $\det \colon M_n(\mathbb{R}) \to \mathbb{R}$ being continuous and surjective, the sets

$$\text{GL}_n^+(\mathbb{R}) = \det^{-1}((0, +\infty))$$

and

$$\text{GL}_n^-(\mathbb{R}) = \det^{-1}((-\infty, 0))$$

are non-empty open sets and

$$\mathrm{GL}_n^+(\mathbb{R}) \cap \mathrm{GL}_n^-(\mathbb{R}) = \emptyset$$

while

$$\mathrm{GL}_n(\mathbb{R}) = \mathrm{GL}_n^+(\mathbb{R}) \cup \mathrm{GL}_n^-(\mathbb{R}).$$

Note also that $\mathrm{GL}_n^+(\mathbb{R})$ and $\mathrm{GL}_n^-(\mathbb{R})$ are homeomorphic, in fact, the map

$$\mathrm{GL}_n^+(\mathbb{R}) \to \mathrm{GL}_n^-(\mathbb{R}), \qquad \begin{pmatrix} x_{11} & x_{12} & \cdots & x_{1n} \\ \vdots & \vdots & \ddots & \vdots \\ x_{n1} & x_{n2} & \cdots & x_{nn} \end{pmatrix} \mapsto \begin{pmatrix} -x_{11} & x_{12} & \cdots & x_{1n} \\ \vdots & \vdots & \ddots & \vdots \\ -x_{n1} & x_{n2} & \cdots & x_{nn} \end{pmatrix}$$

provides a homeomorphism. The claim now follows since it was already proved that $\mathrm{GL}_n^+(\mathbb{R})$ is path-connected, and hence $\mathrm{GL}_n^-(\mathbb{R})$ is path-connected as well.

9.50 For a given $B \in \mathrm{SL}_n(\mathbb{R})$ we construct a path

$$\gamma : [0, 1] \to \mathrm{SL}_n(\mathbb{R}) \qquad \text{such that} \qquad \gamma(0) = \mathrm{id}_n \text{ and } \gamma(1) = B.$$

By the polar decomposition theorem there exist a symmetric positive-definite matrix L and an orthogonal matrix P such that $B = LP$. In particular, since $\det(B) = 1$, it holds that $\det(P) = 1$ and so $\det(L) = 1$. The matrix L being symmetric, there exists an orthogonal matrix C such that

$$L = C^t \Lambda C$$

and since L is positive-definite with determinant equal to 1, the diagonal matrix Λ has the form

$$\Lambda = \begin{pmatrix} \exp(\mu_1) & & \\ & \ddots & \\ & & \exp(\mu_n) \end{pmatrix},$$

for some $\mu_1, \ldots, \mu_n \in \mathbb{R}$ such that $\sum_{j=1}^n \mu_j = 0$. The matrix P being orthogonal, there exists an orthogonal matrix M such that

$$P = M^t K M$$

where

$$K = \left(\begin{array}{c|c|c} \mathrm{id}_p & & \\ \hline & -\mathrm{id}_q & \\ \hline & & S \end{array} \right),$$

with $q = 2r$ and, for $k = n - (p + q)$,

$$S = \begin{pmatrix} \begin{matrix} \cos\alpha_1 & -\sin\alpha_1 \\ \sin\alpha_1 & \cos\alpha_1 \end{matrix} & & \\ & \ddots & \\ & & \begin{matrix} \cos\alpha_k & -\sin\alpha_k \\ \sin\alpha_k & \cos\alpha_k \end{matrix} \end{pmatrix}$$

for some $\alpha_1, \ldots, \alpha_k \in (0, \pi) \cup (\pi, 2\pi)$. For all $t \in [0, 1]$ define

$$\Lambda_t = \begin{pmatrix} \exp(t\mu_1) & & \\ & \ddots & \\ & & \exp(t\mu_n) \end{pmatrix},$$

and define

$$K_t = \begin{pmatrix} \mathrm{id}_p & \\ & \begin{matrix} R_t & \\ & S_t \end{matrix} \end{pmatrix},$$

where

$$R_t = \begin{pmatrix} \begin{matrix} \cos(t\pi) & -\sin(t\pi) \\ \sin(t\pi) & \cos(t\pi) \end{matrix} & & \\ & \ddots & \\ & & \begin{matrix} \cos(t\pi) & -\sin(t\pi) \\ \sin(t\pi) & \cos(t\pi) \end{matrix} \end{pmatrix}$$

and

$$S_t = \begin{pmatrix} \begin{matrix} \cos(t\alpha_1) & -\sin(t\alpha_1) \\ \sin(t\alpha_1) & \cos(t\alpha_1) \end{matrix} & & \\ & \ddots & \\ & & \begin{matrix} \cos(t\alpha_k) & -\sin(t\alpha_k) \\ \sin(t\alpha_k) & \cos(t\alpha_k) \end{matrix} \end{pmatrix}.$$

Hence, define for all $t \in [0, 1]$

$$\gamma(t) = C^t \Lambda_t C M^t K_t M \in \mathrm{SL}_n(\mathbb{R}).$$

Since $\gamma: [0, 1] \to \mathrm{SL}_n(\mathbb{R})$ connects id_n to B, we get that $\mathrm{SL}_n(\mathbb{R})$ is path-connected and hence connected.

Index

© Springer Nature Switzerland AG 2021
C. Alabiso and I. Weiss, *A Primer on Hilbert Space Theory*, UNITEXT for Physics,
https://doi.org/10.1007/978-3-030-67417-5

Printed in the United States
by Baker & Taylor Publisher Services